Intoxicated Ways of Knowing

Intoxicated Ways of Knowing

THE UNTOLD STORY OF INTOXICANTS AND THE BIOLOGICAL SUBJECT IN NINETEENTH-CENTURY GERMANY

Matthew Perkins-McVey

The University of Chicago Press   CHICAGO AND LONDON

The University of Chicago Press, Chicago 60637
The University of Chicago Press, Ltd., London
© 2026 by The University of Chicago
All rights reserved. No part of this book may be used or
reproduced in any manner whatsoever without written
permission, except in the case of brief quotations in critical
articles and reviews. For more information, contact the
University of Chicago Press, 1427 E. 60th St., Chicago, IL 60637.
Published 2026

35 34 33 32 31 30 29 28 27 26   1 2 3 4 5

ISBN-13: 978-0-226-84611-8 (cloth)
ISBN-13: 978-0-226-84613-2 (paper)
ISBN-13: 978-0-226-84612-5 (ebook)
DOI: https://doi.org/10.7208/chicago/9780226846125.001.0001

Library of Congress Control Number: 2025032522

Authorized Representative for EU General Product Safety
Regulation (GPSR) queries: **Easy Access System Europe**—
Mustamäe tee 50, 10621 Tallinn, Estonia, gpsr.requests@
easproject.com
Any other queries: https://press.uchicago.edu/press/contact
.html

Contents

Convergences * 1

I. Vital Substances

1. Pharmacy Bodies * 11
2. Brown, Kant, and the Crisis in Medicine * 33
3. Brunonian *Naturphilosophie* and Intoxicated Knowing * 49
4. A (Brief) Historical Ontology of an Alkaloid * 73

II. Contra Intoxicatio

5. Great Expectations: The Humboldts, Johannes Müller, and the Rise of Neomechanism * 91
6. The "Young" Neomechanists and the Problem of the Brain * 115
7. A Tale of Two Cities: Berlin, Leipzig, and Scientific Psychology * 135

III. The Intoxicated Subject

8. A Postalkaloidal "Golden Age" of Intoxication * 165
9. The Life and Times of Emil Kraepelin: Drugs, Bodies, and Minds * 183
10. Kraepelin's Nosology and an Intoxicated "Physiologie der Seele" * 209
11. Drunken Songs of Tomorrow: Nietzsche, Freud, Weber, and Intoxication * 231

A Horizontal Fall * 263

Acknowledgments * 267
Notes * 269
Bibliography * 275
Index * 303

Convergences

When we turn to these phenomena today, we are concerned on the one hand with their chemistry and on the other hand with their psychological effects. These substances, whose finer structures were until recently hardly known and whose powers had been attributed to the virtues of the plants, do not merely act on our cells, but form new compounds with their molecules. They form chains and rings in aberrations that the body is incapable of producing on its own. The same is true for perception. It is made fertile with images that, without [these intoxicating plants], no thought and no flight of the imagination would be capable of creating them. For our purposes, chemistry is a mode of "transition," and the psyche is an organ attuned to "supervention." (Jünger 1970, 263–264)[1]

Ernst Jünger's *Annäherungen: Drogen und Rausch* is an uncompromising essay penned by a deeply compromised man. At once poetry, history, science, and philosophy, it comes as little shock that a figure so ingratiated with violence, with extreme politics, with the callous abandon of all that is soft and tender in liberal bourgeois life, would discover in intoxicants a rending of worlds, bellicose echoes of the atomic bomb. And yet, for all his personal and political misgivings, this excerpt speaks to something altogether familiar to modern readers. It describes a subject of flesh, blood, and sinew, a cellular being, a molecular being, a fatty brain stewing in a neurochemical soup. But this very same creature is, at once, irreducible to neurochemical chittering; it is experienced and understood psychologically as a being of *psyche*. This particular formation of the modern experience, Jünger contends, is made apparent not in Shakespeare's "'common' dream," in the humdrum awareness of daily life, but in the raucous intercession of ecstatic intoxication. This subject, it would seem, is a biological one. At the sites in which we encounter such a subject, intoxicants leave their trace.

2 CONVERGENCES

What is the nature of this "trace"? How did a subject of mind and body become mingled with, even enjoined in, intoxication? Answering these questions will entail looking to the historical sciences of the body and mind, focusing in on those instances where intoxicants intersect with the creation of knowledge. Let us briefly consider two examples plucked from the annals of nineteenth-century science. In the 1810s, a young pharmaceutical chemist named Friedrich W. Sertürner gathered three local teenage boys. They all agreed to consume half a grain of a new isolate, rendered by the junior chemist, in a weak alcoholic solution. When nothing happened, the group took another 0.5 grains, and yet another still, all in 45 minutes. Soon fading fast, and fearing that this hastily devised experiment may very well prove fatal, the chemist quickly saw to it that everyone drank vinegar, inducing vomiting. This had been the first human encounter with a morphine isolate. Elsewhere, a tad short of 300 kilometers (and 67 years) away, another young researcher—this time a psychology-trained psychiatrist named Emil Kraepelin—has just downed between 7.5 and 60 grams of pure alcohol, barely masking the bitterness with a smattering of raspberry syrup. The subject at hand heard the letter *o*, followed by another, and another still—recording each occurrence with a press of the hand. The besotted were participating in experiments on the influence of various intoxicants on different forms of reaction time.

We can try to disregard the presence of intoxicants in either of these cases as a distraction, a red herring. Each investigation occurred in conjunction with a remote series of underlying research objectives, material conditions, and antecedent conceptual circumstances. But intoxication is not a state of intellectual vagrancy, as some would have it. Nor is it a dream state, out of which the creative unknown ruptures into conscious awareness. Everywhere around us, the presence of intoxicants is in many cases so everyday that they are hardly worth mentioning. And yet, their everydayness is thrown into sharp relief by their potent ability to move us, *to force us to feel*.

What both of those vignettes share is much more than the mere presence of intoxicants. Both exemplify what I have termed "intoxicated ways of knowing." Whether in the laboratory or in one's average, everyday surroundings, the entities and scientific objects that populate the space around us are gathering points for concepts, theories, and associations. Entities themselves, once they pop into existence that is, have a certain degree of durability. One not so easily attributed to the mercurial flux of ideas that surrounds them. Any theory or idea can move a human subject to think, see, and behave in a different way. But intoxicants are unique even among entities for their ability to engender conceptual associations and make them "real" in our perceptions; they intercede directly in our lived experience of embodiment, where "embodiment" refers to the lived experience of being in a body. Intoxication is a form of *tacit knowing*, an inexplicable interface between perceptions, concepts, and scientific objects gathered in the radical particularity of place and time, made explicit in the embodied lifeworld. Far from the totalizing influence of epistemes, intoxication is a dynamic, epistemic attunement of the senses—an epistemic modality. It is in this emotive element of intoxication, this seizing upon perception, that intoxication's epistemological significance lies.

That intoxicants inform perception is nothing new. Few are surprised to hear of mind-altering states featuring in all manner of mystic rites and experiences. But what if intoxication helped constitute who we are as human beings in the twenty-first century, informing our basic scientific understanding of what we are as bodies and minds? It this overlooked role of intoxicants in the constitution of the modern biological subject that this book takes as its object.

To be biological—contrary to perceptions of some biogenetic reductionists—entailed, and still entails, so much more than mere self-identification with a clockwork of flesh and bone (Lewontin 2003; Ruberg 2019). If being biological refers to nothing more than a subject that reductively understands itself in bodily terms, then what is to be made of theories of mind, psychological treatment modalities, and genetic epistemologies? Biologism is a Grecian chimera—its many faces visible in the diverse ways that it is practiced, across competing, even contradictory, subdisciplines. How did this come to be, and how is biologism distinct from mid-nineteenth-century mechanism? "The body" has a history, with the seeming facticity of being-a-body being a relatively young notion. This doctrine of biologism—and, with it, the biological subject—arose in a specific place and at a specific time. It is at the center of precisely this narrative—the historical emergence of the biological subject toward the end of the nineteenth century—that I place intoxicated ways of knowing as having primary significance. Here, substances of intoxication are called upon to testify, and compel testimony, on the ultimate nature of the body and mind. From the development of John Brown's Brunonian system out of opium therapy to the passage of alkaloids into being as objects of scientific inquiry, or the first ventures into "pharmapsychology"—intoxicated ways of knowing hold a crucial place.

It cannot be ignored that there is something unsettling to references of a "biological subject," least of all because it smacks of pretense, of an effort to totalize the rich diversity of lived experiences. How could a "biological subject" possibly speak to the Black experience under Jim Crow or to the

First Nations in Canada? Can it be found in the church, in the shtiebel, at the gas station, or among everyday people in their everyday existence? Is it not simply an abstraction of biomedical scientists and the academics who study them? Such language of the "subject" is not intended in the much-maligned sense of a singular modern subject emergent of a broadly "Western" tradition. The "biological subject" denotes a horizon of many possible ways of being a self, a sphere of negotiations surrounding definitional possibilities, the candidacy of certain claims for truth or falsehood, and potential forms of life. There is not one biological subject but many ever-changing biological subjects. It indicates those spheres of life where biomedical categories give form and voice to everyday experience, just as readily as it marks sites of resistance and divergence. The "biological subject" is much more than the categories of experts, unidirectionally passing from scientific minds to private individuals, whose lived experience is hardly so easily encapsulated. It is a nomenclature for multifarious convergences upon lived, embodied experience, implicating medical clinics, beer gardens, visits to Dr. Google, government health ministries, recreational drug use, and safe-injection sites just as forcefully as concepts, scientific objects, entities, apparatuses, laboratory settings, research programs, and scientists themselves. This is the story of how intoxicants, and the ecstatic states they induce, participated in constituting the shape and focus of just such a "biological subject."

The starting point of this narrative is the appearance of what I have called "vital substances." These intoxicating vital stimulants, therapies grounded in the Brunonian system of medicine, would sweep across the German-speaking world, taking root in *Naturphilosophie* during an age when new and old intoxicants were stepping into daily life like never before. Inextricably linked to the experimentalist reorganization of the materia medica into the European pharmacopoeia, these vital substances were also central to the very principle of what it meant to be a body at the beginning of the German nineteenth century. For the Romantic Brunonians, the vital principle of "excitability" did not merely reflect, but directly flowed from, the perceptional encounter of intoxicated embodiment. Even in Sertürner's identification of morphine as the first alkaloid, the vital substance concept would participate directly in the determination of the alkaloid as a novel kind of thing to be. So entrenched would the notion of vitally stimulating intoxicants be that, as much as he critiqued the Brunonians, Johannes Müller would experimentally validate the vital substance concept as physiologically substantiated.

This factors into the history of the biological subject as much through the tenacious response of its opponents, the organic physicalists (or neomechanists), as it does through its own claims. Through Müller's own students, in particular Hermann von Helmholtz and Emil du Bois-Reymond, the "vital substance" concept would be stalwartly opposed, entangled as it was with vitalism and the Romantic science that Helmholtz and du Bois-Reymond so eagerly sought to supplant. Their opposition to the vital stimulus concept would be so virulent that, in effect, any introduction of intoxicants into the experimental negotiations surrounding the body would come to be met with scrutiny. The conceptual opposition to vitalism, and the vital substance concept, saw the elevation of the methodologies of physiochemical reductionism to a doctrinal status. It would be this shift that led "young" neomechanists, such as Eduard Hitzig, Paul Flechsig, and Theodor Meynert, to identify brain tissue as the locus of "the real" apropos the mind, equating even higher-order mental states with neural states. As a result, the neomechanists would repudiate the validity of experiential empiricism. For psychiatrists like Meynert and Karl Wernicke, this would mean leading a clinical life profoundly integrated with a rapidly growing armamentarium of newly isolated, and newly synthesized, intoxicants that, remarkably, never meaningfully figured into their conceptions of the body and mind. The neomechanist's rejection of the principles and methodologies underlying vitalistic and Romantic science would, thus, entail both a radical turn to the body and a forgetting, or devaluation, of *embodiment*.

It is here that substances of intoxication intervene in the emergence of the biological subject. Hasty conflations between psychological concepts and neurophysiological features would help justify the formation of the separate, novel science of experimental psychology, a science of *embodiment*. At the core of this new science would be parallelistic critique of neomechanism, of biomechanical reductionism in all its forms. For Gustav Fechner, the principle of psychophysical parallelism had followed from his understanding of the duplicitous entanglement of the material and the psychospiritual that lay at the ground of the universe. But Wilhelm Wundt expelled the cosmic, spiritual element from psychological parallelism, instead emphasizing the parallelistic notion that psychological processes were irreducible to physical states. It is in this radically specific context that Emil Kraepelin would bring intoxicants into Wundt's psychological laboratory, in the hopes that experimentation on states of intoxication might be the key to finally understanding both the rudimentary structures of the mind and the nature of mental illness.

Through the intoxicated mind, Kraepelin's experiments would make the physicality of the body real in the psychical—experimentally grounding the psychological parallel in the body, while upholding the independent validity of psychology as a science. Kraepelin would counter the neomechanist's equation of mental states with neural states with his own identification of intoxicated states with mental states as such. In many cases, this would be accomplished through a process of self-experimentation not unfamiliar to Romantic experiential science, but now endowed with the validity, and critical distance, of an experimental science. This would be the emergence of a radical new conception of the body—biologism. Not of the body as a fleshy clockwork, but rather an approach to the body that included an independent science of embodiment, and yet still aligned mental states with bodily events. This occasion did not go unnoticed in the developing philosophies of the late nineteenth century. On the contrary, looking to the burgeoning body philosophies of Friedrich Nietzsche, Sigmund Freud, and Max Weber, one finds that psychological research on the effects of intoxicants on the mind, as well as their own intoxicated lives, figured centrally into their conceptions of the bodily subject. In the end, it is worth pondering if the entirety of the biological subject hinges in some way on intoxicated ways of knowing.

Many will have noted the conspicuous absence of two topics interwoven with the history of intoxication: addiction and pain. The suffering wrought by addiction is typically regarded as so inextricably linked with substances of intoxication themselves that addiction and intoxication are, for some, all but synonymous. Drug use becomes drug abuse. But intoxication was not always warily regarded through the looking glass of addiction. Nor have the later scientific encounters with intoxication of the historic past always entreated addiction as a question of central importance. Intoxicants, and the powerful states they induce, have exerted a profound effect on the historical sciences of the mind and body not merely despite the medicalization of habit, but independent of it. Pain makes for a different case. Pain intercedes in our experience of embodiment. As Karl Marx famously wrote to his daughter in 1881, "there is only one effective antidote for mental suffering, and that is physical pain" (Marx 2010). It, too, makes us feel something. In this sense, is self-experimentation with needles and bodkins really so different from taking intoxicants? Further still, many have taken intoxicants precisely *not to feel*. Chloroform and nitrous oxide stayed the painful dramas of surgery, much as opium and alcohol beat back psychic and physical pain. And yet intoxication, even when marshaled against pain, opens onto avenues of embodied experience otherwise hardly accessible in daily life. There is more in taking chloroform or morphine than the mere absence of pain. As with addiction—pain's prominence in the history of intoxication only further foregrounds the importance of intoxication where pain is not the primary object. Though the histories of pain and addiction science may prove inseparable from the historical advent of biologism, the intoxicated subject has its own story to tell.

This is a history of the biological subject, of the intoxicated subject, one that establishes "intoxicated ways of knowing" as a persistent, critical, and, indeed, constitutive feature of German physiological, psychological, and psychiatric research throughout the nineteenth century. It establishes that the modern human entanglement with intoxicants is fundamental to the sciences by which we have come understand who we are as biological beings. This serves as both a new definition of biologism and a reckoning on its uncertain origins. The concept of "intoxicated ways of knowing" further raises questions surrounding the nature of knowledge production itself, barely scratching the surface of the epistemological significance that intoxicants hold for the biological sciences and possibly calling for a critical reassessment of history as we know it. To examine these elements in tandem means taking a look at the subtle nature of changes in various conceptions of the body, taking seriously the idea of the body as a dynamic gathering point of perceptional phenomena. This entails a historicization of not only the body as a scientific object (and its relationship with the subject) but also of the perceptional experience of intoxication, as a knowing relation given to fluid conceptual associations. Historical ontologies of various substances of intoxication, as they pass into being and immediately change the lives of those around them, will also be an important part of this story. By turning to the historical interactions between bodies and intoxicants, old ambiguities can be resolved, and new questions posed, particularly about where else the formative influence of intoxicated ways of knowing may have gone overlooked in the history of human thought.

I
Vital Substances

I
Vital Substances

❋ 1 ❋
Pharmacy Bodies

Berlin 1748

What are substances of intoxication, really? The answer that forms ready-made in our minds is some variety of psychoactive chemical. If we want to be more specific, we might say a vague grouping of more or less distinct chemical entities that, through some affinity with our own neurochemical interface, produce changes in mood, perception, thought, and behavior. A popular term is "drugs," a word of late-medieval provenance that meant something akin to what is now meant by "pharmaceutical," only coming to denote opiates and other narcotic materials at the close of the nineteenth century. It is hard to ignore just how much this definition leaves something to be desired. Never mind that it makes for a perilously broad and porous natural kind—as readily applied to vanilla, cinnamon, or ginger as to ketamine or heroin. Tugging at our biomedical heartstrings, it seeks to define substances of intoxication by describing what they are in the test tube or at the synaptic cleft, and so ends up saying very little, only to maybe get that part wrong too.

Substances of intoxication, alongside so many other nonhuman actors, have been many things since enlisting in the collective enterprise narrowly referred to as "human civilization." The place of intoxicants in the ancient world is the stuff of modern-day legend. After the biblical flood waters receded, Noah's first task was to plant a vineyard. The Sumerians wrote of *hul-gil*, or the "plant of joy" (Nencini 2022). The Vedas are replete with mentions of *soma*, at once a ritual-drink, plant, and deity, as well as the likely cognate of the Zoroastrian *haoma* (Staal 2001). To this very day, scholars debate the proposed identity of both plants, the identification of *hul-gil* with the opium poppy having taken on a particularly mythic status (Nencini 2022). But what would the ascertainment of *soma*'s chemical structure tell us about *soma*? Not very much. A great deal more can be said about what substances of intoxication are when we look at what they have been, at how they move us to act. It is no secret that intoxicants have acted upon societies in a nigh ineffable diversity of religious, magical, cultic, social, medicinal, therapeutic, recreational, cultural, and identiary capacities. Intoxicants have played emissary of, and to, the gods, illuminating humanity's place in the cosmos, and intoxicants have helped illuminate humankind to themselves—bringing to light the nature of our own bodies and minds. Further still, intoxicants showed themselves to participate in the process of knowledge production. Far from mere epistemically crucial actants, the mind-states engendered by intoxicants interceded directly in the lifeworld of perception, tacitly concretizing some concepts and denying others. Toward the latter half of the eighteenth century, intoxicants could be found fighting on either side of a battle being waged over nothing less than the future of the human soul. They were then, simultaneously, soul-affecting materials and what I have come to refer to as "vital substances." If this is to be a story about the emergence of the modern body, this would seem a natural starting point. Let us begin by looking to one extreme of this conflict, before considering the other.

In the year 1748, König Friedrich II graciously hosted a particularly scandalous Frenchman in his Berlin court, in what Wilhelm Dilthey later characterized as a demonstration of Prussia's "boundless tolerance" (Dilthey 1901/1927, 116). This peculiar visitor was none other than the French physician and philosopher Julien Offrey de La Mettrie (1709–1751). That very year, La Mettrie had anonymously published *L'homme machine*, an ambitiously pithy endeavor to realize the metaphorical equivalence between the human body and machines into an absolute description of the relationship between body and soul (La Mettrie 1748). As the late Ian Hacking put it, La Mettrie "conveniently forms one extreme edge of a framework of which Descartes forms another" (Hacking 2009, 180). Hacking admits this is, of course, a grotesque simplification of a far subtler debate hinging on the eighteenth-century conceptualization of substance, but Descartes and La Mettrie are useful signifiers for dualism on the one hand and strict materialism on the other (Hacking 2009, 180).

Even by the standards of eighteenth-century French mechanism, La Mettrie was a radical. This sentiment perhaps rang truer still in the milieu of Prussian society. Prominent German physician, philosopher, and chemist Georg Ernst Stahl (1659–1734), an influential champion of vitalism—the fundamental distinction between living and "dead" matter—had died in Berlin a mere fourteen years earlier. And while different degrees of naturalistic mechanism were widespread in Germany, devoted mechanism was still something to be guarded against (Rumore 2014). The repeated mention of La Mettrie's raucous hedonism, craziness, and buffoonery, including by would-be fellow French rationalists Voltaire and Diderot, speak to the depths of intellectual, and perhaps also personal, contempt held against him (Hacking 2009, 182). Whether rumors of his debauchery were greatly exaggerated or not, these extravagant assaults on his personal character reflected on the controversial nature of his published writing. *L'homme machine* was sufficiently dangerous in the eyes of physiologist Albrecht von Haller (1708–1777) that he initiated a very public campaign against de La Mettrie's work, echoing concerns raised in the broader community (Knabe 1978, 121; Rumore 2014).

But was La Mettrie's work truly as radical as his detractors claimed? After all, naturalistic and mechanistic attitudes had already taken root across Germany, and Friedrich II's court was replete with Francophiles, then a major source of materialist thinking (Rumore 2014). La Mettrie argues from the perspective of a physician turned philosopher, someone primed to think *gut first*. La Mettrie makes no secret of this: "we should be guided by experience and observation alone[;] they abound in the annals of physicians who were philosophers, but not those philosophers who were not physicians" (La Mettrie 1748, 25–26). Rather than engage with metaphysical arguments against materialism, La Mettrie's argued for materialism from physiological evidence (Wellman 1992, 195). Although there is something of a latent vitalism even in La Mettrie's materialism, it is clear that by "the soul" La Mettrie simply refers to the experience of animated consciousness, understood to be a product of the arrangement of parts of the brain. In making this case in *L'homme machine*, La Mettrie enlists a panoply of interlocutors, among them muscular irritability, brains, intestines, fingers, knives, and (most importantly for our purposes) intoxicants such as opium and spirits.

Writing on the soul, La Mettrie remarks that "opium is too closely related to the sleep it produces, to be left out of consideration here" (La Mettrie 1748, 33–34). Of its effects, where the soul "was prey to the most intense sorrow, it now feels only the joy of suffering past, and of sweetest peace" (La Mettrie 1748, 34). Opium "even alters the will, forcing the soul which wished to wake and to enjoy life, to sleep in spite of itself"; "this drug intoxicates, like wine, coffee, etc., each in its own measure and according to the dose" (La Mettrie 1748, 34). For La Mettrie, that opium, coffee, and wine—not merely material objects but the produce of laboring hands—could have such a direct effect on the soul was a stark affront to any doctrine opposing the physicality of the human spirit.

Now, La Mettrie is not a particularly important figure in this narrative.

But this tale of foreign medical ideas taking root in German soil, particularly ideas about the relationship between intoxicants and the body, beautifully presages what is to come. Here, the ideas are not those of eighteenth-century French materialism, but the vitalistic system of Scottish physician John Brown. It's a story about the cavernous rift between physiology and medicine, spurring physicians and theorists in their search for medicine's scientific foundation. It's a story about how wild notions surrounding the deep significance that intoxicants held for the body came to find an eager audience in the crisis-stricken hearts of German physicians. And, above all else, it's the story of how this very encounter engendered the emergence of what I have provisionally called "vital substances": intoxicating substances that drive the vital forces at the heart of organic matter and ultimately participate in one's historical notion of what it means to be a body. Yet a history of vital substances and their reception will scarcely be meaningful without first considering what these medical substances were to begin with. Let us canvas some of the developments in the study of plant- and animal-based therapeutic remedies that took place over the preceding seventeenth and eighteenth centuries, a narrative that turns on the formation of national and regional pharmacopoeias.

Materia Medica

The seventeenth and eighteenth centuries were periods of tremendous development in the theoretical and material conceptions of plant- and animal-based therapeutics, largely owing to a growing interest in experimentally validating the therapeutic value of existing remedies. There are any number of individual instances that one might call upon in illustrating the experimental approach that was rapidly being extended to therapeutic remedies. Some of these are quite dramatic. Take, for example, the early history of intravenous injection. It was 1659 when none other than Christopher Wren, Thomas Willis, and Robert Boyle used a makeshift contraption, consisting of a goose quill and an animal bladder, to deliver an opium tincture to a dog intravenously (Moon 2021). By all accounts, the dog immediately started running, only to summarily die. This stands out as a particularly pioneering instance of experimentalism shaping the development of plant-based therapeutics. But by far the strongest exemplification of the scope and influence of experimentalism on plant and animal therapeutics is to be found in the sleepier history of the official pharmacopoeias.

In the seventeenth and eighteenth centuries, the pharmacopoeia would emerge as an organizational medium and conceptual locus, a centralizing authority around which therapeutic knowledge was oriented and mobilized. At first little more than a crude repository of known therapeutics, the pharmacopoeia rapidly became the most formative, and conceptually sophisticated, element in the developing science of medicine. Experiments would be conducted so that their findings could be incorporated into the latest edition of the relevant pharmacopoeia. It was the medium by which experimentation with plant and animal therapeutics was integrated as knowledge, setting the foundation for the attempts at "Newtonian" medical system building that would follow. Understanding how this came to be first requires taking a look into the development of the early modern pharmacopoeia.

Physicians of prior centuries had relied on the materia medica, both a collective name for the varied medieval and early modern compendiums of therapeutic remedies and a shorthand for the cumulative knowledge of medicinal substances. The term *materia medica* remained in use to refer to the body of knowledge on medicinal substances until the twentieth century, before it was phased out by the term *pharmacy* (Lytle 1906, 217). The use of that term, pharmacy, to refer to a collection of medicinal substances that have been studied and experimentally verified comes to prominence in the eighteenth century, with the proliferation of city pharmacopoeias. This change began with the sudden increase in the number of new city pharmacopoeias over the course of the seventeenth and eighteenth centuries.

City pharmacopoeias were not without precedent. They sporadically cropped up in various cities throughout early modernity. The first pharmacopoeia was published in Florence in 1498/1499 (Lentacker 2019, 225). This was followed more than half a century later with the publication of a pharmacopoeia in Nuremburg in 1546, after which a number of cities put together pharmacopoeias (Lentacker 2019, 225). Florence published a new edition in 1550, followed by Mantua in 1559, Augsburg in 1564, and Cologne in 1565 (Crawford 2019, 283; Urdang 1946). The objectives behind the publication of these earlier pharmacopoeias varied greatly from location to location. The Cologne and Augsburg pharmacopoeias are at least partially a response to the perceived limitations of the Nuremburg text, and each posed an opportunity for emoluments (Urdang 1946, 46). On the other hand, the impetus for the issuance of the Mantua pharmacopoeia may have been nothing more than the rivalry between Duke Guglielmo Gonzaga of Mantua and Duke Cosimo de' Medici I of Florence, whose pharmacopoeia had already reached its second edition (Urdang 1946, 47). The number of city pharmacopoeias in circulation began to increase in the seventeenth century with Venice, London, Amsterdam, and Brandenburg soon publishing pharmacopoeias of their own (Marriott 2010; Friedrich n.d.).

The emergence of the pharmacopoeia as a discrete artifact was not strictly a response to swelling experimentalist fervor. Some of the earliest pharmacopoeias appear to have been political instruments, extensions of the theopolitical dominion of the state into the chaotic regnum of therapeutics. With respect to their contents, these earlier pharmacopoeias were intended to be little more than "official" prints of the materia medica, telling us a great deal about what subtle changes in the contents of the materia medica were already underway. Though they were conceptually Galenic, these early pharmacopoeias demonstrate an openness not only to iatrochemical remedies but to novel plant remedies from the colonies of the New World, such as "ipecacuanha, guiac, and Peruvian bark" (Maehle 1999, 3). This reflects the successful integration of iatrochemical remedies into the Galenic hegemony. But, above all else, it reveals the fluidity of the materia medica, even if the nature of these additions was not explicitly the by-product of willful theoretical innovation.

Things began to change markedly in the eighteenth century. The number of city and state pharmacopoeias exploded, increasing almost exponentially over the course of the eighteenth century. Cities which were already publishing pharmacopoeias now produced new editions at a greater frequency. Edinburgh produced its first pharmacopoeia in 1699 and waited until 1722 to produce a second version, only to release new editions in 1735, 1744, 1756, 1774, 1783, and 1792 (Redwood 1847). The Austrian Empire produced its first state pharmacopoeia in 1729, but saw six new editions by 1770 (Redwood 1847). Venice produced city pharmacopoeias in 1730, 1781, and 1790. Brandenburg's late seventeenth-century state pharmacopoeia was updated with new versions in 1713, 1731, 1744, and 1781. Württemberg produced its first pharmacopoeia in 1741, noteworthy for its inclusion of various details pertaining to the chemical properties of some of its ingredients (Friedrich n.d., 6; Redwood 1847). Later editions of the Württemberg pharmacopoeia were also printed in 1750, 1754, 1760, 1771, 1786, and 1798 (Friedrich n.d., 6; Redwood 1847).

The sudden interest in developing city pharmacopoeias paints a vivid picture of eighteenth-century pharmacy. It speaks to the existence of a broadly held notion that the available literature on therapeutics wanted revision. Further still, it sheds light on the mindset behind those overseeing the compilation of pharmacopoeias. The rapidity with which new editions were produced suggests that this process of revision was understood as an ongoing process, one whose initial foundations were laid in the prior century. But what were the nature of these revisions? One of the starkest examples can be found in the fourth edition of the *Pharmacopoeia Londinensis*, published in 1721 with a new preface by the President of the Royal College of Physicians, Sir Hans Sloane (1660–1753) (Sloane 1721). Here, Sloane explicitly advocated for the need to put an end to the use of "superstitious" therapeutics and compound remedies (Sloane 1721, 38–39; Earles 1961, 77). Galenic ingredients such as "puppies" and "hedgehogs" were removed from tried-and-true compound remedies, though components containing ingredients such as human fat, skull material, and feces remained (Sloane 1721, 38; Earles 1961, 75). The second edition of the *Pharmacopoeia Edinburgensis* followed suit in 1722, likewise retaining human materials as well as viper fat and a variety of insects as viable ingredients (Anonymous 1722, 20). By the time a new edition of the *Pharmacopoeia Londinensis* was released in 1745, the extent of the changes made to the number of compounds and the lists of ingredients amounted to nothing less than a full-on reformation (Pemberton 1746, 1). The 1745 edition still contained several of the familiar ingredients of medieval Galenism, among them viper flesh and millipede; however, the preface itself contained expressions of doubt concerning the therapeutic value of these components, deferring to common practice in lieu of a better answer (Pemberton 1746).

The crown jewel of eighteenth-century pharmacopoeias, so to speak, may very well be the *Dispensatorium Lippiacum*, or *Lippian Pharmacopoeia*, of 1792–1794, created by Johann Christian Friedrich Scherf (1750–1818). Printed toward the end of the 1700s, it somewhat unsurprisingly embodied the culmination of the patterns of thought developed over the course of the preceding century's unprecedented investment in the continuous production of new and improved pharmacopoeias. It contained notes on the water solubility of chemical ingredients and prescribed rules for the testing of remedies (Scherf 1799; Redwood 1847). Not only did it exemplify the trend toward the simplification of Galenic remedies seen in other pharmacopoeias, but Scherf's pharmacopoeia valued the role of the practical pharmacist as a constant partner in the experimental encounter. The second edition, published in 1799, was also the first pharmacopoeia to use the German language (Scherf 1799).

An interest in reforming the established hegemony of Galenic compound medicine, expressed through the medium of the pharmacopoeia, is indicative of the nature of the eighteenth-century sea change in therapeutics. If the 1600s are to be deemed the period in which the pharmacopoeia comes to the fore as a site upon which medical authority could be gathered and constituted, the 1700s were undoubtedly the period in which the very ground of medical authority was experimentally rebuilt. This was overwhelmingly clear to figures of the period. Famed Scottish physician William Cullen (1710–1790) directly attributed changes in the content of the materia medica to advancements in chemical knowledge (Cullen 1789, 24, 31). The distinguishing factor was the emergence of the experimental apparatus as the crucible through which therapeutic value was independently determined. In this sense, Cullen's observation still rings true, though with an important shift in emphasis: improvements in the materia medica may have affected chemistry, but this is only because chemistry (outside of the iatrochemical sense) had suddenly become relevant to their understanding of therapeutics.

How did the pharmacopoeia—a dozy, practical compendium of medicinals—come to represent the forefront of the empirical study of plant- and animal-based therapeutics? One can imagine several answers. There's the argument from precedence: the pharmacopoeia already existed in a limited form, so avid experimentalists simply expanded on a familiar medium. Or, perhaps, it is because of the pharmacopoeia's constituting function—its capacity as a tool of centralizing claims about the natural world—that it was taken up as with such gusto by those who sought to, in the words of the *Novum Organum*, "establish a new and certain course for the mind from the first actual perceptions of the senses themselves" (Bacon 1844, 2). After all, the pharmacopoeia was a tangible, material embodiment of the systematic unity of empirical knowledge, rendered absolute by the might of regional sovereignty. In this sense, it was a participant in the early modern centralization of knowledge and the rise of the nation state. Nor can one easily overlook the Foucauldian refrain that this method of reasoning was characteristic of the period in which the discursive formation gave way to an impulsive need to dissect, measure, and weigh the world, ultimately so that it could be systematically categorized (Foucault 2005). All are correct in their own right. What can be said is that, with the rise of experimental empiricism and the subsequent investigation into plant- and animal-based medicines, a new assortment of claims gained candidacy for truth or falsehood. Conversely, observations, questions, and methods of inquiry that were once absolutely reasonable—even wise—rapidly became untenable. Nowhere is this more apparent than in the theory of therapeutic action.

Therapeutic Action

Any systematic approach to therapeutics entails a theory of the body, as well as an accompanying theory of therapeutic action. This is as much the case for intoxicants as it is for the most muted tonic. It would be glib to retort that the lotus-eaters of yore wrote about intoxicants in unyielding magicoreligious terms. Our conceptions of therapeutic actions are immanently participatory in how we conceive of our experience of embodiment. Unsurprisingly, then, the reorientation of the ground of therapeutic authority further entailed an explosive effort to reimagine the nature of therapeutic action. Under the previous Galenic model, therapeutic efficacy was determined by a remedy's intrinsic characteristics, supported by the physician's ability to marshal these qualities in their attempts to quell the humoral inequilibrium behind their patient's suffering. An excess of cold, damp phlegm, for example, called for a remedy that was warm and dry, such as coriander, cumin, or holy thistle (Bos 1996, 229). A remedy's curative potential was intrinsic to the real or perceived characteristics of the substances that made up the remedy. But just as many of these Galenic remedies were called into question with the rise of the experimental approach in therapeutics, there was a parallel interest in exploring novel theories of therapeutic action.

The rash of new conceptions of therapeutic action were nearly as varied and multifarious as the fresh prints of the pharmacopoeia that accompanied them. For some, the corpuscular theory of Sir Robert Boyle (1627–1691) provided a possible foundation for explaining the effects of remedies and poisons (Maehle 1999, 4). One such figure was Johann Jakob Wepfer (1620–1695), who cited his study of the ulcers found in the vivisected gastrointestinal tracts of poisoned animals to advance the idea that the cause of the damage was specially shaped, sharp particles found in poisonous substances (Maehle 1999, 4). Alternatively, Friedrich Hoffman (1660–1742) advanced investigations into a neo-iatrochemical theory of therapeutic and poisonous activity at the University of Halle (Maehle 1999, 4). Or let us consider Sir John Floyer (1649–1734), who attempted to establish a theory associating a remedy's outward, sensory qualities with its effects, an idea that gained considerable interest (Floyer 1687). None other than Carl Linnaeus (1707–1778) participated in an attempt to systematize a framework associating therapeutic action and sensory characteristics much later in 1751, a model that reflected the methodological rationale employed in his own classificatory system (Earles 1961, 34). William Cullen spoke in opposition to the value of this idea, primarily on the basis of the apparent subjectivity involved in ascertaining the characteristics of qualities such as taste and smell and the resultant difficulty of classifying substances in this way (Cullen 1789, 26). Cullen did, however, concede to the notion that only those remedies with a pronounced taste and smell were likely active (Cullen 1789, 26). In spite of their differences, each of these models remained paradigmatically aligned with the Galenic conception that therapeutic action is associated with the substance's intrinsic characteristics. Wepfer's theory of toxicity derived from Boyle's corpuscular philosophy was explicitly Galenic, relying as it did on Galenic language (Maehle 1999, 4). Floyer, Cullen, and Linnaeus, meanwhile, were more theoretically inventive, but nonetheless Galenic in their approach.

Most of these newfound approaches gained little, if any, traction. However, some ideas that arose in the midst of these broad debates concerning therapeutic action would be familiar to contemporary readers. The cornerstone of modern pharmacology is the idea that drugs are active principles: discrete, individual substances that engender bodily responses. The active principle concept was likely first proposed by Alexander Monro Secundus (1733–1817) in 1771 via a study of how opium and other "powerful medicines" entered into the body of animals (Monro 1771, 294). Monro endeavored to do so by subjecting his frogs to a battery of tests, in the hopes of ascertaining by what means opium affected the body. Having dosed the frogs with an opium solution through the outer and inner layers of the skin, through muscular nerves, various organs, and finally a pipe directly into the heart, Monro eventually identified two means by which opium affected the body (Monro 1771). The first was through local absorption at the site of the nerve, after which, he theorized, the body responded *sympathetically* (Monro 1771, 325). The second was through absorption into the blood stream, which Monro argued carried opium to nerves surrounding blood vessels throughout the body and in the heart (Monro 1771, 338).

Both suggestions rely on the idea of absorption via the nerves, a concept that Monro was not unique in supporting but which had been subjected to a great deal of contradictory experimental scrutiny (Monro, 1771, 337). Monro himself acknowledged the observation made by others that, in cases where a high dose of opium had killed an animal subject, the opium had usually stayed within the stomach and lost little of its weight (Monro 1771, 339). These experimental findings appeared to deny the possibility of absorption being the cause of the overdose, as the opium remained in the stomach without a significant loss in mass. Monro was quick to dismiss the "erroneous supposition that the stomach is not provided with lacteal vessels" (Monro 1771, 339). It was fully possible, Monro reasoned, for opium to be absorbed via vessels surrounding the stomach. As for the mass of the opium remaining in the stomach, Monro made a tremendous passing remark: "it seems very probable that its active part makes but a small share of its bulk" (Monro 1771, 339). However brief the statement, there is little space to argue that Monro is suggesting anything other than the idea that opium consists of a combination of inert material and an *active principle*. This argument flies in the face of the entirety of Galenic medical knowledge and problematizes nearly all of the classificatory frameworks that formed the bases of the new pharmacopoeias. It particularly undermined the notion that a substance's medicinal effects corresponded with some outward or intrinsic characteristic of the substance itself, and therefore should be classified as such. Yet Monro made little effort to push this point further. It is possible that an awareness of the uncertainty of his claim, as well as its controversial nature, dissuaded Monro from leaving it as anything more than a comment. But the suggestion was made.

These vignettes do a great deal to illustrate the extent to which the latter half of the eighteenth century bore witnessed to tremendous upheaval in the long-standing classificatory and conceptual language of therapeutics. The other side of the coin, as will soon become clear, was an interrelated, and yet distinct, effort to realize a systematic basis for medicine, one that would reconcile anatomy with therapeutic practice. A formative force here, and one that has thus far gone undiscussed, was the heretofore unprecedented eruption in the availability of all manner of intoxicants. Colonial expansion had discovered, invented, and made accessible intoxicants in a radical new way, at the heart of which was the emergence of an economy surrounding the international production and sale of intoxicants for the purposes of mass consumption.

The European enterprise of exploration and colonization saw the appearance of a vast assortment of hitherto unknown intoxicants on the European continent. Coca, tobacco, tea, and chocolate were suddenly the curiosities of an age. Equally as crucial was the flood of familiar intoxicants—alcohol and opium—now available in never-before-seen preparations and in unprecedented quantities.

Let us begin with the greatest of all European narcotics, alcohol. The world's oldest spirit, rum, originated in seventeenth century Barbados (Foss 2012). What began around 1650 as a means of producing alcohol from molasses promptly formed the basis of an industry that spanned the Atlantic basin, with rum distilleries cropping up not merely in the American colonies but in the British homeland (Foss 2012). By 1750, Massachusetts alone supported 63 rum distilleries, producing approximately 700,000 gallons of rum per year (Ostrander, 1956, 83). This was followed closely by Rhode Island, reaching 500,000 gallons annually by the 1770s (Ostrander 1956, 83). As the prevalence of molasses meant there was no shortage of smaller, local distilleries, much of this New England rum was prepared for export, where it went for half the price of West Indian rum on the market (Ostrander 1956, 83). This tremendous flushing of rum into the market was ultimately what saw the Royal Navy transition from brandy to rum, where tots of rum became part of a sailor's daily ration (Tannahill

1973, 273). Rum was never the be-all and end-all of alcohol spirits. But this was the period in which the perennial companions of wine, beer, and other fermented drinks gave way to an ever-increasing variety of spiritous formulations. It would not be long before gin and brandy would be counted among the most popular forms of alcoholic drink available in European cities (Tlusty 1998).

Rum was not the only intoxicant to wind its way across the pegboard of international trade. The opium trade was a similar exercise in imperial commerce, with perhaps even further reaching implications. Before Britain entered into the opium trade, it was already a prized commodity in the portfolios of Dutch and Portuguese traders, who shipped opium from India to the ports of Canton, Java, and Macau, among others (Bailey and Truong 2001, 174). However, such shipments were naturally limited by the relatively minor scale of poppy cultivation, despite consistent demand for the product. The East India Company overcame this limitation, and ultimately gained dominance, by gaining control not only of the sale of opium but of its production, which they accomplished by seizing hold of poppy cultivation centers in Bengal (Bailey and Truong 2001, 174). The year 1773 was the start of the Bengal Monopoly, and by 1797 the EIC outlawed the cultivation of opium poppy for those who did not hold their license (Bailey and Truong 2001, 174). Chinese cultivation of opium did not begin until the 1860s and only became competitive with Indian opium by the 1870–1880s (Bailey and Truong 2001, 174). From the last quarter of the eighteenth to the first half of the nineteenth century, the British had all but total control over the opium trade, and such dominance saw opium elevated to one of the top commodities in the empire.

From its start in 1668, the volume of tea imports grew almost exponentially in the eighteenth century (Derks 2012, 51). At some point, a shrewd outfit of traders put together that Indian opium was one of the few commodities that could be sold to Chinese buyers, thus beginning the triangle trade of Indian opium to China, Chinese tea to England, and British manufactures to India (Derks 2012, 51). The opium trade rapidly showed itself to be far more than a means of covering a trade deficit. The EIC's net profit from trade with China grew from approximately 234,000 pounds in mid-1770 to just shy of 1,000,000 by the year 1800 (Derks 2012, 52). This still only reflected imports of around 200 tons of opium into China. By 1835, Chinese imports of opium swelled to approximately 1,390 tons, seven times the 1800 number, and this would continue to grow almost exponentially across the nineteenth century, eventually covering over a third of Britain's visible global trade deficit (Deming 2011, 11). This made opium the most valuable single-commodity market in the world.

Nor was this "just" the matter of a transformative shift in the volume of known, medically significant intoxicants—though that would have been enough. Colonial commerce also brought in now popular consumables then unseen by Europeans. In addition to opium and spirits, cannabis, tobacco, chocolate, and coca leaves had trickled in from outposts in the New World and the Far East since the earliest vestiges of a lasting colonial enterprise. By the 1670s, Robert Hooke was sporting about London while experimenting with laudanum, chocolate, tea, cannabis, and tobacco, nearly all of which would not have been available in Europe sixty years earlier (James and Withington 2022, 2). But what had been a relatively minor trade in exotic consumables had, by the second half of the eighteenth and into the beginning of the nineteenth centuries, become titanic, socially formative, industries. Of all of the commodities pilfered, proffered, and peddled out of colonial properties, arguably none were as profitable or socially formative as the colonial trades in Chinese tea, sugarcane for alcohol, and Bengali poppies for opium. These were the true spoils of empire. Nor was the influence of these intoxicants merely economic or social. The availability of intoxicants on a mass scale profoundly changed the discussion surrounding medicinal therapeutics. While pharmacy had rapidly developed into one of the most conceptually sophisticated and empirically rigorous aspects of medicine, the connection between physiology, therapeutics, and the practical art of medicine remained murky and fractured. In the theaters of coin and economy, colonial trade had rapidly deepened these ambiguities, through both the introduction of new intoxicating substances and the increased access to those already known.

It becomes clear that the period from the latter seventeenth to the early nineteenth centuries was strongly characterized by a shift in the social and conceptual relationship between substances of intoxication and the people of Western Europe (James and Withington 2022). Intoxicants such as opium, spirits, and tea had risen to count among the most desired and profitable commodities on the planet, more or less freely passing between southern and eastern Asia, Europe, and the Americas, where they spurred human subjects to action. From the genesis of England's tea obsession or the growth of tobacco culture to the emergence of East Asian opium society, many of these influences were far from subtle. They took up space in the world. Not just in warehouses and shipping yards—different kinds of businesses and locales entered into cultural consciousness, destinations where these substances could be consumed in the company of others. They altered the average everydayness of common people around the globe.

Why does this matter for the purposes of the narrative at hand? The creeping scope and efficiency of imperial commerce had made new kinds of therapeutics available and in heretofore unseen quantities. It is not a quirky fluke of history that changes in both the availability and variety of intoxicants coincided with conceptual shifts in therapeutics, or that experimentalists like Robert Hooke were passionate adopters of these new substances. The force with which intoxicants flooded the colonial world only further drove these lines of thought to the foreground of inquiry, as different therapeutics gradually passed through the experimental crucible.

Pharmacy had seen tremendous development and theoretical elaboration in an effort to incorporate both the empiricist priorities of the age and the deluge of novel therapeutic materials. Yet efforts to systematize medicinal therapeutics and make use of this newfound wealth of plant remedies were rapidly making apparent medicine's desperate lack of empirical systematicity. The murky and fractured connection between physiology, therapeutics, and the practical art of medicine perhaps only deepened as the absence of systematicity become increasingly problematic. As colonial trade changed the therapeutic landscape, its products raised questions: What of their application? Or, more pressingly, how did these changes affect theories of therapeutic action, bodily function, or etiology? The questions posed here bordered on nothing less than an inquiry into the very foundations of medical validity—the makings of a true crisis in medicine. For many German physicians and philosophers, the answer to many of these questions came in the form of a Scottish physician named John Brown (1735–1788) (Tsouyopoulos 1990, 107). It is there, in the theories of Brown, that the content of the materia medica was conceptualized in a new way, one in which the spoils of empire—among them opium, spirits, and other intoxicants—were placed at the center of a system that finally unified therapeutics and pathology.

John Brown's Body

Dr. John Brown was sick. A youth of "generous living" had been cut short by a bad episode of gout at the ripe old age of thirty-six (Brown 1795, xvi). Much to his despair, Brown's efforts to secure a lasting reprieve from his condition reaped little in terms of results. Medicine, he felt, was failing him. Finally, deliverance came neither by way of some new therapeutic nor through the efforts of a gifted physician, but in the form of opium. Brown discovered in "opium and some other stimuli, the secret of repelling the fits of the gout as often as they returned, and, at the same time, of reestablishing the healthy state, a secret that has hitherto been so much wanted and despaired of" (Brown 1795, xxiii). Opium, spirits, and other "stimulants" seemed to have unique curative powers, then unrecognized.

Powerful *vital substances* seemed to be the secret to treating all manner of afflictions. With the publication of *Elementa Medicinae* in 1780, John Brown made his new, totalizing system of medicine, derived from these extraordinary insights, available to the European public. Brown argued that the energy that animated organisms was created through the internal and external stimulation of excitability, the basic vital force (Bynum and Porter 1988). This made life, as the historian Guenter Risse put it, "the product of constant reactions between stimuli consuming specific amounts of excitability," an unending balancing act between external inputs and bodily function to maintain a stable amount of excitability (Risse 2003, 165). Any inequilibrium in this vital force, either an excess or shortage of excitability, would result in a pathological state on the physiological level (Tsouyopoulos 1990, 107).

In terms of the theoretical structure of his framework, Brown's system was quite unremarkable insofar as it mirrored other vitalistic medical theories circulating at the time. Brown's system drew heavily on the medical system of his teacher William Cullen and so was more essentially derived from Albrecht von Haller's theory of irritability (Dyde 2015). The Hallerian model shared in the foregoing Galenic-iatrochemical theories in its emphasis on the qualitative disequilibrium of disparate bodily elements, yet distinguished itself through its anatomical bases in the distinction between irritability and sensibility (Steinke 2005). In this way, the tendency is to represent the Cullenian system as built around the nervous system (Barfoot 1993). As Cullen was a leading representative of medicine in the Scottish Enlightenment, there is a presumption of systematicity attached to Cullenian thought, in a clear reflection of the system-building impulse often attributed to Enlightenment mechanists. Michael Barfoot, among others, suggests that the presumptive systematicity of Cullen's medical philosophy may largely be an invention of John Thomson, Cullen's personal editor and biographer (Barfoot 1993). Cullen intended to impart students with a heuristic, rather than merely pass along a received system of medical knowledge. Whatever the actual systematicity of his thought, Cullen nevertheless followed Haller's work in defining all disease as the excess or lack of sensibility, the modulation of which was the objective of therapeutics (Rocca 2007, 94). Brown's system undoubtedly derived the notion of a medical system that linked disease to vital inequilibrium from Cullen, although Brown's points of departure from his former mentor would prove to be the most determinative of his success.

Brown's system engendered several possibilities that were seemingly unavailable to Cullen. As Brown understood it, a logical entailment of the reduction of etiology to a discrepancy in the quantity of a given force meant that illness itself became quantifiable. This leap was possible precisely because of Brown's focus on stimulating therapeutics: it is difficult to quantify vital principles in vivo, but far simpler to quantify external inputs. This allowed Brown's system to aspire to the Newtonian value of mathematical quantifiability while also appealing to therapeutics as the most robustly empirical aspect of practical medicine. Arguably, it is because of these factors that Brown's theory would enjoy a degree of popularity, if short-lived, which vastly outstripped its competitors. His framework was especially impactful in two ways: (1) it wedded pathology to physiology and (2) it shifted the emphasis from qualitative observations to quantitative measurement (Tsouyopoulos 1982, 14).

Brown's vitalistic theory of excitability also provided a simple explanation for the difference between animate and inanimate objects using a single principle, easily applied to descriptions of physiology and pathology (Risse 2003, 165). With its central emphasis on quantifiability, Brown's *Elementa Medicinae* (1780) "represented the culmination of all eighteenth-century efforts at medical system building" (Risse 2003, 166). One is hard-pressed to find remnants of excitability theory in the language of modern medicine, but it is in Brown's attempt to make excitability mathematically quantifiable and, in this way, help medicine become an exact science that we can see his lasting impact (Risse 2003, 166). That excitability could not be concretely defined was not a problem: excitability was no more or less describable than Newtonian gravity. Excitability, like gravity, was an occult force. One lorded over the cosmos, the other over life. Although fundamentally simple, Brunonian theory becomes deeply intricate when used to explain specific phenomena (Tsouyopoulos 1990, 108). Many of these difficulties can be traced to what is easily one of the most rhetorically attractive components of the Brunonian system, which is that excitability was a measurable force. Brunonianism seemed, in some ways, an ideal foundation for a Newtonian revolution in medicine. The obvious problem was the difficulty of measuring this vital force in the body.

Brown's solution was to shift focus away from the body itself and instead track those things entering the body, a perspective that would also shape the nature of Brown's therapeutic approach. Brown's system for grading levels of excitability consisted of a vertical chart with 80 degrees, with a baseline set at 40 degrees (Brown 1795, 27).

The baseline of 40 degrees represented the ideal state of health, while deviation either above or below 40 degrees was a quantification of disease (Brown 1795, 27). "Exciting power," the energy needed for all the activities of life, caused an equivalent deduction in excitability that was then recouped through stimulus (Brown 1795, 27). Given that existence itself puts

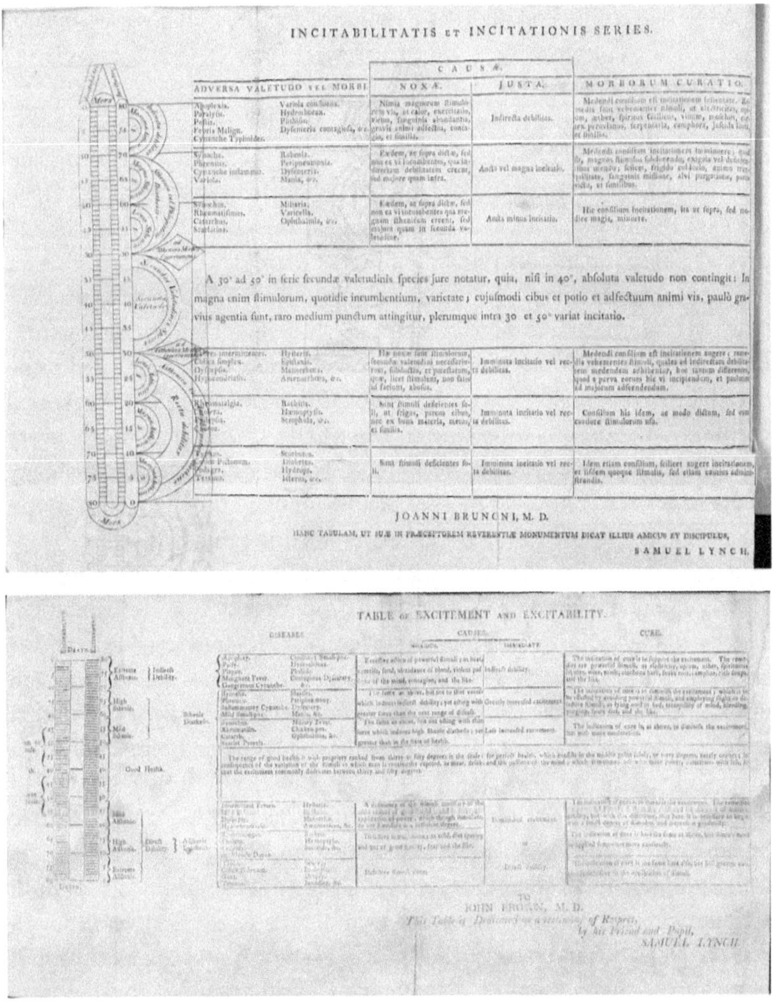

FIGURE 1. Mathematical representations of the Brunonian scale of excitability, prepared by "his friend and pupil" Samuel Lynch as found in the 1790/1791 (*top*) and 1804 editions (*bottom*) of Brown's *Elements of Medicine*. Note the very limited changes in both presentations and the use of vital stimulants at both extremes of the scale. Courtesy of the United States National Library of Medicine.

constant strain on the vital force, excitability could only be liminally observed, being as it was in a state of unending flux. This dialectical opacity of Brunonian excitability meant that it could only be observed indirectly. The task of a Brunonian diagnostician was to rigorously track a patient's activities before and after falling ill, as well as any possible stimuli. Once the relevant factors had been dutifully recorded, the level of excitability could be determined relative to a baseline of 40. Thus, all diseases, regardless of their otherwise perceived causes, were reduced to a singular issue, that of an imbalance in excitability.

This made therapeutics extraordinarily simple: "every disease, that debilitating powers remove, is sthenic, every one, that is cured by stimulant means, asthenic" (Brown 1795, 92). A patient who was "sthenic" (having excitability levels exceeding 40 degrees) merely required a swift and precise reduction in excitability. "Asthenic" patients (having less than 40 degrees of excitability) required stimulant therapy. Treatment for sthenics effectively amounted to deprivation of the body. Cooling, abstaining from meat, eating unseasoned foods, and bloodletting were all alleged to be curative in the sthenic. However, Brown himself considered such cases exceedingly uncommon, reporting an average of 3 sthenic patients for every 97 asthenic patients (Brown 1795, 96). This meant that the vast majority of diseases could only ever be cured with stimulant therapy, calling upon the physician to marshal vital substances in curing the diseased.

Although a healthy level of excitability could be maintained by a stable environment, appropriate exercise, and proper diet, once individuals became asthenic more drastic remedies would be required to save their lives and, for John Brown, there were no stronger medicines than liquor and opium. "Opium," Brown wrote, "is the most powerful of all the agents that support life, and that restore health, and a truly blessed remedy," which "cures any of the diseases depending upon debility" (Brown 1795, 237).[1] Strong spirits were less powerful than opium in their curative potential, but they were still regarded as essential stimulants (Brown 1795, 104). Musk and powerful alkalis, examples of the enduring influence of both Galenic and iatrochemical remedies, were also listed as effective, although it is unclear how much stock Brown actually placed in their curative potential given the amount of attention he allocates to alcohol and opium (Brown 1795, 104). Prescriptions of opium and spirituous beverages were so frequently reported that they came to be synonymous with Brunonian therapeutics. These recommendations on Brown's part were certainly influenced by his own heavy use of opium and spirits in treating his severe gout (Maehle 1995, 52).

That Brown structured his etiological model first in terms of how diseases were cured was not accidental. Nor was it incidental that Brown relied as heavily as he did on powerful intoxicants. After all, the world around him was awash with intoxicants. If William Hogarth's satirical prints are to count for anything, the streets were slick with cheap gin. Imperial commerce ensured that opium and spirits were accessible like never before. Brown was even proximal to many of the ongoing discussions concerning how remedies were to be classified and how they functioned, close as he was to William Cullen. At least in part, it was this timeliness that distinguished Brown's model from many of its competitors. The Brunonian system was open to its position within a greater commercial, cultural, and material zeitgeist that had seen intoxicants become a lively participant in daily life. Further still, it was sensitive to the consequences of these events for any efforts to develop a systematic understanding of pathology, etiology, therapeutic effect, and medical practice. Brown was able to seize upon the foregrounded role of familiar intoxicants and interpolate these elements within established medical models, systematically unifying etiology and physiology while further making therapeutics highly accessible. But the Brunonian system is not merely a totalizing theory of medicine and the body guided by a scientific interest in therapeutic remedies; it is a claim on the *potentia vitalis* of intoxicating substances.

Brown conceived of any given body as the product of a dialectic between the vital force and the world. In this sense, Brown did not treat intoxicants as the singular source of vital stimuli, any more so than roast beef or black pepper. Nevertheless, intoxicants occupied a privileged place. With opium heralded as the vital stimulus par excellence and spirits trailing closely behind, intoxicants are inseparable from Brown's conception of vitalism. They were his powerful therapeutics, and yet they were also his frontline therapeutics (Brown 1795). It becomes tempting to suggest that, in the Brunonian framework, intoxicants themselves become something akin to a physical manifestation of the vital force. It is this conceptual association between intoxicants and the vital force which I have tentatively termed "vital substances."

Where did Brown's theory come from? Why was it capable of making all manner of innovations, from the curious to the critical, where Cullen and Haller had failed? Whereas Haller's and Cullen's medical systems were grounded in the experimental approach of eighteenth-century anatomy, Brown attached his system to a unique claim on its validity: his own conversion story—the tale of a man plucked from death's maw by nothing other than the use of opium and other stimulants. This is made explicit in the preface to Brown's original textbook, in a realization that likened to "the first break of day" (Brown 1795, 1:xvi):

> At first, for the purpose of removing fits of the gout, he went no farther than the use of wine, and other strong liquors, with nourishing food, that is, seasoned meat, and kept the more powerful remedies in reserve. But, for many years past, his surprising success in the use of the latter, has enabled him to find in opium and some other stimuli, the secret of

repelling the fits of the gout as often as they returned, and, at the same time, of reestablishing the healthy state, a secret that has hitherto been so much wanted and despaired of. (Brown 1795, 1:xxiii)[2]

At first glance, this is a familiar song. In equal measure, it recalls bygone quakeries and proffers a tale of shrewd empiricism. Upon further investigation, however, the narrative scarcely follows the Semmelweisian formula. Brown is not merely reporting on the curative effects of opium and spirits. Something different is going on here.

In Brown's recounting of events, the consumption of intoxicants becomes an avenue for a tacit awareness of embodiment. Understanding this requires the acknowledgment that Brown is implying a great deal more than merely that opium and other intoxicants abated the symptoms of gout. Brown is tracing his entire system—from therapeutic effect to his conception of debility—to the *experience* of intoxication. The evidence for vital substances is implicit in the experience of intoxicated embodiment. This is made particularly overt in his inference that the feelings of intoxication are the discernible effects of an extraordinary stimulation of the vital force (Brown 1795). Successful treatment of his illness is simply a demonstration of these principles. The experience of intoxicating stimulation imparted information about what it means to be a body, called attention to the perceived nature of embodiment.

Knowledge, here, is not an association of facts and information, but an experience itself. It is something viscerally felt. In varying degrees, this is always the case. The recollection of knowledge is a temporal experience, one ineliminable from emotions and other perceptions. This is as much about how knowledge makes us feel or its place in memory as it is about clearing the insurmountable epistemic hurdle of discerning thinking from feeling—if they are even separable. Then there is the matter of how recalled knowledge is personally signified. What is belief if not the *experience* of knowledge as truthful? If you believe yourself free, it is because you experience freedom, no matter what we are free to or from.

How does intoxication fit into this? It is understandably popular to conceive of intoxication as a filter. Some temporary lens, a transparency over the senses through which sensations and stimuli are processed and then forgotten. Or maybe like a dream, an irrational space that pours out all manner of coarse novelty and jagged inventiveness. But the example of John Brown suggests something different. Returning to knowledge as an experience, intoxication appears to be unique because it enhances the magnitude of the experience of knowledge, while further making explicit the experience of cognition's entanglement with the perceptual lifeworld. It engenders a tacit awareness of embodiment that has the potential to reorient or reorganize existing frameworks, bringing latent conceptual associations into the foreground of perceptual experience. In this sense, intoxicants are ways of knowing—a distinct manner of making real possible candidates for truth or falsehood in the perceptional experience of embodiment. Possibilities drawn from the space of potential ideas are reified, externalized in the very experience of intoxication.

I have come to refer to examples such as these as intoxicated ways of knowing. Not in the sense of John Pickstone's "ways of knowing," or, for that matter, of Ian Hacking's "styles of reasoning" (Pickstone 2000; Hacking 2002). Examples of intoxicated ways of knowing occur within the perspective horizon of what Hacking means by "style" or what Pickstone means with "way of knowing," yet the term itself designates something that transcends any given "style" or "way of knowing." To this point, intoxication as a way of knowing encompasses a plurality of different "ways" of encountering and knowing the world. Each particular instance is historically circumscribed, specific to its unique context, and yet the occasions for its possibility span human existence. It is a specific kind of comportment, made possible only in concert with discrete material actors—substances of intoxication—when they gather with human subjects in specific settings and contexts. Intoxicated ways of knowing might instead be described as an epistemic mode, though linguists have already seen the over-lofty term "epistemic modality" pressed into service.

In Brown's case, intoxication with opium and spirits revealed these substances to be powerful stimulants of the vital force, capable of righting the bodily inequilibriums that caused disease. Many of the foundational elements were derived from the work of those around him. The experience of intoxication did not engender the development of these background theoretical elements. These structural concepts were available in the environment. Preexisting conceptual frameworks, drawn from Brown's medical education, were made real in association with the use of intoxicants via the experience of intoxication itself. The potential for this event was only made possible by a radically local gathering of concepts, associations, intoxicating substances, and—of course—Brown in a specific place and time. The vital substance concept was the distillate of a particular coalescence of entities, actants, and words in their respective sites, much as the finest single malts discern themselves by the unique mineral signature of some private Highland spring. Intoxicating substances, you could say, became vital substances because Brown was thinking with the body.

As for the man himself, Brown died in a London debtor's prison a trifling eight years after his magnum opus was first published. It appears he initially failed to gain a following in his native Britain. The lingua franca of the "Brunonian moment" would actually be German, the new system soon becoming meteorically popular on the German-speaking continent. By 1795, Brown's system was already known throughout the Germanies, Austria, and Switzerland, crossing borders and engendering discussion in every sphere of educated society. Some of Europe's foremost philosophers, physicians, and experimentalists could soon be counted among its disciples. Far from some zany medical sectarian enthusiasm, the Brunonian system and, through it, the vital substance concept would prove an exacting influence on the course of nineteenth-century medicine and physiology. But these sorts of uncomfortably lofty claims demand context, a context that, like many aspects of the nineteenth-century German-speaking world, begins by taking a look at the role of Immanuel Kant and his role in the German perception of a crisis in medicine.

* 2 *
# Brown, Kant, and the Crisis in Medicine

Among German physicians, physiologists, and natural philosophers, it was already a commonly held notion that medicine was in a state of crisis. As in other parts of Europe, there was a desire to see medicine realized as a science, reconciling medical practice with empiricism and physiology. Yet there was an additional caveat, one which, arguably, uniquely handicapped German aspirations in this endeavor, a barrier looming over any designs on a scientific medicine in the Germanies: the undeniable stature of Kantian thought. Comparing Robespierre (1758–1794) and Kant, Heinrich Heine (1797–1856) once remarked that the men "represented, to the utmost, the type of the provincial bourgeois; nature had destined them to weigh coffee and sugar, but fate determined that they should weigh other things and placed on the scales of the one a king, on the scales of the other a god" (Heine 1834/87, 86). There is little space to doubt who Heine saw as having wrestled with a king, and who with a god. Such was the societal force of Kantian thought, with its exclusionary definition of a proper science, for German medical reformers.

Immanuel Kant's critique of rational psychology had proven tremendously influential in nearly every corner of the German world. In the wake of Prussia's defeat at the hands of Napoleon, Kant's political thought weighed heavily on the minds of Reformers Freiherr Karl vom Stein and Karl August von Hardenberg, weaving Kant's legacy into the very fabric of the social and constitutional order (Levinger 1998, 243). For those whose concerns were primarily of the philosophical sort, Kant's critical works would become defining, foundational texts, against whose arguments new thinkers would inevitably be weighed. So, too, was it for those studying the natural world.

Kant's attempts to place apodictic certainty at the heart of a shared understanding of the self and the world was implicitly significant for scientific practice (Hacking 2002, 3, 24; Aldea and Allen 2016, 1–2). Others were more overt, such as his definition of an authentic science. What then were the principal concerns of those would-be practitioners of scientific medicine? There is no denying that Kant's definition of scientificity was a millstone round the necks of medical reformers. And yet the foremost considerations for German medical theoreticians were undoubtedly theoretical, namely: What is the relationship between the mind and the body and, correspondingly, what did one's definitions of the mind and body mean for any attempt to integrate anatomy into pathology? To all these questions and more, Kant's critical system presented explicit answers. Let us briefly consider how Kant framed the problem before discussing why John Brown provided such an appealing answer.

Kant understood his own work as an extension of a broader Baconian project, invoking Francis Bacon no later than the opening preface to the second edition of the *Kritik der reinen Vernunft* (Kant 1787, B-xii) There are discernible boundaries cordoning off those things that are certainly knowable from those that are ultimately unknowable through the powers of perception and reason. The work of philosophy and science is to define those limits and function within them. Keeping with this basic premise, Kant would develop his own comprehensive, systematic account of the categories necessary for the existence of consciousness, which would form the foundation of his epistemology of science. The cornerstone of this foundation is Kant's *Kritik der reinen Vernunft* (1781), the impact of which reached far beyond its immediate function as a refutation of Descartes and his disciples.

Kant reconfigures the question concerning the Cartesian relationship between the mind and body into an analysis of the extent to which representations can be equated with necessary truths. The importance of this issue for any systematic study of the natural world is so apparent that it is hardly worth stating. How are we to be sure of anything we know about the physical world if we cannot differentiate between representations and necessary truths at the mind-body level? Kant is famously sparse in his direct discussion of the nature of the relationship between the mind and the body. This is to the credit of his critical approach, which assumes the notion that human beings are unable to speak meaningfully about what goes on beyond the realm of human experience. Turning against his rationalist forbears and contemporaries, he recognized that the true objective of rational psychology was to resolve questions concerning the possibility and nature of a gemeinschaft (community or communion) of the human soul with the organic body and that this doctrine posed very real epistemic problems (Powell 1988, 404).

Kant's position on the relationship between the mind and the body is inseparable from his critique of Cartesian dualism. From the outset, the crux of Kant's critical engagement with Descartes rests on his assessment of the viability of Descartes's argument from doubt. At its most basic, Descartes's argument suggests that, while the mind can conceive of not having a body, the mind cannot doubt its own reality as that which is thinking (Descartes 1993, 14). It is on these grounds that he proposes that there is an essential difference between the mind and the body. This deceptively simple argument draws its strength from a position of radical doubt, challenging the reader to set aside all preconceived assertions in the search for the truth. Truth can only be discerned by parting with any and all a priori assumptions that might shape, and thereby predetermine, the outcome of a course of reasoning (Descartes 1993, 13). In this way, Descartes imagined that lasting truths could be developed from the ground up, providing the basis for more complex scientific questioning.

Although he presents a forceful, even revolutionary, idea, Kant is quick to recognize that Descartes's purported argument from doubt relies fundamentally on a confusion of apparent representations with necessary truths. It is apparent to Kant that the notion of thought divorced from external stimulus is impossible, as far as can be proven. For Descartes's argument from doubt to hold water, Descartes must be able to demonstrate that the relationship between thinking and sensing is subject to doubt. Kant undermined this assertion by establishing a connection between thinking and sensing as a foundational principle. Critiquing the tendency to isolate the mind from the body in his "Refutation of Idealism," Kant argues that the apparent temporality of cognitive existence ceases to be explicable if the contents of mental life want for an abiding background with which to temporally order their succession (Winfield 2011, 228). All mental contents are necessarily experienced in a temporal succession. Thought void of external context, if such a thing were conceivable, would be insufficient to provide the frame of reference necessary to phenomenologically temporalize a series of ideas. Rather, the distinction between yesterday, today, and tomorrow is only possible against a non-mental backdrop, something that appears to be external to thought itself (Winfield 2011, 228). This argument alone is sufficient to affirm that many of Descartes's central arguments require further investigation, but it fails to introduce any major claims that are not included in Descartes's meditations. In fact, it would appear the primary thrust of Kant's critique of Cartesian dualism actually focuses on what Kant perceives as a conflation of differences in representation with differences in essential constitution.

Kant readily concedes that "I differentiate my own existence, as a thinking being, from other things outside of me (to which my body also applies)" (Kant 1781, B-409). One experiences themselves as a thinking subject and identifies first and foremost with that subjectivity, over and against the external world. This can also be extended, at least to a certain extent, to one's own body. The obvious temptation of this idea is difficult to deny. Even today, our world is replete with mantras championing mind over matter. We have a tendency to generally identify with our bodily selves, but only insofar as it houses the mind. For Kant, the problem is that this perception is treated as a necessary truth rather than a representation. The mind operates in such a way that it *appears* to be distinct from the physical world. Yet, as La Mettrie reminds us, whatever makes up the stuff of mental life, it is terribly quick to succumb to the effects of hunger and thirst, or, for that matter, a cool, crisp glass of Franconian Riesling. The difficulty in addressing this line of questioning forms the bedrock of Kant's entire methodological approach.

Only that which can be experienced can be spoken about meaningfully. Thus, speculative theories addressing "the pre-established harmony or the supernatural assistance" that makes the mind-body relationship possible are set aside, insofar as they cannot be confirmed through experience (Kant 1781, A-390). Since the self cannot take itself as an object of intellection, the only way left to discuss the relationship between the mind and the body meaningfully is in terms of their respective forms of representation (Kant 1781, A-386). Kant argues that, before we can begin to entertain the Cartesian argument for a distinction between mind and body, we must understand how we experience the mind and body as representations.

The body, or "matter, whose connection with the soul rouses such great concern, is nothing more than its mere form, or a certain style of representation of an unknown object, through intuition that one calls outer sense" (Kant 1781, A-385). Our bodies are deeply personal to us, the ultimate manifestation of our ownmost individuality, personhood, and identity. We are present to ourselves and others in the form of a body. But, as Kant suggests, every interaction we will ever have with our body is strictly a complex representation. The body takes on qualities that render it distinct from the mind because the body can only be represented by way of the outer senses. As a result, the subject finds the mind and the body represented so differently that there is an apparent lack of coherence in understanding their relationship. As long as both mind and body remain mere representations, they cannot be thought of as distinct entities with any certainty. In this way, Kant dismisses the mind-body "problem" as scarcely more than a poorly posed question, acknowledging the appearance of a distinction between mind and body, while affirming that the appearance of a distinction is not evidence of true heterogeneity.

Kant's model has been referred to as empirical psychology, in contrast to Descartes's rational psychology (Sturm 2009, 183–185). This transition is to be understood as far more than a passage from rationalism to empiricism. In Kant, the mind-body problem undergoes a radical transformation, one that is not a reduction of mind to body. Rather, Kant keeps both the mind and body at a distance by retreating into the factic character of the psychological self. The result is the assumption of an agnostic position, one which renders the mind-body problem irrelevant by treating both sides of the conflict within the bounds of cognitive representation.

Kant, Scientific Epistemology, and "Proper Science"

It is precisely this tension that forms the foundation of Kant's scientific epistemology, the likes of which conferred a distinct boundary upon claims to scientific authority. Recalling Kant's invocation of a Baconian spirit at the heart of his critical epistemology, the *Kritik der reinen Vernunft* is intended to represent an experimental analysis of pure reason (Kant 1787, B-xviii). If there were any questions about whether this was merely a rhetorical flourish on Kant's part, Kant writes in a letter to Ludwig Borowski, dated March 1790, that "the only path for scientific research" is to engage in "experimentation and observation, which makes the properties of the object knowable to the external senses" (Kant 1790a). For Kant, the experimental apparatus is a privileged site wherein the unseen becomes sensible and thus knowable. As the study of the natural world cannot rely on synthetic a priori judgments to establish any form of certainty, experiments make knowledge possible by making phenomena available to the senses and understanding.

This does not mean that Kant is unreflectively Baconian. Alberto Vanzo and Larry Laudan have attempted to define Kant's epistemology of science relative to the antihypothetical doctrine of Bacon-Hooke-Boyle. Laudan makes a strong case for the argument that Kant upholds the antihypothetical position of Bacon-Hooke-Boyle (Laudan 1981, 10). Kant does, after all, invoke Bacon directly, asserting that experiment alone can make possible knowledge of the natural world by making that knowledge available to the external senses. With such principled inductivism in hand, one could be indulged in thinking that Kant approached the speculative foundations of hypotheses with the same attitude reserved for Descartes's confusion of representations with necessary truths.

The issue with this reading is that Kant's empirical psychology establishes that a robust cognitive and conceptual framework lies at the foundation of all perceptions. Hypotheses, then, are not merely speculative. From the outset, the validity of a hypothesis is at least partially established by its theoretical conformity with our a priori understanding of the world, which in turn guides an experiment's design. It would seem that, for Kant, experiments themselves are extensions of hypotheses, a position which Kant later assumes for himself in *Weiner Logik*. Vanzo is critical of Laudan's position, instead seeking to present Kant's understanding of the relationship between hypothesis and experiment as a subtle interplay in line with his critical philosophy (Vanzo 2012, 76–77).

Whatever the relationship between Kant's philosophy of experiment and critical thought, Kant provides no definition of an experiment, nor does he provide any sophisticated analysis of how experiments function, beyond a common perception that experiments *intervene* in nature (Vanzo 2012, 76–77). Whether this is a result of Kant's own naïveté or a lack of direct experience, it is clear that Kant is primarily interested in situating experimentation within a broader system of knowledge. Experiments are valuable to Kant because, with a tickling of contrivance, they make otherwise obscure natural phenomena available to the senses.

This is not to say Kant advocated for simple experimentalism. On the contrary, the definition of a proper science that Kant provided in *Metaphysische Anfangsgründe der Naturwissenschaft* (1786) excludes many rigorously experimental sciences from consideration. At its most basic, Kant defined "actual science" (eigentliche Wissenschaft) as a systematic body of ideas, characterized by the relationships of grounds and consequences, apodictically certain, and allowing for the use of mathematics (Kant 1787, IV–V). Experimentation then is conceivably scientific when its underlying premises are rooted in a body of cognition that conforms with Kant's four stringencies. However banal this definition first appears, fraught as it is with concerns surrounding calculability and systematicity, this definition would have a tremendous impact on the course of German science. Kant himself recognizes that this meant subjects such as chemistry would not be considered sciences at all; chemistry might instead be called a "systematic art" (systematische Kunst), as its laws were only empirically substantiated. By the same measure, the classification of natural kinds, geography, natural history, and psychology could aspire to little more than systematic arts (Kant 1787, VIII). Medicine seems to be unworthy of mention, so apparent are its artistic foundations. For those from whom the aegis of scientificity was suddenly retracted, Kant's definition of an *eigentliche Wissenschaft* was a call to action. Chemistry *could* become a proper science, if it were able to establish a basis beyond its empirical foundations. For physicians and medical theoreticians, the possibility of a properly scientific medicine had concievably become more remote than ever before.

I hope that this overlong exposition of Kantian thought can be forgiven. We will have to continuously struggle against the current in keeping intoxicants at the center of this narrative. Nevertheless, it is important to properly establish the very real context in which some concepts became possible and others were relegated to the annals of history. To this end, a rough image of the nature and scope of the hurdle that Kantian thought posed to medical reform should be clear. Only now can we begin to understand how the medical system of a fellow like John Brown, with its bacchanalian reliance on vital substances, came to command the fervor that it would among German physicians, theoreticians, and philosophers of the early nineteenth century.

Kant and the Crisis in Medicine

With most of Kant's most influential works being published from 1780–1790, Kant's definition of an *eigentliche Wissenschaft* came on to the scene in the midst of a transformative moment in the German world. Though in many respects Germany was late to the Enlightenment, the frenetic Newtonianism that had already swept France and England had firmly taken root in Germany by the mid-eighteenth century, even reaching the less educated masses (Rogers 2003). By the time of Kant's *Metaphysische Anfangsgründe der Naturwissenschaft*, there was a widespread, populai sentiment that the world could be understood through properly scientific inquiry. Hence, the impact of Kant's definition of science was not merely theoretical. It struck a chord with those from all walks of life, at a time when scientificity was becoming one of the highest aspirations. The impact of Kant's scientific epistemology was such that his criteria merited at the very least a sincere response. This was particularly true of fields of study for whom Kant's criteria for scientificity called into question their basic validity. Of the latter group, there is no example more prominent than medicine.

During the eighteenth century, the successes of figures like Newton and Boyle breathed life into the notion that the truth of the matter, the secrets of eternity, lay in pursuit of an exact science. The world, accurately measured and quantified, was governed by unbending laws, the names of which could be learned through an empirical study of the universe. In medicine, Enlightenment evaluations of human reason's capacity to apprehend the laws of nature only underscored the imprecision of medical theory and therapeutics (Risse 2003, 165). The eighteenth-century Italian physician Felice Fontana once remarked that one would "be astonished to find that in the eighteenth century there have been Philosophers, Naturalists and *Physicians* who, even in the most important matters have ventured to substitute conjecture for experience" (Earles 1961, 74). Kant's definition of an actual science turned an apparent shortcoming into a definitional failure. In theory, the underlying functions of the body would be discerned and the guesswork at play in the budding life sciences could be put to rest, as long as the system's components and their relations were appropriately measured and understood. The most daunting obstacle was the disconnect between the craft of practical therapeutics and physiology, whose own scientificity was itself up for debate. The difficulty was bringing medicine into a state of logical systematicity, with the caveat that systematicity was only properly scientific if it was grounded by apodictic certainty (Van den Berg 2009, 7). This definitional stringency both undermined the legitimacy of recently established sciences such as chemistry and posed further difficulties for those who desired to see medicine realized as a science.

The German Understanding of Brunonianism

A Bavarian physician by the name of Andreas Röschlaub (1768–1835) presented a solution. For all that Röschlaub did to change the face of medicine, his name receives nary a mention in modern medical histories. Promotion of his approach and the scientific footing it promised would briefly place Röschlaub at the center of the German medical world, attracting a following that counted some of Germany's most famous minds. The solution Röschlaub presented was the medical system of John Brown. Röschlaub argued that Brunonian excitability was an object of apodictic certainty, an a priori principle, and thus could provide a foundation for a properly Kantian medical science (Broman 2002; Tsouyopoulos 1988, 65). Further still, Brown's emphasis on the quantifiability of both vital force and stimuli arguably fulfilled Kant's stipulation that any proper science would make use of mathematics (Broman 2002). However unconvincing this argument might appear to modern readers, it proved remarkably appealing to Röschlaub's contemporaries.

As for the man himself, Röschlaub first encountered the works of John Brown as a medical student, and while Röschlaub was ostensibly a Brunonian he gradually reinterpreted many of Brown's ideas into his own framework with the assistance of Adalbert Marcus, director of the Bamberg Hospital. In 1798, Röschlaub published *Untersuchungen über die Pathogenie oder Einleitung in die medicinische Theorie*, his "big book" and the central argument for what he dubbed *Erregbarkeitstheorie* (excitability theory) (Broman 2002; Röschlaub 1798; Tsouyopoulos 1988, 65). Medical historian Nelly Tsouyopoulos argued that Röschlaub was able to engender interest in Brown's ideas by reframing the continuous interaction between excitability and stimuli as a true dialectical synthesis (Tsouyopoulos 1988, 67). This appears to be in contrast with Brown, whose work, at least in some readings, described life as a forced state in the sense that living beings are perpetually in need of external stimuli to maintain life. Life ceases without stimuli, disease being the low-water mark. In its initial articulation by Brown, this idea did not achieve broad popularity. Writing on Brunonian theory in *Von der Weltseele* (1798), philosopher Friedrich W. Schelling (1775–1854) discounted Brown's system on the basis of precisely this idea, which Schelling understood as relegating living creatures to a vulgar passivity, scarcely more than casual recipients of stimuli from the environment (Schelling 1798, 506; Tsouyopoulos 1988, 67). Röschlaub repackaged this relationship such that the effect of external stimuli was mediated by the existing excitability (Röschlaub 1798, 244). External stimuli were still required to maintain life, with excitability remaining the basic energy that made potential action possible (Röschlaub 1798). But now life could be described as the product of an ongoing dialectic between vital energies and worldly stimuli. In this way, an organism is neither a sponge, passively absorbing the output of the external world, nor an atomized self-contained entity, but rather a dynamic synthesis of self and the world.

The new and improved *Erregbarkeitstheorie* evidently left a significant impact on Schelling. By the writing of *Erster Entwurf eines Systems der Naturphilosophie* (1799), Schelling had already gone from Brown's detractor to a stalwart Brunonian (Schelling 1799b/1858). That same year, Schelling contributed a defense of *Erregbarkeitstheorie* to the first edition of Röschlaub's 1799 journal (*Magazin zur Vervollkommnung der theoretischen und praktischen Heilkunde* (Schelling 1799a, 255)). There was even something of a brief friendship between the two of them, with Schelling making the effort to visit Röschlaub in Bamberg. However, Schelling's efforts to further adapt the *Erregbarkeitstheorie* to *Naturphilosophie* quickly materialized into irreconcilable differences in how Schelling and Röschlaub understood the Brunonian system, a situation that later led to a total split by 1805 (Tsouyopoulos 1988, 65). The act that consummated this split was Schelling and Adalbert Marcus's decision to establish their own journal in order to distance themselves from Röschlaub, which they aptly titled *Die Jahrbücher der Medizin als Wissenschaft* (Tsouyopoulos 1988, 65). This severance of the movement swiftly led to the formation of two distinct camps, with Schelling leading the physiologists (such as Lorenz Oken, Franz von Walther, Ignaz Döllinger) and Röschlaub heading the pathologists (Johann Schönlein, Ernst von Grossi, Johann Nepomuk von Ringseis) (Tsouyopoulos 1988, 65).

Evidently, Röschlaub's facelift on Brown's medical system was extremely effective at transplanting the model into a new context, so much so that it was naturalized in the form of its own heresy. Which is to say, Schelling's split with Röschlaub over doctrinal minutiae, and the separation of their followers into different camps, more than anything demonstrated the depth of the German acceptance of Brunonianism's foundational concepts. Whatever their differences, both exalted Brunonianism as the savior of scientific medicine. How different really was Röschlaub's *Erregbarkeitstheorie* from its source material? Röschlaub altered the form in which the theoretical components of Brown's system were understood, and thus fundamentally altered the mechanisms by which excitability was modulated. However, regarding the application of Brunonian theory, Röschlaub staunchly upheld Brown's therapeutic methodology. Röschlaub, in particular, reaffirmed the Brunonian emphasis on the use of stimulant therapy in curing diseases. Vital substances remained central to *Erregbarkeitstheorie*, and so they remained central to Schelling, Röschlaub, and their followers. Röschlaub arguably further elevated the vital substance concept by guarding it through the crucible of translation and modification.

Regarding the more general openness to Brown's ideas, Tsouyopoulos argues that the sudden receptiveness to Brunonianism among German thinkers after 1795 was not coincidental, suggesting that the phenomena can be traced back to the publication of *Wissenschaftslehre* by Johann G. Fichte (1762–1814) (Tsouyopoulos 1988, 70). The issue at stake here is how excitability could convincingly be presented as an a priori principle if *Lebenskraft* (life force) could not (Tsouyopoulos 1988, 67). Published in 1794, *Wissenschaftslehre* dominated the academic discussion of the following year, with the poet Novalis (1772–1801) being the first to propose that there were significant conceptual similarities between Brown's and Fichte's systems (Tsouyopoulos 1988, 70). Though compelling, it would seem the inverse is the case. Taking into account Schelling's initial perceptions of Brunonianism as a mechanical philosophy, it would seem that Röschalub's success might largely be the product of his efforts to bring Brunonian excitability into agreement with the *Wissenschaftslehre*. This is evidenced by Röschlaub's inclusion of a public appeal to "Dr. Johann Gottlieb Fichte" to give consideration to, and ultimately pass judgment on, *Erregbarkeitstheorie* in the introduction to the second volume of the first edition of his *Magazin* (Röschlaub 1799, xi).

Even if this is taken to be nothing more than a publicity stunt, it demonstrates Röschlaub's intention to bring intoxicated excitability, and the Brunonian model in general, into alignment with Fichte's authority. Röschlaub brought the Brunonian system into agreement with what he would eventually come to understand as Fichte's antiformalist, dynamist ontology, focused as it was on relationality rather than substance or categories (Grant 2008, 103). The Brunonian system operated on the notion that excitability was a quantifiable force, at least *in principle*, and, in this way, it was able to formally satisfy the Kantian definition where simple *Lebenskraft* would not suffice. Once again, this is owing to the binding and enmeshing of excitability to vital therapeutics, without which there was no practical means of quantifying excitability. Thus, Brunonianism's candidacy for consideration as a rationally valid medical model hinged on its conformity with Kant's emphasis on quantifiability, though its enthusiastic adoption was ultimately a product of Röschlaub's efforts to realign Brown with Romantic thought.

German Brunonianism in Practice

How far afield did thoughts about vital substances spread? Although the partitioning of the movement threatened the legitimacy of Brunonianism, it would receive sporadic support over the next five or so years, culminating in a revival that reached into the 1820s. This already leaves us with somewhere around two decades of intentional tinkering with Brunonian ideas and vital substances. Some examples of this are quite noteworthy.

Brunonian ideas formed the basis of a neurological theory of alcohol addiction created by Baltic German physician C. von Brühl-Cramer (?–1821), using a model of Brunonian stimulus dependence derived from Brown's system (Perkins-McVey 2023). Published in 1819, "Ueber die Trunksucht und eine rationelle Heilmethode derselben" arguably represented the first systematic study of alcohol addiction in a clinical setting (Perkins-McVey 2023). Parallel representations of chronic drunkenness as a disease of Brunonian stimulus dependence further figured in the works of Thomas Trotter and Benjamin Rush, likewise credited with the advent of the disease theory of alcoholism (Perkins-McVey 2023). In addition to being one of the earliest scientific commentaries on fetal alcohol syndrome, von Brühl-Cramer's work outlined alcoholism's pathogenesis, etiology, and different sequelae, such as dementia (Kielhorn 1996). Notably, the introduction was written by celebrity doctor Cristoph Hufeland (1762–1836), who, though one of Brunonianism's earliest critics, "himself wrote several articles about John Brown, in 1819, 1822, and 1829, and compared Brown with Galen" (Tsouyopoulos 1988, 66).

Eduard Löbenstein-Löbel, of the University of Jena, relied on

Röschlaub-Brown in the creation of what is likely the first formal medical manual on wine therapy in 1816 (Löbenstein-Löbel 1816; Paul 2001, 28). Löbenstein-Löbel came out against his fellow Brunonians in elevating wine above opium as the pinnacle of all vivifying therapeutics, going as far as identifying different wines for the treatment of different illnesses (Löbenstein-Löbel 1816; Paul 2001, 30). Champagne, for example, was particularly suited for the treatment of extreme vomiting among pregnant women, depression, chronic gout, and kidney stones, while a red Bordeaux was more effective in cases of constipation, general inflammation, and hepatic diseases (Löbenstein-Löbel 1816, 55–56, 65). Somewhere between a medical textbook and a wine catalog, Löbenstein-Löbel goes as far as extolling the virtues of particular wineries for specific treatments—a Château Margaux for patients with a slower pace of life, but a Libourne for those accustomed to hard, outdoor work (Löbenstein-Löbel 1816, 65). This is an excellent exemplification of the relative diversity of Brunonian thinking from the moment it gained popularity in the German world. Brunonianism had stretched beyond the boundaries of the Brunonian system itself, giving rise to an assortment of individual models. Even as conceptual elements were altered, these disparate systems were united in their shared understanding of vital substances and their therapeutic value.

Brunonian thinking further shaped the emergent Romantic approach to psychiatry. Ernst von Feuchtersleben's influential 1845 *Lehrbuch der ärtzlichen Seelenkunde* is remembered for popularizing Karl Friedrich Canstatt's language of "psychosis" and developing the language of "psychopathy," in addition to pioneering psychiatry as a science of mental pathology, distinct from neurology (Ban 2006; Feuchtersleben 1845; Mason 2006). For von Feuchtersleben, the nature of psychosis, of psychopathy, was inextricable from the naturphilosophic conception of Brunonian excitability. A Schellingian *Naturphilosoph*, von Feuchtersleben understood all mental phenomena as emergent of a dialectical interplay between self and world (Beer 1995). The *Seele* itself being incorruptible, mental disturbances were understood to be *Persönlichkeitskrankheiten* (personality illnesses), as the personality was the experiential space within which the dialectic of mind and world became actively manifest. The nature of mental disturbances was, in this sense, understood in parallel to the broader conception of illness as the product of sthenic and asthenic states of excitability. This rationale translated over into von Feuchtersleben's approach to intoxicating vital substances. Brunonian stimulants such as wine and opium "operate psychically in that they promote psychical function through a feeling of uninhibited organic vitality" (Feuchtersleben 1845, 365). The curative, but also dysregulating, function of vital substances corresponded to the nature of mental pathology as the product of a disruption in the balance between bodily vitality and the mind (Feuchtersleben 1845).

Von Feuchtersleben would even commit his interest in vital substances to a body of ecstatic prose, first published in 1843/1844 alongside illustrations by Mortiz von Schwind (Feuchtersleben and Schwind [1875] 1978). Outside of acute cases of interest to psychiatric doctoring, vital substances occasion an awakening of the vital senses, as he writes in a poem:

> Dreams like to seduce,
> feigning at the appearance of lawlikeness—
> impressing forms
> and yet they are mere swindles

FIGURE 2. Schwind's illustration accompanying Feuchtersleben's poem 9.

As Nirot's fragrant plant
And the brown juice of Arabia
In an enigmatic dance
Create colorful fantasies. (Feuchtersleben and Schwind 1978, 9)[1]

The "brown juice of Arabia" delivers the intoxicated into an imagined palace of the "Sultan's delight," "twisting the smoke into images which only the artist's sense recognizes" (Feuchtersleben and Schwind 1978, 9, 10, 11). Drawing on his own experiences consuming hashish, opium, wine, and tobacco, von Feuchtersleben mused of a "wine- and smoking- philosophy" (Feuchtersleben and Schwind 1978, epilogue). Vital substances, among them wine, opium, and—now—hashish, enable the realization of a waking dream, an immediate reflection of the psychiatric significance von Feuchtersleben would accord to intoxicants in 1845. Of course, in typical Brunonian fashion, this is also to the peril of those who consume them. Extraneous vital stimulus was equally as capable of engendering states of physical or mental illness, even if natural cases of asthenia were considerably more common.

The scope of the Brunonian influence on German Romantic philosophy and science was so great that it began taking root among the English Romantics, many of whom were Germanists themselves. Remarkably, though Brown was a Scot, English interest in Brown was almost exclusively through Schelling and Röschlaub. Among the most famous examples of this conceptual reverse migration was the Brunonianism of England's highly storied opium-eaters, Samuel Taylor Coleridge and Thomas De Quincey, both of whom engaged with Brown's ideas through German publications.

For Coleridge, his initial exposure to Brown came through his good friend Thomas Beddoes, a physician and Germanist who had been the first to publish translated passages of Kant's *Kritik der reinen Vernunft* and *Kritik der Urteilskraft* into English (Vickers 1997, 47, 59). Beddoes had become aware of the German Romantic Brunonianism of Röschlaub-Schelling in the late 1790s, having already translated and edited Brown's *Elements of Medicine* from the Latin in 1795, and it was through him that Coleridge ultimately became a disciple of Schellingian medical *Naturphilosophie* (Brown 1795:1; Vickers 1997, 59). By 1819, Coleridge placed Brown alongside the likes of Luther, Milton, Cicero, and Wordsworth as one of history's greatest geniuses (Vickers 1997, 48).[2] A similar pattern is seen in the case of Thomas De Quincey. His *Confessions of an English Opium-Eater* awoke the unaware to the phenomenon of English opium-eating in the 1820s (Gao 2020, 6). There, De Quincey used Brunonian language to describe the effects of opium intoxication, musing about opium's excitatory capacity as a bodily stimulant (Gao 2020, 9). Not only was De Quincey a Germanist himself, but he met Coleridge in 1807, at which time they struck up a friendship, and it was likely by way of their relationship that De Quincey was converted to Brunonianism himself (Morrison 1997, 27–28).

In strikingly Brunonian fashion, De Quincey places opium at the center of a radical reconception of the body and mind. Youngquist argues that *Confessions* was—in large part—a jab at Kant's transcendental philosophy (Youngquist 1999). Rather than subordinate the body to the mind, bodily incorporation to mental representation, opium pulled De Quincey from his own dogmatic slumber, underscoring the sense in which cognition turned on bodily consumption (Youngquist 1999). Where transcendental philosophy undermined the centrality of embodied existence, opium was a reconstructive force; it placed physiology at the center of mental life (Youngquist 1999). Coleridge's and De Quincey's enthusiastic givenness to opium-eating can be framed as a particularly Romantic case of nineteenth-century addiction. Yet we might further understand it as a personal practice supported by the Brunonian system of medicine. Both parties certainly understood Brown's philosophy as the secret to not only personal, professional, and national health, but also individual excellence (Morrison 1997; Cooke 1974).[3]

The Brunonian system even figured into the discovery of nitrous oxide by Humphry Davy (1778–1829). Beddoes had a hand in this development as well. Davy had been a pupil and collaborator of Beddoes, who participated directly in Davy's discovery (Golinski 2011, 17). It was Beddoes who identified the Brunonian system as the correct hermeneutical framework with which to understand the new discovery, suggesting nitrous oxide was a vital stimulant (Golinksi 2011, 19). Beddoes went as far as inviting Coleridge to experience its effects (Golinski 2011; Jay 2009). The lines of demarcation between experiment and experimenter, research and play, blurred as an aristocratic microculture of nitrous oxide parties emerged around Davy, Beddoes, and the novel intoxicant (Jay 2009). It is clear that Davy himself ascribed to the Brunonianism of his mentor, though with perhaps less overt enthusiasm, and went on to describe nitrous oxide as a stimulant affecting the balance of excitability (Bergman 1991, 535–536, 538). Notably, Davy himself described the effects of nitrous oxide by comparing it with those of opium, perhaps *the* vital substance (Golinski 2011, 19). A compelling piece of experimental evidence supporting this view was the observation that nitrous oxide supported life. Davy and Beddoes noted that nitrous oxide was respirable without producing the expected effects of hypoxia, or even shortness of breath (Hermann 1864).

In a Brunonian context, this was an expected result. Nitrous oxide was a vital substance; it stimulated excitability. That continued respiration of it eventually terminated in death only further proved this view. Too much of any vital stimulus resulted in an acute state of sthenic shock, just as readily a cause of death as complete vital depletion.

German physiologists and advocates for scientific medicine, as well as their followers elsewhere in Europe, took to Brunonian ideas, using them to classify and categorize novel substances, techniques, and illnesses. In each and every case, the juncture of their encounters with Brunonian ideas was mediated by intoxication. This was not only because Brown's notion of excitability initially flowed directly from the experience of intoxication, out of intoxication as a way of knowing. From Davy to von Brühl-Cramer, Brown's initial development of the "excitability" concept out of opium intoxication was repeatedly reconsummated with the establishment of new therapeutic intoxicants and methods. The state of intoxication served as a way of knowing that formed the basis for meaningful perceptual understanding of health and what it means to be in a body. As they were taken up by different physicians, the finer points of Brown's system even shifted in light of individual experiences with intoxication. Löbenstein-Löbel's elevation of wine over opium was, in this way, a profound testimony to Brunonianism not merely as a theory of medicine but as an embodied praxis. Löbenstein-Löbel relied on his ownmost experience of wine intoxication to empirically verify its superior therapeutic value. At the center of the process of mobilization and realignment remain vital substances, a concept that was now able to reach a broader audience than Brown alone could have feasibly accomplished.

✳ 3 ✳
Brunonian Naturphilosophie *and Intoxicated Knowing*

The Intoxication of Philosophy: Kant

Vital substances, Brunonian medicine—it is easy to write these off as curious abstractions, theory utterly divorced from the lived experience of those who championed it. Given the ubiquity of Brunonianism in late eighteenth- and early nineteenth-century German medicine, it is worth assessing the extent to which Brunonian ideas truly affected the broader discussion of intoxicants in the thought of the period. In doing so, it is important to take a look at the idea's reception in the midst of two of the most influential Germans in the early nineteenth century: Kant and Schelling.

Was Immanuel Kant a Brunonian? The simple answer, the one limited to his published works, is that it is difficult to say anything for certain. The honest answer is a bit more complicated. His tragic death in 1804 in the throes of cognitive disarray blocked Kant from participation in the different waves of Brunonianism that would continue into the 1820s. Kant was alive and at least somewhat active, if in the midst of mental decline, during Röschlaub's efforts to interpolate Brown's ideas, as well as during his friendship with Schelling. For Kant to have encountered Brown via Röschlaub, like so many others, would preclude any Brunonian influences from all but Kant's last and least-known works.

But this is far from the end of the story. It turns out that Kant did have the opportunity to engage with Brunonian theory, even before it was sublated into Röschlaub's *Erregbarkeitstheorie*. Nor is this merely circumstantial. It is said that the primary factor in determining the terminal condition of stellar gravitational collapse is the star's mass. With the death of Immanuel Kant, the immense gravity of this star attracted some form of commentary from almost all who knew him. Ehregott Andreas Christoph Wasianski (1755–1831) was one such figure, a particularly noteworthy one in light of his status as the dying Kant's personal assistant, support, secretary, and confessor. Titled *Immanuel Kant in seinen letzten Lebensjahren*, Wasianski's memoir furnishes posterity with a wealth of testimony concerning Kant's condition in the years preceding his death in 1804. For our purposes, it is a source of explicit assertions concerning Kant's personal support for the Brunonian system of medicine, the means by which he came to know Brown's ideas, and even the extent to which Brunonianism influenced Kant's daily life.

By Wasianski's account, "Kant was very heterodox in medicine" (Wasianski 1804, 189). This interest was at least partially theoretical, as Kant had long followed new developments in the arts and sciences. Kant's "care for maintaining his health was another reason why he was so interested in new systems and inventions in medicine," and "he saw the Brunonian system as a major invention of this sort" (Wasianski 1804, 42). He was initially introduced to Brunonian thought through the work of Melchior Adam Weikard (1742–1803) (Wasianski 1804, 42). Weikard was actually one of the first Germans to promote Brunonianism, arranging for a German print of Brown's *Elementa* in 1794, to be followed by the release of a German translation the next year alongside Weikard's outline of Brown's system (Broman 2002, 144).[1] At some point following the publication of Weikard's translation and commentary, Wasianski suggests that Kant established a firm belief in the Brunonian system and organized his life in accordance with its principles (Wasianski 1804, 42).

The timeline presented by Wasianski's testimony proves instructive for several reasons. For one, it explicitly states that the source of any alleged Brunonianism on Kant's part preceded Röschlaub. The roots of Kant's entanglement with Brown might even extend into the years 1794–1795. It also broadens the scope of possible influence exerted by Brunonian thinking on Kant's later corpus. Were it not for this prior engagement, their tenuous rivalry might have chilled Kant's responsiveness to the Fichtean bent of Röschlaub's interpolation, which had swiftly become synonymous with Brunonianism as such.

Considering the language of trust and devotion implicit to Wasianski's portrayal of Kant's Brunonianism, it is worth asking to what extent Brunonianism affected Kant's unusually fixed schedule. The fixedness of Kant's daily regimen is the stuff of legend—ready fodder for quips about a man seemingly alien to disorder or spontaneity. Kant had likely bound himself to a rigid routine for most, if not all, of his adult life, something many scholars have acknowledged (Clark 1999; Kuehn 2001). However, this general tendency seems to have been in keeping with Kant's philosophical commitment to operate in accordance with established maxims.

Some have even questioned whether Kant's itinerary was truly as unbending as has frequently been suggested (Kuehn 2001).

Wasianski, on the other hand, leaves no such ambiguities about the nature of Kant's post-1795 lifestyle. From orderly drams of rum to measured glasses of wine, Kant's regularly consumed vital stimulants "à la Brown," and many if not all of Kant's other habits, appear to align with Brunonian proclivities (Wasianski 1804, 189). Let us briefly consider an average day. Kant started his day at five in the morning with a single, brisk cup of continuously refilled tea, quickly chased followed by a pipe of tobacco (Wasianski 1804, 40). The remainder of the morning—until precisely fifteen minutes before one—was allocated to writing or teaching. A quarter to one was when his cook carried out a container of wine, and sometimes spirits (Wasianski 1804, 40). At this point, it was time for lunch and a walk, after which Kant committed the afternoon to less intensive work. Kant tended to enjoy animated, libatious dinner in the company of friends, albeit followed by a sprightly walk without accompaniment. Until dusk's last crepuscular rays, Kant spent his evening in a book from his library. By ten in the evening, Kant was lying in bed, where he could be found swaddled in his linens "just like a cocoon" (Wasianski 1804, 29–31).[2]

Taking Wasianski's account at face value, it is difficult to deny the points of congruence between Kant's schedule and a Brunonian prescription. Waking up early; scheduled, moderate meals; and consistent exercise are all confirmedly square with Brunonian methods of palliating surging excitability. Kant's commitment to the scheduled consumption of wine, spirits, tea, and tobacco likewise served as powerful vital stimulants that acted to maintain a steady level of excitability. Though perhaps individually unremarkable, the structure of each day tells a story. Rising in the morning calls for stimulus, succeeded by moderating activity, followed by further stimulus, ad infinitum. An exhilarating lecture calls for a dram of liquor, much as wine with friends demands a speedy jaunt. Every element—even Kant's nightly chrysalis—fits within a greater effort to see the Brunonian system realized as life praxis. But the most demonstrative argument in favor of the suggestion that Kant was a confirmed Brunonian, and not just an old kook in his twilight years, stems from the measured character of each component. Food, exercise, and—of course—intoxicants are pursued with exacting regularity. Kant never abstains and never consumes in excess. It would seem that Kant's lifestyle is in fact a treatment plan—wine, tea, tobacco, and spirits his prescription.

Wasianski goes as far as identifying the Brunonian system with the Kantian project of enlightenment, attaching the Brunonian system to Kant's philosophical project:

[Kant] considered it a significant advancement, not only for medicine but for all, finding it in agreement with the common entirety of humankind: [for humanity] to return, after many detours, from the compound to the simple meant, among other things, a good thing for the patient in the economic regard, who is prevented by poverty from using expensive compound remedies. [Kant] therefore ardently wished that this system would soon receive more followers and be brought into general circulation. (Wasianski 1804, 42–43)

Wasianski alleges that, for Kant, the advent of the Brunonian system represented not merely a radical change in medicine, as it with Weikard or Röschlaub. Much as the central impetus behind Kant's *Was ist Aufklärung* was a desire to rend the rational subject from the oppressive condition of social *Unmündigkeit* that they might realize their rational duty to *Freiheit*, the popularization of Brunonianism was nothing less than the body's emancipation from tyranny of medicine past (Kant [1784] 1999). The exorbitant expense associated with the Galenic compound remedies of yore, a yoke upon the necks of the common people, would be replaced by the comparative simplicity of Brunonian "vital substances," an economic as well as a material liberation. It embodied a revolution in the population's capacity to attend to their health, tantamount to an Enlightenment in matters of the body, and an extension of Kant's philosophical project.

It is natural to have major questions about the credibility of Wasianski's statements. Some of these uncertainties are allayed by better picture of their relationship. Wasianski had been Kant's amanuensis while a student at the University of Königsberg from 1772 to 1780 (Kuehn 2001). After reconnecting by chance at a wedding in 1790, Wasianski—then a deacon—soon found himself invited to dinner parties at the philosopher's home, quickly becoming one of Kant's closest friends and confidants (Kuehn 2001). To this point, Kant biographer Manfred Kuehn paints Wasianski in a highly favorable light, further asserting Kant's absolute trust in his companion and legal executor (Kuehn 2001). Wasianski aside, there was also simply little by way of impediment to Kant's adoption of Brunonian ideas. Kant had long espoused that the existence of organized beings likely relies on a preharmonizing principle, emergent of a sophisticated conception of vitalism (Shell 2013). Brunonianism was, in a certain sense, simply an expansion upon, and a specification of, the conceptual grammar and syntax that already patterned Kant's understanding of organized beings. A change to the Brunonian approach hardly constituted something akin to a conversion from unyielding materialism.

Active Play

What are we to make of Wasianski's lofty claims? For all the biographical intrigue, the question of Kant's Brunonianism matters little if it never translated into a tangible influence on Kant's broader philosophy. Of Kant's post-1795 work, there are two significant publications that, in considering the nature of intoxicants, are likely candidates for allusions to Brunonian vital substances: *Die Metaphysik der Sitten in zwei Teilen* and *Anthropologie in pragmatischer Hinsicht abgefaßt*. Though neither explicitly mentions Brown, both contain striking allusions to a heretofore unrealized interest in the potential represented by intoxicating substances, one that is worth considering in light of Kant's possible, even likely, Brunonianism.

Many of Kant's starkest remarks about intoxicants and their effects are found in *Die Metaphysik der Sitten in zwei Teilen*. It is here that some readers have might found opportunity to giddily muse about a different Kant—less the Old Man of Königsberg and far more at ease with a bottle in the crook of his arm. Kant admittedly devotes a relatively substantial body of discussion to the matter of stupefaction, through food as well as through alcohol and opium. Here, Kant identifies both intoxication and gluttony as examples of vices derived from "misuse of the means of nourishment which limits and exhausts our ability to use them intelligently" (Kant 1797, 427). Scholarship has generally treated Kant's remarks concerning intoxication and gluttony as a more or less singular condemnation, suggesting that the problem for Kant is really one of stupefaction (Hoffman 2021).[3]

Though at first Kant does seem to treat intoxication and overeating with identical disgust, closer examination would suggest that Kant conceives of intoxication as a discrete mental state, or experience. While gorging oneself "brings the senses into a passive state," the state of intoxication engenders "active play of representations," stimulating one's imagination (Kant 1797, 427). Both are undeniably associated with a stupefying influence on the mind. It is nevertheless clear that Kant attaches significance to the specific way in which different avenues of stupefaction affected the rational mind, even if a person is diminished to their basest form. More interestingly still, intoxication appears to do so through acute stimulation of the imagination, a sentiment echoed in Kant's *Anthropologie* (Kant 1798, 165).

Much of this distinction is borne out is Kant's elucidation of the moral problem of intoxication. Is intoxication a failure in one's moral duty to oneself? Can it can be justified? For those seeking simple answers, Kant is characteristically restrained. Drinking to excess, Kant argues, constitutes a debasement of oneself, undermining the regular function of the senses and atrophying an individual's resolve to carry out their affairs. It is in this sense that Kant proposes that overindulgent use of intoxicants violates one's duty to oneself, an infraction against our ownmost nature as rational subjects (Kant 1798, 427; Timmermann 2006, 517). The same argument holds for overeating. And yet—in the same breath—Kant appears to propose that intoxication, *unlike* overeating, cannot be reduced to a mere enjoyment of the senses, instead distinguishing it through its effects on the imagination. The language employed by Kant in describing what takes place in the intoxicated mind is that of *play*, terminology strongly associated with the activity of the imagination in Kant's *Kritik der Urteilskraft*. In spite of this, a discussion of intoxicants never materialized in Kant's critical period—save his famous dig at the purported judgment of sommeliers (Kant 1790b).

While it first appeared that the moral dilemma posed by intoxication was reducible to a question of excess, there is clearly more to Kant's reasoning. Working through the Kantian argument for the moral consumption of drink and drugs will further furnish us with a clearer conception of how Kant conceived of the nature of intoxication and what it meant for the body. In furtherance of these objectives, let us consider two basic lines of inquiry: (1) What is it that leads Kant to associate the imagination with intoxication to begin with, when it could just as easily be aligned with the senses? (2) How does intoxication amount to a failure in one's duty to oneself, considering the apparent benefits discussed by Kant himself? In seeking an answer to the second question, we will have to begin with the first, convergently bringing to light the unique position accorded to intoxication in Kantian thought.

Imagination

It has already been suggested that the language Kant uses to describe the state of intoxication bears a striking resemblance to his discussion of aesthetic judgment. In *Kritik der Urteilskraft*, Kant explains that "the cognitive powers brought into play by this presentation are in free play because no particular concept limits them to a specific rule of cognition" (Kant 1790b, 217). Imagination generally conforms to the concepts applicable to a given situation as presented through the senses. However, the aesthetic brings the imagination and the understanding into *freien Spiele* (free play) by introducing sensory material without constraining imagination to any particular rule of cognition. As a result, the imagination and the understanding participate in a certain conceptual freedom, which follows the laws of cognition without conforming to a specific law. The importance of this for Kant's critical system is that aesthetic judgment appears to be a demonstration of the noumenal self's capacity for free will, thus engendering the possibility of moral duty. Importantly, this experience is accompanied with a pleasant sensation for the subject (Kant 1790b, 217–18).

Intoxication, comparatively, is described as eliciting "active play of representations" through the imagination (Kant 1797, 427). Although he contemptuously refers to intoxication as an animalistic enjoyment of the senses, the Vigilantius notes on Kant's *Metaphysik der Sitten* confirm the impression given by the text that Kant does not regard the primary object of intoxication to be sensory enjoyment but rather the play of the imagination (Vigilantius 1997, AA. XXVII, S. 527, Z. 27). The result of this is not only pleasure, but also a livening of spirits and even (in the case of opium) a dreamy inwardness (Kant 1797, 427–28). A possible explanation for how Kant understands this is given in *Anthropologie*, where Kant states "drunkenness is the unnatural state of inability to classify your senses according to the laws of experience" (Kant 1798, 165). This is because of the excessive activity, or play, of the imagination. A secondary effect is an anesthetization and disorientation of the sensory faculties, which Kant likens to when one has quickly woken up from a deep sleep (Kant 1798, 166). Drugs and alcohol seem to have this effect, as Kant tells us, by stimulating the vital force (Kant 1798, 169–70).

On the finer points of Kant's understanding, the absence of a direct, systematic account leaves some aspects of this explanation ambiguous. It remains unclear from the text alone whether *active play* occurs because intoxicants stimulate the vital force, artificially overwhelming the senses and thereby freeing up the imagination, or because artificially stimulating the vital force holistically invigorates both the powers of imagination and the sensory faculties.

Just as the aesthetic experience brings the understanding and the imagination into free play, intoxication brings the imagination, and with it the understanding, into active play. Parallels can even be found in Kant's explanation of sensory enjoyment, for—if intoxication were merely about enjoyment of the senses—it would only be comparable to judgments of agreeability ("the wine tastes good"), which are always subjective. Instead, both the aesthetic judgment and the state of intoxication are associated with a distinct activity of the imagination. This might appear to be an unlikely comparison. But aesthetic judgment and intoxication remain the only instances in Kant where the imagination is brought into a distinct state of play, unbound from compliance with a specific rule of cognition. To this point, it is worth considering if Kant is suggesting that intoxication can be morally significant in a fashion similar to aesthetic judgment. If this is the case, Kant leaves little trace of it, and Kant does make repeated reference to the dangers of intoxication, in particular the possibility of developing a need for the pleasurable experience it imparts.

This brings us to the second question. Kant makes a point of asserting that the moral difficulty of intoxication "is not judged here by the damage, or the physical damage (such as illnesses) that man gets through [excessive intoxication]" (Kant 1797, 427). Disease caused by excessive intoxication can only ever establish a rule of prudence, rather than a moral duty (Kant 1797, 427). Kant himself freely admits that intoxication, including the occasional minor excess, is not without its benefits, including a virtuous candidness, social limberness, elevation of the spirits, and even reprieve from the worries of day-to-day existence (Kant 1798, 170–71). This is all in addition to, or a result of, the effects on the imagination just discussed. What then is the basis for the moral duty of restraint in all matters of intoxication? The answer it appears is not reducible to the debasement of befuddlement, but rather the risk of dependence.

To understand this argument, it is important to recall Wasianski's testimony that Kant was a committed Brunonian by the time he wrote *Metaphysik der Sitten*. With that in mind, Kant's initial connection of intoxication with illness can be taken quite literally, as excessive stimulation is an "actual" cause of disease. Brown's model, which establishes the bodily need for outside stimulus and in turn opens the door to behavioral feedback loops between vital force and stimuli, has elsewhere been attached to the concept of drug and alcohol dependence. In fact, Brühl-Cramer's 1819 text "Ueber die Trunksucht und eine rationelle Heilmethode derselben" used Brown's system to establish a clinical theory of alcoholism (Kielhorn 1996, 121). Kant himself recognized that a real possibility of using intoxicants was the need to use them again in the future, and at higher doses (Kant 1798, 427). In such a case, excessive consumption entails a real risk to one's bodily autonomy, subordinating rational freedom to an irrational cycle of bodily dependence. Far from the contextual and subjective question of debasement, the state of dependence on alcohol and drugs fits the very definition of a failure in one's duty to oneself (Timmermann 2006, 517).

Relying on Wasianski's account, it becomes overwhelmingly clear that Kant perceives intoxicants as what I have earlier described as "vital substances." There is little reason to doubt Wasianski's reckoning of events, though when the subject is a figure as historically consequential as Immanuel Kant there is an understandable hesitance to rely on second hand accounts. Kant's approach both to the matter of intoxicants and to the concept of intoxication, as presented here, nevertheless supports the argument that Kant treats intoxicants as vital substances. Even if Wasianski's memoir is taken out of consideration, Kant still conceives of drugs and alcohol as having a privileged relationship with the vital force, far removed from other medicines, and uniquely significant to the human body and mind. But Kant was not the only foundational figure of the early nineteenth century for whom vital substances played a crucial role. As previously discussed, Schelling, too, was deeply invested in the notion of vital stimulus, to such an extent that it figured centrally into his *Naturphilosophie*.

The Intoxication of Philosophy 2: Schelling

However determined Kant may have been in his dealings with Brunonian vital substances, it is advisable, indeed prudent, to remain circumspect about the influence of Kant's Brunonianism on those around him. As much as intoxicants suffused the material, discursive, and praxic culture that surrounded Kant and his followers, the difficulties involved in assessing even Kant's personal proximity to Brown's work leave all the more uncertainty surrounding its potential reception, if it was ever known outside of an expressly limited group of individuals. The same cannot be said of Schelling, a man for whom vital substances left an indelible mark on the grand philosophical and scientific project of *Naturphilosophie*.

How did Schelling come to embrace vital substances with such ardor? The story is inextricable from Schelling's sweeping encounters with some of the most prominent scientific minds in the German world at the close of the eighteenth century, a period of intellectual gestation whose immediate impact can be seen in the development of the system of *Naturphilosophie* and Schelling's associated departure from Fichtean philosophy. From 1796 to 1798, a couple years before his rise to great fame (and, at times, infamy), monetary need found Schelling in Leipzig. The twenty-one-year-old Schelling had moved from Tübingen to Leipzig in order to take a job tutoring a pair of young barons. His job was twofold. Schelling was first to attend lectures on an array of scientific subjects and then later reconvey their contents to his pupils (Risse 1976, 322). This arrangement proved itself enormously beneficial for Schelling. Effectively paid to attend class, the young Schelling was afforded an opportunity to interact with experts on an assortment of subjects concerning the study of the natural world, organic life, and medicine.

It is difficult to conceive of the existence of Schelling's *Naturphilosophie* outside of this substantive engagement with the extant tradition of

Romantic science. This is not to detract from Schelling's originality. But the likes of Wolfgang von Goethe (1749–1832), Georg Stahl, Alexander von Humboldt (1769–1859), Karl Kielmeyer (1763–1844), Johann Blumenbach (1752–1840), Johann Reil (1759–1813), and Johann Ritter (1776–1810), to name just a few, helped pave the way for the later acceptance of Schelling's idea by first uncovering "Romantic" nature as a dynamically unfolding holism. Many of these individuals were counted among the foremost minds in their subjects and in their time, although the depths of their influence have long since been wiped from popular scientific narratives.

Blumenbach, as well as his acquaintance Johann Reil, saw an immanent purposiveness in the organization of life (Lenoir 1982; Watson 2010). Organisms and their constituent parts were evidently subordinate to a lawful teleology that they understood in terms of a formative vitality. Blumenbach's pupil from 1786–88, Karl Kielmeyer further developed on these ideas in his attempt to derive the fundamental laws of organic form by way of comparative anatomy (Lenoir 1982; Watson 2010). Kielmeyer's eventual argument for morphological affinity across animal forms led him to ultimately suggest that many, if not all, living species had in fact come from other species (Watson 2010). There was also Goethe's famous 1790 scientific poem "Die Metamorphose der Pflanzen," which analyzed plant morphology as a dynamical unfolding of an archetypal concept or species (Goethe [1811] 1989). Timothy Lenoir and Peter Watson credit Romantic biologists Blumenbach, Reil, and Kielmeyer with shaking up hegemonic classification schemes and later paving the way for Darwinism (Watson 2010; Lenoir 1982).[4] Watson particularly emphasizes the sense in which the underlying principles of universal holism and dynamic relationality that pervade Romantic thought de-essentialized the classification systems that Linnaeus and others had erected throughout the natural world.

One of the lectures that Schelling attended during this formative period was that of the chemist Johann Ritter. A gifted but lavishly indulgent young experimenter, Ritter's lodgings were reportedly ridden with spent bottles of alcohol, and he was known to consume large doses of opium, possibly to help treat pain caused by extreme electrophysical self-experimentation (Strickland 1998, 456).[5] With the help of an introduction by his friend Novalis (whose own fondness for wine and narcotics features in "Hymen an die Nacht"), Ritter rose to a position of prominence and authority within Romantic circles (Snelders 1970, 201; Novalis [1800] 1988, 14). Schelling was introduced to Johann Ritter in 1798 when he saw Ritter lecture on his book *Beweis, daß ein beständiger Galvanismus den Lebensprozeß in dem Thierreich begleite* (Ritter 1798; Weatherby 2016, 186). Schelling quickly took to Ritter's work on galvanism and Ritter, in turn, would come to see Schelling's *Naturphilosophie* as a viable basis for a systematic grounding of the physical sciences. Ritter expanded upon Alessandro Volta's observation that heterogeneity between conductors (polarity) was a precondition for both inorganic and organic (galvanic) conduction (Ritter 1798; Weatherby 2016, 186). Ritter argued, and attempted to demonstrate experimentally, that galvanic conductivity arose not only out of the heterogeneity of muscle and nerve but from within nerves themselves (Ritter 1798; Weatherby 2016, 186). For Ritter, this raised deeper questions about whether a satisfactory degree of heterogeneity itself was all that was required to generate an electric impulse, both within and without the organism (Ritter 1798; Weatherby 2016, 187). With the discovery of the electric pile in 1800, Ritter went on to conduct a number of extravagant experiments on the effect of electricity on sensory nerves. Unfortunately, his promising, if peculiar, career was cut short. Johann Ritter died in 1810 at the age of thirty-three, leaving behind open questions about the impact of his risky experimental practices on his overall health.

Sometime in this period, Schelling was also introduced to Brunonian medicine, an idea that Schelling initially rejected on account of the system's perceived mechanism. Of course, Schelling underwent a sharp reversal of this position in 1798–1799, when Schelling became acquainted with Röschlaub's *Erregbarkeitstheorie*.

Elements of all of these ideas made up the foundation of Schelling's thought at the end of the eighteenth century, owing to the informal scientific education he received during this period. In many cases, the influence of early Romantic life science can be seen directly, such as in Schelling's references to Kielmeyer and Blumenbach in *Von der Weltseele*. In addition to impressing a preoccupation with the role of galvanic activity, Ritter similarly helped Schelling develop his *Naturphilosophie* by introducing differential polarity as a possible ground for the occasion of not merely electrical impulse but all manner of force and activity. Röschlaub's interpretation of Brown provided Schelling with a conception of vital phenomena that upheld the individual organism as an independent entity while affirming the fundamental unity of self and world. Whether by generating a community that would eventually be receptive to Schelling's thinking or through Schelling's own efforts to mirror these ideas, the conceptual tendencies that pervaded these earlier Romantic thinkers made Schelling's fusion of natural science and post-Fichtean speculative philosophy a viable model within strong sectors of the German scientific discourse.

Unfortunately, further information on what Schelling studied during the period between 1796–1798 is sorely lacking, due to a temporary lapse in the correspondence between Hegel and Schelling (Fuhrmans 1962, 73).

Whatever it was that he studied, its impact on Schelling was tremendous. In a flurry of productivity, Schelling published *Ideen zu einer Philosophie der Natur als Einleitung in das Studium dieser Wissenschaft*, *Von der Weltseele*, and *Erster Entwurf eines Systems der Naturphilosophie*, all before 1800.

Schelling understood his system of *Naturphilosophie* as both furthering and clarifying Fichte's philosophy by overcoming some of its limitations and taking it into the next phase of its development. Kant had sought to limit philosophy to what we, as subjects, could know. Fichte meanwhile was concerned with the practical, ethical dimension of the transcendental subject. Nature could only be recognized as an idea emergent of the synthesis of necessarily limited perceptions. Both the philosophies of Kant and Fichte had, in this sense, reduced nature to mere mental projection. This greatly narrowed the nature of the interface between self and world. Schelling instead hoped to develop a real, concrete connection between nature and the mind. To accomplish this lofty ambition, Schelling's *Naturphilosophie* shifted focus away from the limitations Kant imposed upon the subject and instead asked how nature must be in order for the human mind to be as it is. This was an exercise in turning Kant on his head: where Kant quested after the categories necessary for consciousness to be possible, Schelling sought to uncover the "categories" in nature that made the human mind possible. Beginning with factical nature as an a priori, Schelling characterized primordial nature as a struggle between an infinite force of expansion and an infinite force of contraction. This dialectical struggle between opposed and yet infinite forces gradually gives rise to quantities such as space and time before generating natural forces and finally the physical world as we understand it. Thus, the entirety of nature is imagined as a self-organizing totality fundamentally determined by conflicting forces. The human mind is nothing less than the culmination of the teleological evolution of spirit, first unconscious in nature and later realized as conscious in the human mind. In this way, Schelling overcame the problems inherent to Kant's subject-object distinction by discovering nature as the dynamic ground of human consciousness, which is distinct from and yet essentially an unfolding of nature itself. Each entity was both a reflection of the dynamic forces behind all that is and a self-contained whole: the many-in-one and the one-in-many.

At the heart of Schelling's optimistic project were the scientific ideas he had been exposed to by teachers, friends, and colleagues. Chief among them were the Brunonian concepts of Röschlaub and the galvanic principles of Ritter, both of which would shape Schelling's conception of living beings. Rather than immediately seek answers in Schelling's published work, it is worth considering one of the greatest scandals of Schelling's life:

the death of Auguste Böhmer. Here, Schelling's support for the Brunonian system, and the vital intoxicant therapy it promoted, figured into a very real life and death struggle with holistic, Galenic medicine. Most of all, it demonstrated the concrete, practical nature of Schelling's belief in the curative potential of intoxicating stimulants.

The Death of Auguste Böhmer

Only fifteen years old at the time of her death on July 12, 1800, Auguste Böhmer was the stepdaughter of August Wilhelm Schlegel and the daughter of Caroline Schlegel (Wiesing 1989, 275). Likely due to the scope of the scandal that surrounded her early death, many aspects of Auguste's life have been obfuscated by centuries of academic debate, at times bordering on scholarly gossip. Officially, she was the daughter of Caroline and Johann Franz Wilhelm Böhmer, although biographer Walter Ehrhardt has made a case for the suggestion that Auguste's biological father may have been Johann Wolfgang von Goethe (Ehrhardt 2006, 277). As for her connection to Schelling, Auguste has been represented as Schelling's fiancée or, at the very least, the daughter of Schelling's lover (Caroline divorced Schlegel and married Schelling later in 1803) (Steinkamp 2002, 478). Whatever the nature of their relationship, the burning question then, as today, was deceptively simple: Was Friedrich Wilhelm Schelling responsible for the death of Auguste Böhmer (Wiesing 1989, 275)?

In May 1800, Caroline, Auguste, and Schelling traveled together to Bamberg, with the stated intentions that Schelling would attend some of Röschlaub's lectures on Brunonianism while Caroline could retreat to the Franconian baths (Wiesing 1989, 277). It was hardly a secret that the secondary purpose of the trip was to allow Caroline (then still Schlegel's wife) and Schelling to spend time alone together (Wiesing 1989, 277). Schelling left Caroline and Auguste to visit his parents, only to be called back in July when Caroline fell ill with dysentery. Caroline recovered speedily enough, only for the young Auguste to be cast into the throws of illness shortly after. Initially, Auguste's treatment was solely overseen by a surgeon from Bad Kissingen named Büchler. But Schelling, having just attended medical lectures from Röschlaub, soon interceded. Though there was tension between the two concerning the proper course of treatment, Schelling officially conceded to the positive prognosis given by Büchler until as late as two days before Auguste's death. This can be seen in Schelling's letters to August Schlegel, although it is worth asking if Schelling was masking a deeper concern for the sake of Auguste's stepfather (Fuhrmans 1962, 196). When Auguste ultimately did not recover, Büchler rushed to place the blame for her death on Schelling's interference, publicly and privately spreading this opinion among the officials and dignitaries of Würzburg (Wiesing 1989, 278).

The primary dispute over the proper course of medical treatment centered around opium. Büchler and Schelling not only disagreed about the appropriate dosage but also the particular preparation that ought to be used (Wiesing 1989, 280). Büchler, consistent with the Galenic compound medicines of the period, prescribed the young Auguste opium mixed with rhubarb tincture and gum arabic (Wiesing, 1989, 280). In a letter to Schlegel, Schelling remarked that the rhubarb tincture and gum arabic was—at best— diluting the opium and—at worst—a laxative that actively undermined Auguste's recovery, instead supporting a prescription of smaller doses of pure opium (Fuhrmans 1962). After all, Brown identified colonic illnesses such as dysentery as typical diseases of asthenia, and pure opium was the remedy most capable of rectifying any radical disequilibrium in vital force (Brown 1795, 182).[6] It is impossible for modern researchers to determine the extent to which Schelling's involvement helped or hampered Auguste's progress. It is clear, however, that perceptions of culpability were inevitably impacted by Schelling's role as an ambassador for the Brunonian system, as well as Büchler's symbolic stance against it. Büchler and Schelling were caught in a real life-and-death struggle over the very concept of disease and the body, with Auguste tragically paying with her life.

Two years after the death of Auguste, an anonymous reviewer in *Allgemeine Literatur-Zeitung* criticized the new philosophy of a medical student named Joseph Reubein, writing ominously about their hope that Reubein not befall the same fate as Schelling by killing those he hoped to heal (Anonymous 1802, 329). This anonymous reviewer, generally identified as either Franz Berg or Christian Schütz, passed off Schelling's guilt as an open secret. Yet Brown's work would continue to exert periodic influence for roughly another twenty years. As for Schelling, this tragic story makes one thing overwhelmingly clear: Schelling's alignment with the Brunonian system was not an association of convenience for the purposes of fleshing out his *Naturphilosophie*, nor a passing interest. Schelling vehemently believed that the Brunonian system would save Auguste's life. In her final days, it was Andreas Röschlaub, then the greatest German Brunonian, that Schelling called upon (Wiesing 1989).

Opium and Bodily Causation

So it comes as no surprise that vital substances featured centrally within Schelling's philosophical system. Powerful intoxicants were not merely potent curatives relegated to the purview of medical theory and practice. They were intercalated within the very structure of Schelling's understanding of organized life, and—by extension—Being itself. An appreciation of what such a lofty attestation could possibly represent merits some further context. Schelling upholds the Brunonian notion that "[t]he vital activity goes out [would go out] without object, it can only be excited by external influence" (Schelling 1799b/1858, 82). But the organism, Schelling contends, is not merely "a machine that winds its own springs" (La Mettrie 1748, 34). Because it is a self-organizing body and not merely a dependent, an organism itself should have some role in constituting itself in relation to external influences or, as Schelling puts it, "this external influence [on the product] is itself determined by organic activity" (Schelling 1799b/1858, 82).

The organism is the primary agent in its own construction—but it can only do so by interacting with the outside world. The external world constantly constructs the organism as object and the organism, in turn, develops itself as subject. For, Schelling reasons, if the organism as subject was determined solely by the input of the external world, then the organism would possess no capacity for independent activity (Schelling 1799b/1858, 82). Because of the organism's *being a subject* it constantly constructs itself through its engagement with the external world, which takes the organism as object. The dynamic unity of activity and reactivity, this duplicity of organism as subject and object, is what Schelling understands Brown to have intended by the concept of excitability. This accounts for the lived experience of the subject as well as, in the spirit of Schelling's doctrine of the all-in-one and one-in-all, the microbiological activities of organic life itself. Thus, the body is not something *acted upon*, as though it were dead material; it is both a body and *productivity*, with the potential for an external influence (such as opium) being made possible only in and through the living body (Schelling 1799b/1858, 144). Schelling frames this as an agonal dialectic where, just as the body warms itself in cold weather, the body constantly strives for equilibrium in the face of constant external influences. In this sense, Schelling firmly establishes that external influences cannot affect the entire organism chemically, as the effects of the external influences are only ever mediated by the organism (Schelling 1799b/1858, 82–83).

In an overt allusion to the central importance attributed to vital substances in this dynamic, opium emerges as the prime example of Schelling's principle of causation in the *Erster Entwurf*, as he states:

> That opium has an exciting effect is explained by its chemical, or, what is the same, its electrical nature (that is why it also works in galvanism), but

its indirect effect, that is, the effect mediated by the activity of the organism itself, is narcotic, and this effect is of course chemically unexplained: because it is indirect. Thus, on the whole it is shown that the very same materials which cause the most violent excitability (which must be explained from their chemical and electrical properties) indirectly exhaust excitability (which, of course, can no longer be explained from their chemical properties). It's no wonder chemical explanations go no further than this. The ultimate effect of external causes on the organism can no longer be explained chemically. (Schelling 1799b/1858, 83)

Opium's potent capacities as an excitatory vital stimulant are emergent of opium's implicit chemical and galvanic properties, while the subjectively dominant effect of narcosis reflects the basic dynamic of excitability. The state of narcosis is not mediated chemically, nor are such effects explainable galvanically. Narcosis is instead the perceptional manifestation of the body's mediation of oversufficient stimulation, nothing less than the occasion of the intersection between the atomized organism and nature. Excitability, understood in this sense, signifies the continuous process of sublation at the ground of all life, the collapse of the duplicitous nature of the organism, as something that acts and is acted upon, into a higher unity. For Schelling, "the entire secret rests upon the on the opposition between inner and outer, which must be conceded if one concedes to anything individual in nature overall" (Schelling 1799b/1858, 84). Opium is a critical exemplification of this idea in that, when a person consumes opium, this principle becomes immanently experienceable.

By merit of the reflexive structure of Schelling's *Naturphilosophie*, the case could be made that Schelling's use of opium here could be replaced with practically anything else. Sleeping, eating, marital relations—excitability is the phenomena at the foundation of the organism, and it mediates all forms of bodily activity. Yet, from the wording alone, it is evident that the reference to opium specifically is fundamental to Schelling's argument. Opium is not merely opium, but the chief representative of a discrete class of substances: "materials which cause the most violent excitability" (Schelling 1799b/1858, 83).[7] Opium represents the pinnacle of what I have termed "vital substances," defined by their unique purchase on the organic, and thus becomes the crucial exemplification of Schelling's concept of the body. Citing anything weaker than opium in the hierarchy of vital excitation leaves space for counterexamples and creates a need for further explanation. The soporific effects of opium intoxication, framed by Schelling's conception of reciprocal activity, themselves become a reflection of the dynamic forces that underlie the foundational conception of the body expressed in Schelling's *Naturphilosophie*, and thereby the world.[8]

Of course, there were other concepts of undeniable significance for the naturphilosophic theory of bodily activity. A great deal of ink has been spilled on the subjects of mesmerism and galvanism—two bygone infatuations of Romantic medicine (Gantet 2021; Montiel 2009; Schelling 1799b/1858). Nor should their import for Schelling be denied. A demonstration of the interplay between the interiority of self and the exteriority of world, the mesmerist-somnambulist dialectic embodied vital action itself (Montiel 2009). The principle of galvanic electricity likewise conditionalized organic function upon the dualistic participation of both receptive actors and a positing environment. Both galvanic energy and animal magnetism were expressions of the elementary symmetry of the cosmos, organic manifestations of the inorganic forces of magnetism and electricity.

Those seeking to critique the claim that vital substances were particularly important for Schelling's *Naturphilosophie* might question the priority of vital substances with respect to these other noteworthy conceptual influences. It cannot be denied that mighty galvanism, which stirred the muscles and marshaled the flesh, was critically significant to Romantics and *Naturphilosophen* alike. Galvanism, and mesmerism, was a phenomenon that made life possible. It was current, rather than opium, that roused Frankenstein's monster from its eternal rest. And yet organic phenomena, among them galvanism and mesmerism, were nevertheless subsidiary to the broader phenomena of excitability. Brunonian excitability referred not to a circumscribable phenomenon, but to the very relationality at the heart of the organic world. Where galvanism and mesmerism were organic reflections of the symmetry realized in the Romantic cosmos, Brown's excitability was the constituting basis of the organic as such. It was intoxicating vital substances that held the greatest power to affect it.

Naturphilosophie, *Lorenz Oken, and Romantic Physiology*

Schelling provided a systematic basis for the integration of experimental science into the post-Kantian conception of nature. These efforts to unite and systematize the disparate Romantic tendencies of many late eighteenth-century scientists through transcendental philosophy would shape much of German scientific thinking for the next thirty years. In the life sciences, the impact of the Brunonian unification of physiology, etiology, and therapeutics was difficult to ignore, and the swelling ranks of the *Naturphilosophen* demonstrated that this was trend was only further propelled by the philosophical supporters of Schelling's system. But Schelling never reached the point of participating in any directed study of the natural world for himself. The self-conscious diffusion of *Naturphilosophie* into the scientific process fell squarely on the shoulders of his followers, of which there was no shortage. Few were as ardent in their championing of *Naturphilosophie* or as influential in their time as Lorenz Oken.

Of all of the students of the natural world who built upon Schelling's *Naturphilosophie*, none is more deserving of the moniker of arch-*Naturphilosoph* than Lorenz Oken. Born Lorenz Okenfuß (1779–1851), Oken was one of the earliest scientific adopters of Schelling's *Naturphilosophie*, demonstrating his commitment from the outset with the 1802 publication of *Grundriss der Naturphilosophie, der Theorie der Sinne, mit der darauf gegründeten Classification der Thiere*. Like Schelling, Oken saw Brunonian excitability at the heart of the organic world and, like for Schelling, that beating heart pumped vital substances. But before considering how Oken further developed the notion of vitally stimulating intoxicants, it is important to first establish Oken's relationship with vitalism, morphology, and Brunonian excitability.

As the title suggests, *Grundriss der Naturphilosophie* was nothing less than a radical attempt to simultaneously deconstruct the Linnaean system of animal classification and establish a novel classificatory scheme that was consistent with the fundamental understanding of nature reflected in Schelling's *Naturphilosophie*. Oken proposed that the number of animal classes could be reduced to no more than five, with each class being associated with the historical development of a particular sense organ (Oken et al. 2007). The most rudimentary of the classes was that of the invertebrates, followed by fish, reptiles, birds, and, finally, mammals. Invertebrates were counted as being one of the oldest and least sophisticated life forms. Fish distinguished themselves through the development of not only a spine but a tongue—realizing the sense of taste (Oken 2007). Reptiles brought with them a true nose, giving animal life the capacity to draw breath from the open air, and birds were associated with the appearance of proper ear openings. Mammals were the last and most developed of the classes, holding all known sense organs in addition to movable eyes with lids. The boldness of this position embodied the revolutionary spirit of *Naturphilosophie*. A system that identified human self-actualization with the teleological fulfillment of nature's grand historical development, *Naturphilosophie* was more than a total theory of the natural world; it was a call to action. Thought of in this way, all life could be understood as progressive elaborations upon an originary theme, which Schelling understood as a mirroring of nature's own developmental process from a primordial interplay of forces. It is to be expected that the social fabric of human society also reflected this pattern, as both Fichte and Hegel (and company) famously argued.

By 1805, Oken's *Die Zeugung* argued that all organic structures consisted of vesicles and cells, which were themselves derived from the collection of *Urschleim* (protoplasma) (Oken 1805). This meant that special differentiation was a matter of disparities in the procedural organization of the protoplasmic masses. The next major development in Oken's work was the publication of the first edition of the *Lehrbuch der Naturphilosophie* in 1810, which would see further editions in the years that followed. While *Grundriss der Naturphilosophie* and *Die Zeugung* focused exclusively on the life sciences, the *Lehrbuch* extended Oken's system into the inorganic world, identifying in the entirety of nature an a priori mode of classification founded on the principles of evolution from more basic forms. As discussed, the blurring of classificatory distinctions was already present in the works of earlier Romantic physiologists like Kielmeyer, Riel, and Blumenbach. But it was the addition of Schelling's *Naturphilosophie* that provided a framework through which Oken could subsume the entirety of the natural world into a central developmental model.

Schelling understood excitability to be the first property of the organism (Schelling 1799b/1858, 144). This conception of the underlying duplicity of the organic body as both a holistic account of the experience of organic bodies and a description of the underlying biological activities that make life possible had a profound effect on Oken's thinking. Unsurprisingly, Schelling himself brought the concept of excitability as an underlying tension or duplicity—life itself—into his understanding of morphology and the historical development of a species. It's clear that Schelling perceived the comparative anatomy of Blumenbach and the associated preoccupation with morphological characteristics as inimical to the possibility of a more essential project of comparative physiology (Schelling 1799b/1858). This is because, for Schelling, the outwardly visible components of organic bodies are actually expressions of the proportional intensities of organic forces inherent to the organism (Schelling 1799b/1858, 171). The oppositional interactions of these organic forces give rise to florid sequences of biological activities, the likes of which determine the physical form of a given organism as well as its specific nature. Thus, any classification scheme established on the basis of anatomical features can only ever point toward a more natural system of speciation on the ground of the variable intensities of organic activities. Schelling identifies the organic forces in question as sensibility, reproductive force, and irritability (Schelling 1799b/1858, 171–72). All of these organic forces, whose entangled conflicts are expressed in the organism as a discrete entity, are ultimately derived from Brunonian excitability, as the first property of all living things (Schelling 1799b/1858, 144).

It is in this way that Schelling succeeds at excising any materialist notions from the Brunonian concept of excitability in order to find in it a reflection of the essential interplay of forces that lie at the feet of Being as a whole. Speciation, thus, is analogous to the originary differentiation of the universe from a primordial unity, with excitability being the overarching principle of organic existence. Species, properly identified, would then be defined by the relationships of underlying organic forces that are derived from the organizational principle of excitability. This is on the basis of the implicit identification of excitability as both that which makes life possible and that which, through the relative fixedness of its derivations in a given organic body, engenders the appearance of species forms.

This becomes a crucial point when considering Lorenz Oken's reception of Schelling, his position as a vitalist, and the influence of vital substances. Oken brought *Naturphilosophie* into the study of the natural world, into the lab, and realized *Naturphilosophie* as a full-fledged research program rather than merely an ad hoc framework. Further still, Oken understood *Naturphilosophie* as the avenue through which the speculative work of natural history might become truly scientific, ultimately combining it with chemistry, anatomy, and physiology into what Oken called *Biologie* (Gambarotto 2018b, 62). Oken's *Grundriss*, *Zeugung*, and *Lehrbuch* were all examples of physiological research conducted with *Naturphilosophie* forming its initial premise, guiding its methods, hypotheses, and interpretations of the evidence.

In spite of Oken's defensible reputation as the quintessential *Naturphilosoph*, there are points of at least seeming divergence. If only superficially, Oken appears to differ from Schelling on a crucial point by assuming an antivitalist position. This is, at the very least, how Hans Driesch generally interprets Oken in his *Geschichte des Vitalismus* (Driesch 1922, 92). Oken does lend his voice to the opposition against the more rudimentary concept of *Lebenskraft* found in Riel and elsewhere, unambiguously asserting that "galvanism is the principle of life [and] there is no other life force than galvanic polarity" (Oken 1810, 10). Yet, both Oken and Schelling prepare distinct modes for understanding the nature of organic activity, not as a vital force but nevertheless as a distinct vital activity. For one, although galvanic energy is closely associated with electricity, Oken does assert the uniquely organic nature of galvanic force. For, "magnetism has a basis: it is metal[,] chemistry has a basis: it is salt[,] so galvanism has a basis: it is organic matter" (Oken 1810, 11). Oken takes this notion as far as the assertion that the "organism *is* galvanism in a homogeneous mass" (Oken 1810, 10).[9]

At such a point it becomes exceedingly difficult to uphold the suggestion that the particular association that Oken establishes between galvanism and organic life can be understood as anything other than vital in nature. Although Oken, like most *Naturphilosophen* and many Romantics, argued for the essential unity of the organic and inorganic world, the objective of Oken's project was explicitly not to reduce life to a by-product of physiochemical processes. Rather, all of Being is likened to an organism, borne out of a primordial oneness (Oken 1810). Organisms, in this way, are elevated to a position of distinction as the culmination of a teleological unfolding of the universe's own self-awareness. Galvanic force may not be a form of energy or force that is incommensurate with the inorganic world, but it does have a distinguishing relationship with organic bodies and, in this sense, galvanic force might be considered "vital."

Perhaps most important in making the case for Oken as a vitalist is Oken's position concerning Schelling's conception of excitability. As Oken himself states, "excitability is the most general phenomenon of organic matter, and it belongs to plants and animals" (Oken 1810, 134). Oken's "excitability" [*Erregbarkeit*] is "the ability to assimilate nature" (Oken 1810, 134). On this topic Oken's thinking seems to be at its absolute closest to Schelling's, as he states in the *Lehrbuch*: "In feeling the animal always goes beyond itself. So it's just excitability. In motion the animal remains in itself. The sense of self emerges only from both states [as both nature and differentiated being]" (Oken, 1810, 135). Galvanism is likened to the vital force in the sense of *Lebenskraft*; it is the animating force. What galvanism is not is the first principle of vital activity, for, as Oken clarifies, the *first truth* of the organic is excitability, as the very possibility of animation.

All of these aspects of Oken's thinking, the continuity between Schelling's and Oken's formulation of vital activity, the underlying principles governing morphology, and the nature of excitability, make up the foundation of any analysis of the place of substances of intoxication in Oken's system. Though Oken does make sparing remarks in his earlier publications that demonstrate that Oken understood opium to be a highly effective and relatively safe therapeutic, there is an unfortunate scarcity of direct remarks on the function and nature of opium, or any other therapeutics, in Oken's corpus (Oken et al. 2007, 48). This is not on account of a judgment of insignificance on Oken's part, but a result of the broad and systematic approach embodied in the understanding of *Naturphilosphie* and *Biolo-*

gie that guided all of his endeavors. That said, it is possible to assess the extent to which the legacy of "vital substances" was perpetuated through his work.

As a starting point, it is worth recalling that Schelling himself understood the excitatory potential of opium and alcohol as pertaining to their apparent galvanic properties. In fact, Schelling is so certain of opium's galvanic properties that he goes as far as, quite confusingly, suggesting that opium, as a plant principle, depresses a plant's potential for stimulation (because Schelling holds the irritability of plants to be a negative form of galvanism) (Schelling 1799b/1858, 82). This is also the case for Oken, as will soon become clear.

All galvanic processes rely on polarization, the unresolved differential of which is the basis for further movement, development, and activity. Substances of intoxication would thus be expected to affect the body through some process of polarization. With respect to opium or other plant-based intoxicants, Oken's *Lehrbuch* does make mentions of medicinally active plant principles, where he classifies them alongside plant poisons and elsewhere refers to alkali plant bodies as "betäubend" (stupefying/narcotic) (Oken 1843, 209). This classification on its own is hardly remarkable. What is significant is the implicit sense in which Oken makes use of the term "poison."

Oken argues that poisoning of the organic body is destruction by some means of excessive polar equalization (Oken 1843). In acute cases or irregular contexts, a disruption in the galvanic process, comparable with *Lebenskraft* no less, could threaten the life of the organism. Yet Oken's concept of poisoning is in actuality considerably broader that this initial definition suggests. Saliva, for example, is identified as a poison that breaks down organic matter (food) into the most basic organic components so that they can be taken up (Oken 1843, 350–51). Localized and controlled "poisoning" is thus part of the essential maintenance of any organic being. Reframing the earlier definition, it could be said that poisoning is a disruption in the galvanic process (Oken 1843, 350).

It is worth questioning whether Oken's categorization of medicinally active plant principles alongside vegetable poisons was not merely coincidental and suggests that Oken's medicinally active plant principles function via this neutral concept of poison. Oken suggests that the difference between chemical or animal poisons and vegetable poisons is that plant derived poisons are nervous poisons—they affect the nervous system (Oken 1843, 350). This sheds light on Oken's use of *betäubend* to describe alkali plant products: Oken's stupefying plant alkalis, medicinally active principles, and (some) vegetable poisons are one and the same, referring to substances of intoxication such as opium. To be intoxicated is to be poisoned. More specifically, Oken says that "vegetable poisons return the animal to the vegetative state" (Oken 1810, 340). The animal nervous system is overtaken by a disruption in the galvanic polarity before finally the animal's senses are pulled into a stupor.

It is clear that Oken's concept of poisoning serves as a model for intoxication. Consistent with that most storied Paracelsian maxim "the dose alone makes the poison," the separation between medicinally active plant principles, alkalis, and dangerous poisons are largely matters of degree for Oken (Paracelsus 1538/1965, 510). After all, vegetable and animal poisons harm or help the subject by intervening in the balance of galvanic polarity, thereby undermining the vital process itself. This leaves an open question about how exactly this process takes place. On this point, Oken offers two possible modes of action: "Neutralization or excessive polarization" (Oken 1810, 340).

Oken provides very little by way of direct explanation of the activity that gives rise to poisoning either by way of neutralization or by way of excessive polarization. He does, however, identify animal poisons, and particularly saliva, as blood poisons. Saliva, which Oken likens to all animal poisons, poisons organic substances through equalization of their galvanic polarity and thus neutralization of the organic substance into mere infusorial mass (Oken 1843, 350–51). This leaves vegetable poisons, including substances of intoxication, to account for poisoning by means of excessive polarization.

As with Schelling, the potent effects of substances of intoxication are a direct reflection of their latent galvanic potential. On the basis of what has already been adduced on the topic of poisons, Oken appears to suggest that opium and other stupefying plant alkalis affect the nervous system not by neutralizing bipoles but by pushing the state of polarity to an extreme, by overcharging its galvanic potential and pushing the vital process to the point of annihilation. However concealed substances of intoxication may be by Oken's prose, the significance they held for Schelling appears to have remained largely intact. The Schellingian infatuation with intoxicating vital substances, and their potent effects on the vital process, was more than a passing whimsy. The vital substance concept endured translation into the praxic study of the natural world. Ceasing to merely describe nature, to coordinate objects, instruments, methods, and concepts, the vital substance concept was now encountered in the world, in situ.

Oken functions in this story as one of many bridges between physiology and medicine, a bridge built at least initially upon the concept of vital substances and their determinative role in the body. He was hardly the only impactful scientist who found themselves cleaved to *Naturphilosophie* or even simply the vital substance concept. Oken also fostered a number of particularly influential pupils, among them Johannes Müller, who would exercise a profound influence on the course of late nineteenth-century physiology. But these details are yet to come. The Romantics did not "discover" the biological subject. Although Oken and the other *Naturphilosophen* were fixated on the notion of the holistic unity of the organic and inorganic worlds, bodies themselves remained vital in their teleological wholeness. The "golden flood of grapes—the almond trees' miracle oil, and the brown juice of the poppy," in the words of Novalis, vitalized the living body—the body remained a vital entity, a stranger to modernity's chimeric biologism (Novalis 1800/1988, 14). No, the "biological subject" would not materialize for nearly another century, and such a discussion will first entail passing through an epoch of deceptively familiar physicalism in scientific thought. But before we get ahead of ourselves, we must address the breaking point of a quietly earthshaking revolution from within the domains of Romantic science, one that continues to shape our world: the discovery of morphine and the dawn of the age of alkaloids.

✷ 4 ✷
A (Brief) Historical Ontology of an Alkaloid

In Search of the Principium Somniferum

The Paderborn of 1804 was scarcely more than a quiet bishopric nestled around the source of the spring-fed Pader river, from whence it got its name. The city was then still a Prussian subject, yet to fumble into the hands of the French vassal Kingdom of Westphalia in the years between 1807 and 1813. This provincial locale would seem an unlikely setting for one of the nineteenth century's most transformative developments—far in every sense from the hallowed halls of the Académie des sciences—and yet it was. It was here that a young pharmaceutical chemist by the name of Friedrich Wilhelm Adam Sertürner would isolate morphine from opium, simultaneously giving rise to a novel classificatory scheme that included "the most important of all the substances of the materia medica": the alkaloids (Hay 1882/1883, 719).

The novelty of morphine's debut on the world stage was not merely a matter of historical precedence. It was the very consummation of the concept of the "active principle," the theory that the effects of medicinal plants and animals stem from the ingestion of a particular substance, or principle, which can be distinguished from its source material. Simultaneously, this was the genesis of an uncountably large category of heretofore unrealized scientific objects, engendering the mobilization of pharmacy into a field of sincere experimental inquiry. The discovery of morphine, and with it the active principle concept, would fundamentally alter the viability of the extant classificatory systems of medicines and their corresponding theories of action. Classification of plant therapeutics on the basis of external characteristics becomes more complicated when the "active" plant principle can be extracted, distilled, and purified, thereby erasing the origin plant itself within the equation. It was, to put it simply, the advent of modern pharmacy.

In 1805, the German pharmacist Friedrich Wilhelm Sertürner (1783–1841) isolated a precipitate from opium for the first time, identifying it as the primary active ingredient (Patil 2012, 165). It was not until a later 1816 paper that Sertürner would not only claim to have extracted the active constituent of opium in the form of what he called "Morphium," but also imply that the substance likely belonged to a new class of organic matter. "By the use of ammonia," Sertürner rendered a precipitate "from an aqueous opium extract," which he could then "crystallize from ethanol" (Phillipson 2012, 4). He published an article on his peculiar findings titled "Darstellung der reinen Mohnsäure," with little impact (Sertürner 1806). After the publication of his 1816–1817 paper, the subsequent recreation of Sertürner's method led Pelletier and Caventou to the isolation of "quinine, strychnine, veratrine, caffeine, and emetine" (Patil 2012, 165).

This was the beginning of something new, coursing, and immense. Yet, "before 1817 the existence of alkaloids was scarcely suspected" (Lesch 1981, 305). What set the stage for this rupture in the rational hegemony? Prior to the emergence of the alkaloid as a new kind of thing to be, the rise of experimentalism had normalized the expectation that traditional therapeutics achieved medicinal validity through a process of empirical investigation. Concurrently, new discoveries in plant chemistry helped define the conditions under which the investigation of plant- and animal-based therapeutics took place. In the late eighteenth century and early nineteenth century—the world Sertürner inhabited—the most potent expression of this project was that of the Lavoisierian research program of organic principle analysis.

There is likely no single field of scientific inquiry that contributed more to the appearance of alkaloids at the beginning of the nineteenth century than plant principle analysis. In the decades preceding Sertürner's publication, the work of diligent experimentalists in plant analysis produced a wide array of methodologies for classifying various plant components. Despite productive efforts, Antoine Fourcroy (1755–1809) and his contemporaries complained about the lack of understanding concerning plant chemistry and dreamed of narrowing the gap between natural history and chemistry (Klein and Lefèvre 2007, 240). One of the principal difficulties faced by researchers in the field, which likely impeded the sense of steady progress, was a complete lack of agreement regarding how plant principles were to be separated.

Organic principle analysis outwardly appears to wed eighteenth-century classifactory and taxonomic impulses with the theoretical-methodological innovations of the Lavoisierian school. It could, in this sense, be understood—as Heilbron suggests—as an expression of the consummate "quantifying spirit" at the heart of the Enlightenment impulse to measure, weigh, and quantify (Heilbron 1990, 2). Or, as John Lesch understands the phenomenon, it might instead be understood as an extension of the model realized in Enlightenment natural history (Klein and Lefèvre 2007; Lesch 1981; Lesch 1990). Ursula Klein and Wolfgang Lefèvre question the validity of both hermeneutical frames, pointing to the disarray that proliferated the various classificatory approaches employed in organic principle analysis. Firmly rooted in the tumult of late eighteenth-/early nineteenth-century Paris, what can be said is that proximate principle analysis embodied a revolutionary spirit. Each distinguishable component of an organic entity was to be dissembled and organized in accordance with its rudimentary elements. A sprig of rosemary ceased to be comprised of leaves, stems, and oils, but became instead lists of vaguely associated components, each available for further deconstruction into their most basic elements. The ancien régime of discrete organic holisms was no more.

At the most basic level of classification, it was generally agreed upon that organic substances were distinguished from inorganic substances. From there, Fourcroy systematized plant principles according to those components that were separable without chemical alteration (Foye et al. 2002, 2). Keeping with his mentor Jean-Baptiste-Michel Bucquet, "Fourcroy distinguished between the common 'humours' or 'sap' of plants, and particular juices" (Klein and Lefèvre 2007, 240). He then "further distinguished juices separated from plants 'by mechanical means' without any alteration, which he defined as their 'proximate principles'" (Klein and Lefèvre 2007, 240). Thomas Thomson, for comparison, created four classes of plant principles, of which three "were defined primarily on the basis of the relative solubility of their members in water, alcohol and ether" and the fourth according to the presence of trace minerals (Lesch 1981, 307).

Evidently, the nature of plant bodies and their components was available as an object of consideration and negotiation, but "whatever the basis of classification, plant alkalis did not appear as a category" (Lesch 1981, 307). Of course, the existence of plant acids was already firmly established: Carl Wilhelm Scheele identified citric and malic acid in the wake of Antoine Lavoisier's development of the oxygen theory of acids in the eighteenth century, finding that natural plant acids played a significant role in vegetative chemistry (Kremers and Sonnedecker 1986, 359). The question faced by those studying plant acids was no longer if they existed, but rather their nature, their function, and their role in the rightful categorization of basic plant principles. In spite of this, there is little to no speculation on the possible existence of plant alkalis in the period before 1816/1817. For all of the energy churned into the careful consideration of the nature of plant bodies and their structure, for all of the variety realized in the classificatory approaches taken to the separation of plant principles, there was no space for the idea of plant alkalis. Thomson "discussed alkalis as a group only under mineral chemistry," suggesting that an alkali presence in plant materials stemmed from mineral contents absorbed from the environment rather than from a distinct plant principle (Lesch 1981, 307). Potash seemed a probable explanation, as it was believed that plants could readily take up potash found in the soil.

Thomson's assumption makes one thing overwhelmingly clear: within the thinking at the turn of the nineteenth century there is no apparent rational necessity that organic and inorganic chemistry have an analogous relationship. Even if the relationship between acids and bases in mineral chemistry could be given analogous consideration, this relationship did not translate into the reality of plant bodies. Such notions rubbed against the Romantic principle, just discussed, of the originary unity of all matter, and the corresponding symmetry of Being. However, this assumption went well beyond a skeptical empirical attitude, which refused to recognize plant alkalis on grounds of insufficient proof. There was, after all, evidence of an alkaline presence in plants. In the face of such contradictory results, scientific chemists could point "to the work of Thénard and Chevreul that had shown that many plant bodies could enter into close combinations with acids" (Lesch 1981, 319). The widespread consensus on the importance of Thénard's and Chevreul's work, and the subsequent blackboxing of the "fact" it supported, provided further grounds on which to disregard the consideration of a substance's taxonomic identity on the basis of its perceived acidity or alkalinity. It was not that this was the work of poor scientists or sloppy chemistry. The question of whether plant alkalis could or could not exist was simply not eligible for consideration within the ruling domain of the rational. The conceptual foundations necessary to make possible the question of whether or not plant alkalis existed were simply not there. Pivotal to understanding this claim is a consideration of the basic a priori structures that supported the epistemological validity of competing systems of plant principle classification.

Eighteenth-century research into plant principles operated within its own distinct series of historically structured classification schemes. The specifics of these classification systems were still fairly fluid at the beginning of the nineteenth century. Plant principles could be differentiated according to what was observable in a plant's "natural" state (meaning unmanipulated), as Fourcroy attempted, or according to their solubility in water and other solutions, as Thomson argued. Debates would rage over the proper means of classifying proximate plant principles (Klein and Lefèvre 2007, 241). While these details were more or less up for negotiation, the most foundational assumptions that underlay these classification systems enjoyed almost unanimous agreement among those involved with plant principle analysis in France and England. In Germany, Friedrich Gleb, Rudolph Vogel, and Johann Wiegleb raised their voices in support of the French model, decrying the Romantics and *Naturphilosophen*, and agreed upon the essential difference between organic bodies, meaning plants and animals, and inorganic bodies, meaning minerals (Klein and Lefèvre 2007, 241). In England, this sentiment was shared by William Lewis and in France by Pierre Macquer, Jacques François Demachy, Jean-Baptiste-Michel Bucquet, Jean-François Derosne, and Antoine-François Fourcroy (Klein and Lefèvre 2007, 241). Whatever the system of plant analysis, every approach was foundationally structured according to the assumption of a fundamental, natural distinction between organic and inorganic matter. It would have seemed irrational, absurd even, to overturn this basic notion and consider an equivalence between the chemistry of life and the chemistry of matter.

Although the rise of Romanticism in Germany had done a great deal to advance the argument that there was in fact a fundamental similarity that underlies the structures of the organic and inorganic worlds, by the time of Sertürner's surprise appearance, conventions in the chemical classification of organic substances, and the theories that supported these classifications, were still primed to resist Sertürner's suggestion that the substance he would call "Morphium" was a salifiable base. Prior to the isolation of morphine from common opium, there is evidence of an unconscious resistance to the possibility of plant alkalis. Why might this be and how is it overcome? It is very clear that the approach taken to the analysis of plant principles and their chemistry was conditioned by the assumption of a fundamental difference between organic and inorganic materials. The theoretical assumption of this difference determined the range of rationally permissible classification schemes in such a way that the chemical relationships seen everywhere in mineral chemistry would not easily be assumed in the case of plant principles. This can be seen in the work of some of Sertürner's predecessors and contemporaries.

Sertürner was not the first to isolate a substance from opium and declare it the *principium somniferum*: Derosne, Armand Jean François Seguin (École Polytechnique), and Louis Nicolas Vauquelin (Fourcroy's assistant from 1783–91) each yielded white, odorless powders from opium, but failed to give it a proper name or identify it as a salifiable base

(Hay 1882/1883, 719). Derosne, as discussed, might have isolated morphine before Sertürner and even published a paper on the subject in 1803 ("Mémoire sur l'opium") that was translated to German in 1804 ("Über das Opium"), but failed to name or classify his substance (Hay 1882/1883, 719). Derosne recognized the alkaline-like quality of his precipitate but overlooked this anomalous consideration because his primary concern was the isolation of the *principium somniferum* (the soporific principle) (Derosne 1804). Seguin had also isolated a similar type of substance from cinchona bark, while Vacquelin, among the first to suggest that singular substances might account for the effects of medicinal plants, produced an isolate of *Daphne alpina*. Derosne and Sertürner both isolated their substances in the form of a precipitate from an aqueous opium solution and both turned litmus paper blue green, suggesting the presence of an alkali (Phillipson 2012, 4). Derosne simply could not, or would not, concede to the existence of a plant alkali; it had to be potash contaminant. Contrasting Derosne, Sertürner "drew the conclusion that the basic nature [of the precipitate] was an integral property of the substance," despite his editor's protestations (Phillipson 2012, 5).

The scope of Sertürner's work and research expands considerably in the transition from his relatively novice 1806 publication to his 1817 paper. In his 1817 publication, Sertürner does a comparison of the crystalline structure of his precipitate and that yielded by Derosne. In differentiating between the structure of the crystals formed by both substances, Sertürner puts forward the suggestion that Derosne's isolate was contaminated by meconic acid, thereby accounting for his failure to identify morphine as a plant alkali (Sertürner 1817). Years later, it was decided that Sertürner was wrong about Derosne: Derosne's *Sel Nacrotigue de Derosne* was found to be a different alkaloidal constituent of opium, having no recognizable narcotic effects (Kapoor 1995, 14). Then, so confident in his findings that the protestations of his editor and friends could not dissuade him, Sertürner's 1817 publication moves away from the agnosticism of his 1806 claim to firmly make a case for the discovery of the new plant alkali (Phillipson 2012, 5).

This hesitance on the part of chemists, pharmacists, and medical experimentalists baffles the modern mind. Derosne's shortcoming was not the result of any methodological misapplication. On the contrary, it becomes clear that Derosne's method was more than sufficient to achieve Sertürner's results. Even an alkali that failed to produce soporific effects could still be made to *perform* as a plant alkali. This pattern of encountering strange substances in the process of plant principle analysis and uneasily forcing them to fit within the preexisting categories pervaded the pre-

1816/1817 isolations of active plant principles. In this case, it appears that chemical "language was insufficient to describe the new reality" (Lesch 1981, 12). Here the language of the scientist succeeded in describing their world: with the existence of plant alkalis already excluded on a theoretical level, language rightly described the world to correspond with the basic conditions of rational viability.

If there was a discontinuity to be found here, everything seemed to point toward an unseen error on the methodological level. The assumption of a potash contaminant affecting Derosne's results was not a product of any demonstration of the presence of potash, nor was it a negative heuristic erected in defense of a primary hypothesis. The detection of alkalinity being attributed to potash in the plant body was an a priori judgment derived from established theory, part of the background noise of theory and methodology that makes any claim about the natural world possible. In this sense, the anomalous alkaline white precipitate only becomes anomalous retroactively.

Sertürner the Romantiker

How is it then that Sertürner came to a different conclusion? One explanation championed by Reinhard Löw, and very much in line with a familiar notion from Thomas Kuhn's *The Structure of Scientific Revolutions*, suggests that Sertürner's youth made him more susceptible to the influence of the Romantic idea of universal polarity (Löw 1977, 310; Kuhn 2012).[1] Löw's argument suggests that the dynamic polarity that the *Naturphilosophen* understood to lie at the very foundations of nature, that had been so crucial to Ritter, Schelling, Oken, and their allies, led Sertürner to consider that plant alkalis ought to exist if plant acids did as well (Löw 1977, 310). John Lesch, however, finds the evidence behind this claim to be lacking. Perhaps rightly so, as there is no mention of polarity in any of Sertürner's writing from this period. If he was familiar with the concept, it does not appear to be foundational to the arguments presented in any of his publications. Instead, Lesch proposes that, as a practical pharmacist, "[Sertürner] was not committed to an exclusive distinction between organic and inorganic," freeing him up to consider the results from a different perspective (Lesch 1981, 319).

Of course, this narrative has its own problems. Sertürner's 1806 publication "Darstellung der reinen Mohnsäure," the work of a youth both in age and in chemical experience, does not concede to the existence of a new class of alkali plant principles, but rather remains true to the findings of his French contemporaries in remarking that his isolate was simply alkali-like

(Sertürner 1806). Lesch's proposal is further belied by evidence to suggest that, despite having a less formal education in chemical theory, Sertürner was acquainted with the work of some of his contemporaries (Sertürner 1806; Schmitz 1985, 65). Sertürner was initially unaware of Derosne's work, and it is only *after* encountering Derosne's paper that he makes the claim that morphine is an alkaline base in his 1817 publication "Über das Morphium" (Sertürner 1817). It should be admitted that this is hardly conclusive, and the case can be made that Sertürner believed his precipitate to be a true alkali in 1806, as among his concluding remarks is the suggestion that similar substances might be at play behind other poisonous or medicinal plants as well as alkali plant principles (Sertürner 1806).

While Lesch's argument is compelling, it ultimately performs one of the historian's cardinal sins in seeing too much of the future in the past. It is all too easy from a modern perspective to perceive the discovery of alkaloids as an implicit rebuke of vitalism, perhaps a crucial one, but this was not the context in which Sertürner conducted his experiments. On the contrary, Sertürner's work was a testimony to the reality of vitalism. Even though the classificatory distinction between organic and inorganic matter arose on the basis of vitalistic concepts, it was fully possible in early nineteenth-century Germans to uphold vitalism and advocate for an allegorical similitude between all physical substances. For Sertürner was not only a chemist but a Romantic chemist.

Friedrich Sertürner's knowledge of chemistry was largely self-taught, entering the profession of pharmacy as both a means to hone his craft as well as a way to make ends meet. In the German world of the eighteenth and early nineteenth centuries, "the only possibility young chemists and pharmacists had of doing experimental work in a laboratory was in a pharmacy," and this seems to be the rationale behind Sertürner's own choice of profession (Schmitz 1985, 63). Early on, it is clear that Sertürner was inspired by the work of Erfurt pharmaceutical chemist Christian Friedrich Buchholz, who had been educated by his uncle Wilhelm Buchholz, a Romantic philosopher and science adviser to Goethe (Partington 1962, 581). Being no older than twenty-two at the writing of his first publication on opium's somniferic principle, it comes as no surprise that Sertürner's initial experimental approach and writing style is heavily indebted, even derivative, of C. Buchholz's own publications (including his work on opium). Sertürner's first publication on opium was actually a cursory text in Trommsdorff's *Journal der Pharmacie für Aerzte, Apotheker und Chemisten*, titled "Säure im Opium" (1805). "Säure im Opium" is both stylistically and methodologically similar to Buchholz's 1800 "Versuche die Zerlegung des Opiums," a work on Buchholz's own experiment in identifying the volatile narcotic principle in opium (Sertürner 1805; Buchholz 1800; Maehle 1999, 190). Sertürner's now famous "Darstellung der reinen Mohnsäure" also appears to proceed from Buchholz's instructions in "Versuche die Zerlegung des Opiums" regarding the separation of plant principles.

The place of Christian Buchholz in the mind of the young Sertürner provides key context concerning Sertürner's own scientific perspective, especially as a self-taught chemist. The importance of this influence becomes clearer in light of Sertürner's later self-proclaimed identification with Romantic theories of science. Between 1820 and 1826, Sertürner would identify himself with the project of *Naturphilosophie* in "System der chem. Physik" (1820/22) and "Annalen für das Universalsystem der Elemente" (1826). Who is to say how early Sertürner began to personally identify with *Naturphilosophie*, but the foundations of this position can nevertheless be traced back to Sertürner's earlier interest in Buchholz. Though Buchholz never openly donned the moniker of *Naturphilosoph* himself, his work, his perspective, was unquestionably intertwined with those of the Romantics and the *Naturphilosophen*.

Nor is the suggestion of a direct Buchholzian influence the only evidence of Romantic and naturphilosophic influences on Sertürner's early years. Around the time that his earliest work on opium was published in 1805/6, Sertürner had also participated in a competition held by the Institut de France on the phenomena of galvanism, a topic of extreme importance to *Naturphilosophie* as well as their varied conceptions of vital substances (Schmitz 1985, 62). This is all to say that Sertürner was not merely some brilliant young mind who, left to his own devices in a backwater bishopric, happened upon one of modern medicine's greatest discoveries. He was evidently aware of, if not directly involved in discussions surrounding, the principal theoretical considerations at work in naturphilosophical science and that such an engagement took place even from the earliest stages of his career. Further still, this early interlocution between Sertürner and the Institut de France casts further doubt on Lesch's thesis that Sertürner's capacity to overcome entrenched classificatory boundaries was a function of his relative ignorance as a practical pharmacist.

Creatio ab Experientia

At this point, it becomes possible to understand why morphine's genesis as a scientific object occurred at the hands of Sertürner in the Germanies between 1805 and 1816 rather than in France at those of Derosne. We can (fairly) safely assume that the entity in question, the substance now known as "morphine," was present in the opium obtained by Derosne for use in his experiments. Few—even Bruno Latour—would deny the discrete *thingness* attributable to the entity later intended with reference to "morphine," as much as the terminal realization of morphine as an alkaloidal opioid-receptor agonist is sustained by a fragile historicality (Latour 2000). The experimental apparatus and its associated methodology were almost identical in the cases of Sertürner and Derosne; the underlying chemical principles implicit to the method of extraction were not up for debate. It was the existence of a narcotic plant principle as the cause of opium's physiological effects that was up for candidacy as true or false, and so the existence of the entity in question was theoretically permissible. In this respect, Derosne was enormously successful: he extracted a white crystalline substance from opium that he believed to be the essential "salt" of opium. The case can thus be made that it was ultimately Derosne who first identified an "active principle" in the field of plant principle analysis, in spite of his failure in identifying the precipitate as a distinct substance, a new *kind* of thing in the world. But when it came to the alkalinity of this "essential salt of opium" it was the particular theoretical foundations of Deronse's scientific encounter with that which precluded him from understanding it as a base.

Derosne's barrier was a matrix of historical a prioris, things which, though not strictly physical beings, are inextricably rooted in a time and a place, through institutions, books, social ties, and so forth. The categorical exclusion of plant alkalis as a possible thing to be, even as a candidate for consideration as true or false, is inextricable from the profoundly institutional influence of Parisian chemistry on a gentleman-pharmacist like Derosne, brought up in the field by his famous cousin Louis-Claude Cadet de Gassicourt. Meanwhile, Sertürner's pharmaceutical practice in Paderborn could not have been further from Paris's community of lively chemical research. The scientific atmosphere of Germany sustained the conditions for its own historical environment, characterized by its own categories of historical possibility. It is perilously easy for modern thinkers to conceive of a scientific monoculture—a fantasy now and even more so back then. Hence, when entities, experimental apparatuses, and theoretical frameworks gathered in Paderborn, Sertürner was open to the possibility of plant alkalis, while also acknowledging the novelty of this claim.

It was Sertürner's famous 1816/1817 publication, "Ueber das Morphium als Hauptbestandteil des Opiums," that actually saw the genesis of alkaloidal chemistry. In many ways this paper was a rehash of the paper published ten years earlier, the primary difference being Sertürner's willingness to see through conclusions he had earlier only alluded to, aided by the inclusion of further experimental methods. It appears that, outside of the apparent rhetorical necessity of mustering the most current methods, many of the changes to the overall methodology of Sertürner's earlier paper serve to directly usurp, as well as sublate, Derosne's work into his own. One such method is Sertürner's description and illustration of the crystalline structure of his precipitate. Sertürner eagerly points out that Derosne included a description of the crystalline structure of his own extract, which "shoots to one side in a prismatic shape at an angle of 30 to 40 degrees" (Sertürner 1817, 64).[2] This is markedly different from Sertürner's description of pure morphine crystals as "completely colorless and in completely regular, horizontally lying parallelepipeds with sloping sides," turning litmus paper blue (Sertürner 1817, 64). After reviewing morphine's appearance in the form of various acid salts, Sertürner concludes that Derosne had not succeeded in extracting a pure extract of morphine at all (Sertürner 1817, 67).

An interesting aspect of Sertürner's painstaking analysis of the varied structure of morphine and its salts is how little it brings to bear on the search for the *principium somniferum*. It demonstrated that there was some difference between what he and Derosne described, as well as allowed later researchers to know if they had reproduced Sertürner's results. It was, to borrow such language, rhetorical. Determinations of parallelepipedic structure as prismatic or otherwise was immaterial to what Sertürner sought to establish. Such methods were effective at sublating prior efforts to obtain opium's somniferic principle, albeit only by differentiating Sertürner's study within a shared field of scientific practice. Where then did the dam break? It has been established that Sertürner operated within spaces at least partially conditioned by a distinct set of a prioris, but this is merely—to misappropriate an *angsty* Kierkegaardianism—the possibility of [a determined horizon of] possibility. The task of the historian is to ascertain the conditions that precipitated in the concretization of a particular possibility in the world. What engendered the realization of morphine out of a range of other possibilities at the hand of Sertürner?

There was an additional methodological plane through which Sertürner was able to discern his work from Derosne's, yet undiscussed and far more consequential with respect to the *principium somniferum*: the means of establishing physiological effect. Derosne tested his isolate on animals, with no narcotic effects of note (Derosne 1804). Sertürner meanwhile disparaged the value of animal tests in ascertaining the effects of a pharmaceutical product (Derosne 1804; Sertürner 1817, 68). Instead, Sertürner enlisted the help of three teenage boys to partake with him in a bout of dangerous self-experimentation. Sertürner and the lads consumed a dose of half a grain of morphine in diluted alcohol, followed thirty minutes later by an additional half-grain and a further half-grain fifteen minutes

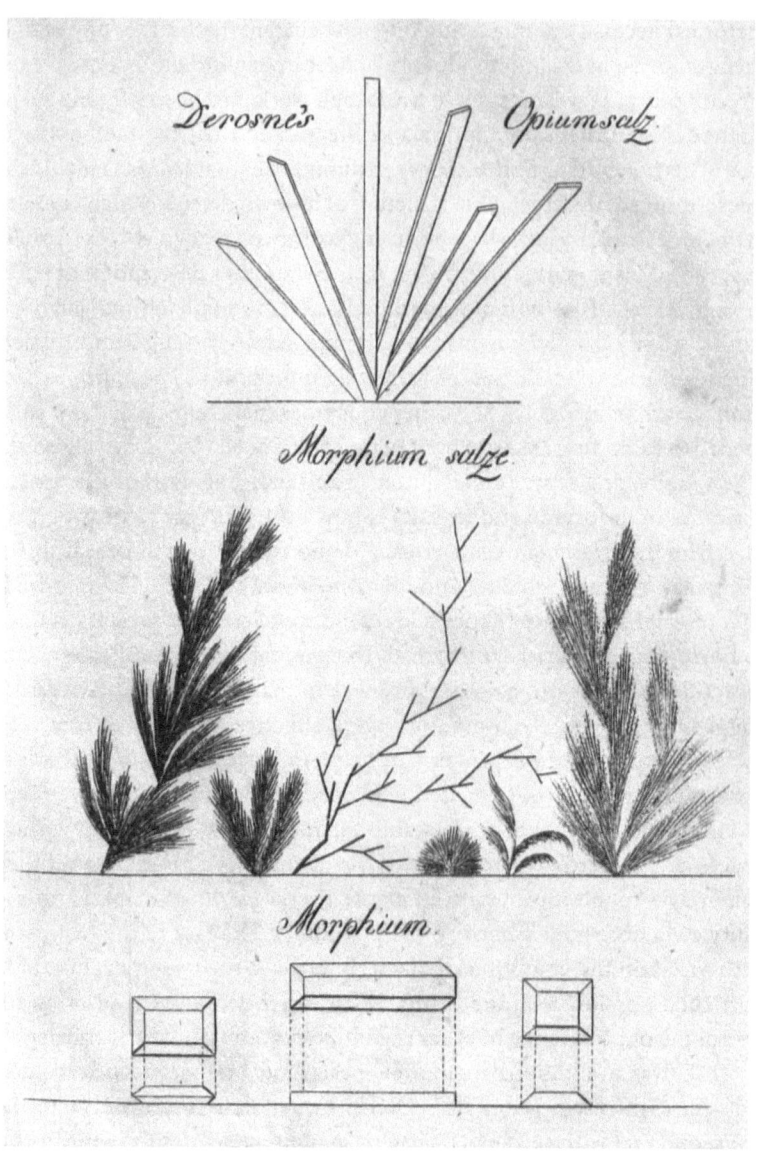

FIGURE 3. Sertürner's drawings of Derosne's opium salt (*top*), Sertürner's morphine salt (*middle*), and morphine (*bottom*) (Sertürner 1817). Note the differences in crystalline structure. Courtesy of the National Library of Israel.

thereafter. The effects were so pronounced that by Sertürner's own account he soon worried for the mortal well-being of both himself and the young men, forcing them all to quickly down strong vinegar to induce vomiting (Sertürner 1817, 69).

Harrowing and ethically dubious as Sertürner's account may be, it is in this process of self-experimentation that the vital substance concept and what I have called intoxication as a way of knowing rears its head. Just as the influence of intoxication as a form of tacit knowledge was allegedly the inspiration behind the Brunonian system and evidently influenced its reception among the Romantics, Sertürner's experience with *Morphium*, the god of dreams, takes center stage during the most crucial step of Sertürner's analysis. Here, this tacit "feeling" of intoxication is so much more than mere feeling; it is a form of knowing, of understanding—it imparts information about the world and the body in that world. The very first words Sertürner musters in recounting morphine intoxication, beyond a flushing of the eyes and cheeks, is that "the vital activity seemed to increase overall" (Sertürner 1817, 69).

No more than a few words. Yet in the context of the 1810s in Germany (coupled with Sertürner's Romantic and naturphilosophic background), it becomes perfectly clear that Sertürner understood the newly realized morphine to bear vital "effects." This is further demonstrated by the battery of experiments Sertürner conducted on the galvanic properties of his novel precipitate, a major point of discussion in the naturphilosophic understanding of opium's vital properties (recall, if you will, Schelling's discussion of opium's galvanic properties). It is unclear the extent to which Sertürner was exposed to a "learned" conception of vital substances, as most of his work on naturphilosophic theories of science were written later in Sertürner's career. Nevertheless, the very first time in history that a morphine isolate was consumed by a human being it was associated with a vital response in the body: morphine was a vital substance.

When Sertürner, opium, reagents, solvents, and pharmacy glassware gathered in the setting of Romantic Germany in the mid-1810s, a novel kind of thing edged its way within the ontological horizon of scientific study. Sertürner testified on the nature of this new scientific object with his own body. Contrary to the arguments made by Lesch, this gathering of entities saw the emergence of morphine not in spite of vitalism but because of a rudimentary belief in the fundamental unity of worldly phenomena on the basis of an organic unity, of which the concept of vital substances was an active component. Morphine, the first time it was identified and named for the god of dreams, passed onto the horizon of scientific consideration as a "vital substance."

What does it matter how Sertürner understood morphine, or the emergent class of alkaloids, as alkaloids are entities about which our knowledge is constantly evolving? There are what we will provisionally call upper-tier theories. They're only upper tier because of their historical prominence in the work of identifying a given theory's candidacy for truth or falsehood. There are also what might be called second- and third-tier (and beyond) theories, which are also attached to a scientific object but are less prominent. All of these designations are in a state of constant flux and are specific to each encounter with a given scientific object. Many of these lower-tier theories are in every sense subject to ruptures in the prevailing domain of reason, passing out of candidacy for truth or falsehood almost unnoticed. There are also far more amorphous, almost unrecognized, associations. All of these theories and associations persist in our relationship with scientific objects because scientific objects are entities themselves, characterized by a nigh irrevocable "thingness" that presses up against our theoretical encounters with them. For the most part, these associations can be traced to the emotive and perspectival reality of embodiment and have a funny way of clinging to our perceptions of the world. They persist in light of their discreet emotional significance to our experience as individual subjects. This is also why these associations rarely apply to theories in a historically significant way; they are the inescapably personal manifestations of the interface between perceptions, theories, and things. But this is the peculiar thing about intoxicants: they intercede in our experience of embodiment. *They make us feel something.*

Just as the tacit knowing involved with intoxication helped establish the vital substance concept, this emotional aspect of intoxicants, which is bound up in human perception, allows the lower-tier theories concerning intoxicants to persist in the form of associations. This is owed to the unique relationship between the embodied and intoxicants, which complicates the expected nature of one's encounter with intoxicants as scientific objects. Returning to the case of Sertürner, the simultaneous generation of the novel class of plant alkalis and emergence of morphine represents a turning point in modern history. If later scientific encounters with intoxicants carried an association derived from the vital substance concept, morphine's emergence is a prominent bottleneck. This is particularly because the emergence of morphine was not a singular development in the history of pharmaceutical chemistry, but the spark behind an explosion in the number of known alkaloidal medications and intoxicants that would send shockwaves throughout the nineteenth century.

Uptake was remarkably swift. Many of the last cries of resistance can be found appended to Sertürner's paper itself. Ludwig Wilhelm Gilbert, editor of *Annalen der Physik*, only accepted Sertürner's study with a footnote qualifying that Sertürner was likely unaware of Thénard and Chevreul's observations that acids could produce true compounds with plant principles (Hanzlik 1929; Sertürner 1817). Yet, once presented for consideration in the form of Sertürner's 1817 paper, Joseph Louis Gay-Lussac, the illustrious French chemist, had little choice but to consider not merely Sertürner's immense discovery of the *principium somniferum* but all of his findings, leading many chemists, such as Vauquelin, to retroactively claim the discovery for themselves (Jurna 2003, 281; Bache 1980, 490). The first effort to formally define alkaloids as a class of things was put forward by Carl Friedrich Wilhelm Meissner in 1819. In the aftermath of Sertürner's 1817 publication, Meissner isolated sabadillen and coined the term "Alkaloide" to distinguish plant alkalis from mineral alkalis (Phillipson 2012, 5; Fattorusso and Taglialatela-Scafati 2008, 74). Many competing definitions soon entered the fray. Chemists like Joseph König reserved "the name alkaloid for plant bases related to pyridine," while others, in keeping with one of Sertürner's observations, saw nitrogen as a necessary component (Fattorusso and Taglialatela-Scafati 2008, 74). Interest in defining plant alkalis as a novel category directly corresponded with an exponential increase in the number of newly identified alkaloids. Pierre Jean Robiquet identified narcotine (1817) and codeine (1832), Friedlieb Ferdinand Runge derived caffeine from Goethe's coffee beans (1819), Joseph Bienaimé Caventou and Pierre Joseph Pelletier isolated strychnine (1818) as well as caffeine (for the second time) and quinine (1820), Meissner and Caventou found ceratrine (1818) and colchicine (1819), nicotine was found by Wilhelm Heinrich Posselt and Karl Ludwig Reimann (1828), theobromine by Aleksandr Woskresensky in 1842, and cocaine by Albert Niemann in 1860, just to list a few examples (Huxtable and Schwarz 2001). The collective impact of alkaloidal ontology borders on the ineffable. Quinine alone may be responsible for turning the tide in Europe's halting efforts to colonize India and Africa (Porter 1999, 465).

At the heart of it all, intoxication as a way of knowing—the very experience of inebriation—emerged as directly constitutive of burgeoning conceptions of the natural world, and the organism within it. Investigating subjects, simultaneously the object of empirical intervention, were enveloped in the perceptional experience of intoxication. From Brown and Schelling to Sertürner, intoxication made the external materiality of the bodily object internally available to the process of encounter. The inquisitorial character of empirical instrumentalism was instead mediated by the subjective experience of intoxication, as the crucial site in which subjects, scientific objects, theories, and experimental apparatuses orient themselves in marshaling testimony about the natural world. Intoxicated ways of knowing saw the ontologies of new kinds of things to be—not only "vital substances" and "alkaloids," but novel entities, such as morphine. Most of all, it proved a constitutive force in the realization of a particular historical conception of what it meant to be an organism, and testified to the ultimate place of the organism in the world at large.

But, of course, this was barely the beginning. It has already been intimated that Romantic vitalists cannot be credited with realizing the modern biological subject. Nor is this the end of the road for intoxicants. Alkaloidal preparations would come to represent only a fraction of the new intoxicants available for consumption and, even then, the bulk of the intoxicants consumed over the course of the nineteenth century were the tried-and-true alcohol and opium. Intoxicating substances, too, would soon see remarkable social changes, a wholesale transformation in both the scope and nature of their consumption. What is to come will not be yet another clandestine elegy for noble, dynamic Romanticism, so rudely supplanted by vulgar materialism. Rather, it will tell of how the bitter quarreling, even enmity, between those polar positions, locked as they were in a struggle over the basic nature of life itself, would help engender the biological subject's conditions of possibility. The vital substance concept would soon meet its match in a revolutionary shift in the study of the body and, with it, a change in the meaning of embodiment.

11
Contra Intoxicatio

* 5 *
Great Expectations

THE HUMBOLDTS, JOHANNES MÜLLER,
AND THE RISE OF NEOMECHANISM

Berlin and the Modern University

While novel alkaloidal plant principles were being discovered by leaps and bounds and new intoxicating medicines stepped on to the world stage, hints of a comparable shift could be seen in the budding sciences of the body. Throughout the first half of the nineteenth century, the world of German physiology was still a messy place. More so than medicine, physiology was empirical; however, it still relied on theoretical and philosophical frameworks in inferring the clinical significance of its findings. Though it arguably stood on drier ground, it lacked the systematicity required to identify itself as a proper science and was thus mired in the very same morass as the other aspiring German sciences of the nineteenth century. The ensuing revolution would make names for some of the nineteenth century's most widely lauded scientific figures: Helmholtz, du Bois-Reymond, Meynert—all of whom rose on the shoulders of their teacher and mentor, Johannes Müller. I tentatively refer to members of this camp as "neomechanists," a moniker that may remind some of Oswei Temkin's language of "mechanical materialism." Some will quibble over the use of "neomechanism" as opposed to simple physicalism or "organic-physics." This is understandable, as it could be argued that the principal nature of the conflict between the *Naturphilosophen* and the physicalists centered on the issue of consciousness. Here, I use "neomechanism" as a tentative nomenclature to contrast the intoxicated vitalism of their predecessors with their own comparatively radical physicalism.

It would be their legacy that set the stage for the pioneering efforts in the study of the body and mind that took hold in the middle of the nineteenth century, fundamentally transforming not only how the body was studied, but how it was understood. But such dramatic developments can hardly be attributed to the sweat of any one person's brow. The neomechanist, or physicalist, revolution could just as readily be identified as the product of organizational changes in the German university, political pressure, or the influence of outside benefactors. Of all the individuals who contributed to the circumstances that made possible this shift in the physiological sciences, few are as significant as the Humboldt brothers.

The hands of Alexander and Wilhelm von Humboldt both lay heavily upon the nineteenth-century development of the science of the body, although the nature of the influence outwardly exerted by each brother could not have been more different. Alexander von Humboldt canoed the Amazon and scaled Mount Chimborazo. Wilhelm von Humboldt was a sober-hearted linguist, educator, and government functionary. Unsurprisingly, Wilhelm's contribution to history of the body and its study was institutional in nature. In 1801, the Prussian cabinet began planning the development of a new university in Berlin, which at the time was almost devoid of scientific learning (Rüegg 2004, 16). The task fell on Wilhelm von Humboldt who, with the assistance of Ferdinand Schleiermacher and Johann Gottlieb Fichte, established the University of Berlin in 1809 (Broman 2002; Rüegg 2004, 16). Located in an eighteenth-century palace, the newly founded university was to embody the *Einheit von Forschung und Lehre* (unity of research and education), a true center of cutting-edge inquiry into the hidden wonders of the natural world (Broman 2002; Rüegg 2004, 16). In sharp contrast with the practical-learning model that informed education reforms in Bavaria and Southern Germany, holism in scientific research and education were the focus of the new university from its very inception (Broman 2002).[1] These high ideals were well suited to the Prussian academic reforms initiated back in 1799, which restricted academic rights to graduates whose master's or doctoral thesis was recognized by the scientific community (Broman 2002; Klinge 2004, 137). This emphasized independent research as a core value of the university educator, and promotion within the academic hierarchy increasingly depended on the publication of new research. The University of Berlin also instituted the habilitation, a piece of independent, postdoctoral scholarship that was a required to attain the status of a full professor (Klinge 2004, 137).

With the University of Berlin as its proven model, the emerging concept of the German university realized scientific research as a professional activity, one patterned and officiated by bureaucratic and civil regulation (Rüegg 2004, 17). The German university system thus gave rise to the first professional scientists. As Rüegg suggests, by the middle of the nineteenth century, "practically all researchers in the natural sciences and medicine in Germany were active either as heads or collaborators of institutes or university laboratories" (Rüegg 2004, 17). This development was largely a product of the emergence of scientific research as a new type of profession embedded in the structure of the university. This was also the period in which the inter-university academic conference and the academic journal both gained currency as a means of sharing, policing, and consolidating new knowledges, further entailments of professionalization (Klinge 2004, 130). This is not to suggest that one should necessarily expect a greater share of scientific insights to emerge out of institutional, academic science. The annals of science are fraught with the smashing successes of those who lacked either formal degrees or the institutional support of a research university. French physiologist Claude Bernard made many of his discoveries in a cellar (Rüegg 2004, 18). Louis Pasteur did experiments in a pair of attics (Rüegg 2004, 18). What the emergence of the modern research university produced was a space wherein local encounters were collected and "legitimized," having passed through the crucible of professional science. The research laboratory recapitulated the logic of the factory given expression in the Industrial Revolution. Far from merely spaces of scientific production, research labs became loci of standardization, formalization, and constitutionalization. Wilhelm von Humboldt was instrumental in elevating the German university to just such a privileged, if not exclusive, site of knowledge creation and translation.

If Wilhelm von Humboldt is to be credited with paving the way for organized academic science, Alexander von Humboldt is rightly recognized for doing much to lead intrepid young minds on to such a path. Alexander von Humboldt had been a naturalist, botanist, and explorer in his own right. He had voyaged around the globe, describing all manner of phenomena, and pioneered fields of science such as biogeography, meteorology, volcanism, and geomagnetism, to name but a few (Dettelbach 2001; Jenkins 2007; Echenberg 2017). *Kosmos*, Humboldt's bold stab at a holistic sketch of the natural world, repeatedly sold out, was translated, and was reprinted, his publisher proclaiming the demand to be "epoch-making" (Rupke 2008; Sachs 1995, 28).[2] From his home base in the Prussian capital of Berlin, Alexander von Humboldt's influence even extended into the halls of the Prussian court, where he occasionally served as a foreign emissary in service to the Prussian crown. Rather than marshal his favor in service of personal enrichment or political intrigue, Humboldt exercised his influence to promote Prussian interest in scientific discovery (Dettelbach 2001). This would often be as direct as Humboldt personally reading a draft written by a given protégé to the Prussian king (Finger et al. 2013, 28). Questions surrounding the nature of Humboldt's scientific project have since come to lay at the center of debates concerning "the nature of science itself" (Dettelbach 2001, 11). In the case of Johannes

Müller, it was the enthusiastic support of Alexander von Humboldt that encouraged him in his work. This mentorship spanned Müller's early days in Berlin to his latter days as a widely lauded, seasoned professor, with a cohort of protégés all his own. When Müller died in 1858, it was Alexander von Humboldt who rallied to raise funds for his widowed spouse. Supported by the Humboldt brothers, Berlin would become one of the nineteenth century world's greatest centers of physiological study, largely due to Müller's great commitment as an educator and his tremendous productivity as a synthesizer of knowledge.

But Müller, as will soon be discussed, was more of an end than a beginning—a transitional figure between the dynamic vitalism of the *Naturphilosophen* and the physicalistic neomechanists, who sought to reduce the body to physiochemical processes. For the purposes of this story, Müller will simultaneously be a primary instigator in the shift toward an approach to the science of physiology that singularly privileges anatomical and experimental investigations, as well as a forceful advocate for the experimental validity of a post-Okenian vital substance concept. And it will be precisely this dual significance of Müller's that will make his ideas the object of alterity at the hands of his physiochemically-minded students. They, intent on overcoming the perceived shortcomings of their mentor, will seek to undermine every remaining vestige of the theory of vitalism and, with it, the vital substance concept. As far as the history of intoxicants and the body is concerned, the effort on the part of the neomechanists to divest scientific physiology of the vital substance concept will result in a broader hesitance to study the effects of intoxication at all. Indeed, physiological interest in the effects of intoxicants was sustained, albeit to a limited extent, in France, otherwise falling within the purview of a distinct, emerging field of pharmacy. By the time that their students are entering into the scientific community, among them the "brain psychiatrists" Eduard Hitzig, Theodor Meynert, and Karl Wernicke, research on the effects on the mind would only be conceivable as an investigation in neurophysiology. The anatomical body would become the sole locus of the real concerning a meaningful understanding of the body and mind. This is crucial, for it will be the difficulty in reducing mental processes to brain matter that lends credulity to an otherwise novel approach to understanding the nature of the mind.

Johannes Müller

Prior to Johannes Müller's appearance on the world stage, German physiology was in a state of disarray, despite persistent Romantic efforts at systematization. Oken brought a fresh empirical spirit, but the theories underlying physiological relationships between anatomical features remained reliant on speculative principles. Medicine and pharmacy had seen a flurry of development, and yet many of these changes had only raised more questions. But this was all fertile soil for Müller's calls for an approach to physiology as a science that was concerned primarily with empirical phenomena, rather than grand cosmological questions. Yet the legacy of Müller as reformer is as much a reflection of the revolutionary aspirations of his brightest students as of his actual efforts to change the course of German physiological science.

Müller was the great reformer, the famous father of scientific physiology. He was also a vitalist with Okenian impulses and a toe or two in the eighteenth century. It was through this heritage that Müller carried forward the vital substance concept. Despite openly disregarding Brunonian ideas, a vitalistic conception of intoxication, and intoxicants, was clear for all to see in Müller's bodily concept. Further still, Müller's role as a reformer in the field of physiology meant bringing the vital substance concept into the anatomical lab, watching it unfold beneath the microscope. The very nature of his concept of vitalism would be made clear through his encounters with intoxicants.

It would be these vitalistic, and intoxicatingly Brunonian, influences that would become a point of contention for several of his key students, among them Emil du Bois-Reymond, Hermann von Helmholtz, and Ernst Brücke. For them, Müller's methodological reformation of physiology had not gone far enough. The nature of the body—indeed of the entire world—could be understood exclusively through the mechanistic relationship between empirically identifiable physiochemical processes. These "neomechanists" were in all-out revolt against vitalism and everything it stood upon, including the vital substance concept. They were also fabulously successful, for a time. But before discussing the theoretical shift represented in the anti-vitalistic critique leveled by the neomechanists and the implications for the study of the body and mind, it is worth addressing some of the institutional and societal shifts that made a so-called neomechanistic revolution possible, before discussing the foundation set by Johannes Müller.

Johannes Petrus Müller was born in 1801 in French-occupied Koblenz. After a year of military service, Müller began studying medicine at the University of Bonn in 1819 (Zimmer 2006; Otis 2007). These studies concluded in 1822 upon the acceptance of his thesis on the motion of several animals (*De Phoronomia Animalium*) (Zimmer 2006; Otis 2007). Müller continued his education in 1823 when he received a scholarship to study comparative anatomy with Carl Asmund Rudolphi in Berlin, who had also been a mentor to Moritz Romberg (Otis 2007; Zimmer 2006). His early research, while centered around topics within the scope of physiology and anatomy, was more speculative and philosophical in style, given the enduring impact of Lorenz Oken. It was not until Müller began researching comparative anatomy with Rudolphi that Müller committed himself to relying solely on observable phenomena in the discernment of physiological principles (Zimmer 2006; Lohff 1978, 247; Otis 2007).

Müller was swift to make a number of essential discoveries in the study of bodily function. With the publication of *De glandularum secernetium* Müller established the role of glands in regulating bodily functions. In 1831, he experimentally demonstrated Charles Bell's and François Magendie's unproven theory that dorsal roots carried sensory nerves and ventral roots carried motor fibers on frog spines (Steudel 1963, 570). In total, he would publish 267 titles over the course of his life. From 1833 to 1844, he gradually consolidated his own research and the available work in physiology into his highly influential *Handbuch der Physiologie* (Handbook of Physiology) (Lohff 1978, 247). During his lifetime, Müller was already regarded as the foremost expert on physiology and his *Handbuch* its primary treatise (Rachlin 2005, 41)

The *Handbuch* was one of the most circuitous textbooks of its time, both incorporating the products of Müller's own experiments as well as bringing together disparate publications in French, English, and German (Zimmer 2006). Far-reaching in the scope of its detail, Müller provided extensive macro- and microscopic description of various tissues, nerves, and fluids in different organisms, all of which was organized into nine books and a lengthy prolegomenon (Müller 1834, III). During the process of writing the *Handbuch*, Müller also founded a journal in 1834 with the intention of increasing communication within the field of physiology. Titled the *Archiv für Anatomie, Physiologie und wissenschaftliche Medicin* (Archive for Anatomy, Physiology, and Scientific Medicine), his journal quickly became highly respected among researchers in the natural sciences (Otis 2007). In contrast to journals started later in the nineteenth century that trend toward increased specialization, Müller's periodical was relatively broad in focus. It instead embodied the principles of a new approach that affirmed the importance of cooperation between anatomy, physiology, and "scientific" medicine.

Müller as Reformer/Müller and Vital Substances

For all these reasons, Johannes Müller is sometimes "considered the main reformer of physiology in the late 1830s, because he was instrumental in liberating the science from the romantic spirit that had prevailed for several decades," or so has been argued (Zimmer 2006). Such assessments focus on how Müller "emphasized the comparative and developmental aspects in the study of physiology and stressed the importance of meticulous observation" (Zimmer 2006). But on a theoretical level, Müller was in many ways a product of his Romantic education. Müller asserted as early as 1826 that the methodologies of physiology needed to approach the body physiochemically, while simultaneously professing that, in actuality, physiology could never be reduced to physiochemical processes (Müller 1826, 19). For one, Müller's commitment to limit his approach to observable phenomena did not preclude him from remaining a vitalist, affirming the essential difference between organic and inorganic matter and lending his voice as an opponent of rigid empiricism. In 1824 Müller gave a lecture titled "Ueber das Bedürfnis der Physiologie nach einer philosophischen Naturbetrachtung" (On the Need of Physiology for a Philosophical Contemplation of Nature) (Rothschuh 1973, 197). Here he expressed not only that philosophical contemplation of nature was meaningless without scientific observation, but also that empirical rigor was blind without philosophy.

Beyond merely advocating of the importance of philosophical analysis, it is apparent that Müller's physiological work was deeply indebted to the likes of Leibniz, Kant, Schelling, and Oken. The clearest example of this can be found in one of Müller's most impactful physiological insights: *the law of specific sense energies*. An idea that had come to guide much of Müller's work was the notion that "sensation is an attribute not of some real object out in the world but of the sensory nerves" (Fullinwider 1991, 22). Müller's "law" was that sensory nerves were not neutral conductors: some sensory nerves represent energy from the sun as heat and some as light, but not both at the same site (Fullinwider 1991, 22). The law of specific sense energies is first explicitly introduced in Müller's 1826 *Zur vergleichenden Physiologie des Gesichtssinnes des Menschen und der Thiere* (On the Comparative Physiology of Vision in Men and Animals). This deceptively simple idea held tremendous significance for the way the body and mind were to be understood. The law of specific sense energies not only seemed to confirm that our knowledge of an external reality was actually a reflection of the nervous system itself, but it suggested that the stuff of perception, like sound or color, was innate, rather than passively received (Fullinwider 1991, 22).

Many of Müller's conclusions surrounding what the law of specific nerve energies told us about the nature of perception mirrored ideas central to Kant's philosophy of empirical psychology. Müller understood his law of specific sense energies as a physiological parallel to Kant's categories of thought (Hergenhahn and Henley 2014, 222). Just as Kant's categories of thought are necessary for the construction of tangible, conscious experiences, Müller's specific sense energies are physiological a prioris that construct and shape our perceptions of an external world (Kant 1781, B-161). While Müller provided a physiological explanation, he was one with Kant in the idea that we have no awareness of external objects as things-in-themselves (Müller 1834). True knowledge of objects "outside" ourselves is impossible because our perception of them will always be comprised of limited inputs from the nervous system. In this respect, Kant's psychological framework and Müller's physiological model were functionally identical: our awareness is only of our nervous system.

Although Kant made a great impression on Müller's work, Müller was not without criticisms of Kant. In drawing a physiological parallel between the law of specific nerve energies and Kant's categories, Müller did away with Kant's transcendental forms. At the same time, Müller acknowledged the existence of something phenomenologically akin to the unity of apperception. Ultimately, Müller felt the solution was similar to what Leibniz called "pre-established harmony," an idea Kant considered for himself in *Kritik der reinen Vernunft* (Critique of Pure Reason) (Fullinwider 1991, 22). The basis of this "pre-established harmony," Müller suggested, was an unconscious "ruling idea" or "organizing principle" whose action regulated the physical and mental structure of each biological species (Fullinwider 1991, 22). Here, at the core of his physiological Kantianism, Müller presents like a good *Naturphilosoph* (Finkelstein 1996, 78). Like Schelling, Müller saw the apparent harmony in the organization of extremely complex organic systems as evidence of an underlying idea specific to each species, of which the individual was but a particular unfolding. The unconscious action of this organizing idea was what Müller understood to be a "vital principle" (Fullinwider 1991, 23).

Sharing in the vitalistic thinking of his day, it was in this sense that Müller was earlier described as less of the beginning of the "new" physiology than the end of the "old" physiology. But in Müller's case, vitalism was not merely an explanation of the difference between organic and inorganic bodies. It was a solution to fundamental questions concerning the possibilities for knowledge as embodied subjects. The vital force itself is a fact of organic life, and while speculation about its origins belongs in the domain of philosophy, experimental physiology should strive to understand the properties of the vital force (Müller 1834, 18).

The primary distinction between organic and inorganic matter, Müller suggests, is best understood as the difference between beings that are *organized* and those that are not. This notion is explained by the existence of a purposive ruling idea at work in organic life, as Müller writes in the *Handbuch*:

> The organic bodies are not only distinguished from the inorganic by the nature of their composition of elements, but by the constant activity which is working in living organic matter. [This active force] works with purposefulness according to a rational plan, such that the parts are arranged for the purpose of a whole. This is what characterizes the organism. (Müller, 1834, 18)

Organisms consist of many separate components working harmoniously toward a common goal. Müller understood this in terms of purposiveness. Separate components, each with their own purposes, serve to perpetuate the existence of the entire organism. Furthermore, this purpose, or ruling idea, appears to guide the development of the organisms throughout its entire existence. The germ, however different in appearance, is always equitable with its more developed form *in potentia* before it realizes its completed form in *actuality* through development (Müller 1834, 24). In this sense, "Stahl's [idea of the soul as the animating force] is the organizational force itself which manifests in accordance with a rational law" (Müller 1834, 24).

It is the organizing principle that ultimately distinguished living and "dead" matter. This was plainly observable in the peculiar chemistry of organisms, where chemical entities that might normally form simple compounds readily form considerably more complicated bonds: "Before the properties of that force [of generating compounds] are known, one can truly recognize it as a life principle or organizing force, albeit in an undetermined quantity" (Müller 1837a, 285). Pondering the nature of this vital principle, Müller conceded that it was still impossible to describe it materially as a force or an energy, but remarked that the same could be said of important phenomena in physics (Müller 1834, 27). Still, Müller clearly understood the vital force on a physical level, describing at length the reanimation of tissues deprived of vital stimuli (Müller 1834, 27). In this, Müller appears to diverge from his primary description of vital phenomena. Where even Oken rejected *Lebenskraft* as an animating force in favor of galvanic energies potentiated by the underlying principle of excitability, Müller appears to be making a case for a vital force that is at once animating and organizing. An explanation to how and why Müller sustained this seemingly paradoxical understanding is found in Müller's reception of the concept of vital force in relation to substances of intoxication.

Before going to talk about potent vital intoxicants and the energizing effects on the body, it should be acknowledged that Müller did not—unlike the naturphilosophical mentors of his youth—identify as a Brunonian. Müller does make reference to John Brown in his *Handbuch*, a brief mention occupying scarcely more than a couple pages (Müller 1834, 60). He recognized Brown as an intrepid, but crude, discoverer of a number of the laws of excitability and a bold, but dangerous, progenitor of the earliest forms of scientific medicine (Müller 1834, 59–61). However, Müller's commentary here is not solely of a historical nature, and he is quick to make a series of pointed criticisms of Brown and his followers. For one, Müller describes Brown's notion of sthenic diseases as contradictory (Müller 1834, 60–61). All diseases, Müller asserted, are defects in the bodily composition of tissues comprising the "Organsystemen," characterized by a diminishing of vital powers as opposed to overstimulation (Müller 1834, 66). This, in itself, is not a particularly harsh criticism; insofar as sthenic conditions were exceedingly rare, the practical application of Brunonian theory almost always treated diseases as asthenic, with bodily exhaustion being described as a product of overstimulation. But Müller's firmest critique of Brown was a hit squarely on the vital substance concept that guided Brunonian therapeutics. Brown, Müller asserts, erred in the conflation of apparent vital stimulation with *actual* vital stimulation (Müller 1834, 61–62). This is because substances of intoxication produce an effect on organic function, the symptoms of which create only the temporary appearance of vital stimulation (Müller 1834, 61–62).

Müller's anti-Brunonian critiques appear to establish a conclusive break from the vital substance concept as received from Brown. The buck—or so it would seem—stops here. Yet Müller's extensive and diverse remarks on the influence of opium and other intoxicants on the body complicate this position and perhaps tell a different story. Müller's *Handbuch* frequently relays the conclusions of physiological experiments with opium, both conducted by Müller himself and found in others' work, asserting that opium stops the motion of the heart (Müller 1834, 181 and 184), suppresses feelings of hunger (Müller 1834, 466), and locally abolishes nervous activity in the limbs of frogs (Müller 1834, 233). Müller even attempts to reproduce Humboldt's findings that chloric acid, opium, and alcohol (among other substances) have an excitatory effect at the nerve, though Müller singles out alcohol, chloric acid, and opium as halting all excitatory activity in the nervous tissue (Müller 1834, 596). The poisonous effects of narcotics are actually a recurrent theme in Müller's *Handbuch*. Above all, this seems to be a reflection of the unique effects that Müller understands narcotics to have on the body. Unlike other stimuli, the group of substances known as narcotics "alter the material composition of the nerves" (Müller 1834, 607), or, as he states elsewhere, "it follows that the opium itself alters the nervous substance" (Müller 1834, 233).

How does this alteration of nervous material occur? Müller is very clear in stating that it is not owing to any discernible chemical or mechanical destruction of the nervous system but through the actual alteration of nervous material (Müller 1834, 607). Müller mostly emphasizes the capacity of narcotics to abolish nervous activity, although it is important to note that this is not the full extent of the picture. In regard to how narcotics enter the body in order to engender an effect, Müller remarks that they are generally taken up in the blood (Müller 1834, 609). Under experimental conditions, narcotics can affect nerve material locally, which Müller suggests is a result of the narcotic's destruction of the excitability of the nervous material (Müller 1834). As has just been established, Müller asserts that all narcotic substances "when given in large doses deaden [excitability] through alteration [of nervous matter]" (Müller 1834, 607–8). However, "some, such as opium, are stimulants and less depressive at low doses" (Müller 1834, 607–8). So what is Müller really saying here? Do they deaden excitability through alteration or are they stimulating? The answer casts doubt on the extent of Müller's critique of Brown, or at the very least clarifies it and suggests broader implications for Müller's understanding of organized life.

By Müller's own account, opium does produce excitation—it stimulates, at least up until the point that it brings about an alteration in nervous tissue. It is hard to imagine a more vivid echo of Oken's own description of the action of vegetable poison, which, as discussed earlier, brings about the destruction of the nervous system by supercharging its polarity. Müller is more agnostic with respect to understanding the technical process by which this change in nervous tissue occurs, though the effect is the same. Both Müller and Oken describe narcotics, and alcohol in Müller's case, to be initially excitatory before bringing an end to vital activity in the tissue, almost as if the nervous tissue is being overstimulated to the point of neutralization. Müller goes as far as relating the corresponding loss of muscle receptivity at the sites where narcotics have been locally applied to the destruction of the excitability of nervous filaments (Müller 1837b,

15). A further demonstration of this idea is found in Müller's remark that nitrous oxide—a substance recognized by Davy as a vital stimulant—can initially sustain life in place of air, before creating a state of intoxication and eventually death, where other gases merely kill through irrespirability (Müller 1837a, 290). Like his mentor, Müller evidently does in fact understand substances of intoxication, in particular narcotics such as opium and alcohol, to negate the excitability of nervous tissue through excessive vital excitation.

Returning to Müller's remarks on the Brunonian system, Müller's position appears to be more critique of what might be called "classical Brunonianism," as opposed to the Romantic and naturphilosophical reception of *Erregbarkeitstheorie*. This is made clear through Müller's Okenian understanding of narcotic poisoning, though it can also be seen in Müller's own conception of excitability, of which Brown was recognized for having discovered its basic laws. Müller understands excitability to be the *distinct property* of organized bodies (Müller 1834, 592–93). This "property" refers to the constant process by which organized bodies assimilate external vital stimuli to bring about changes in their bodily form in making life possible (Müller 1834, 37). In this sense, organisms are, at once, dependent on environmental stimuli and yet irreducible to a passive mechanism. This is, of course, clearly derived from Oken-Schelling's understanding of excitability as interpreted through Röschlaub's *Erregbarkeitstheorie*, for whom excitability is not a substance but a process, a dynamic relationality that forms the basis of all vital activity.

The turn, however, was Müller's efforts to bring the vital substance concept into the anatomist theater and validate the idea by overtly experimental means, and with limited open recourse to speculative philosophy. By taking the Schellingian-Okenian concepts of excitability and reading them through the investigatory methods of scientific physiology, the formulation of intoxication as the overstimulation of galvanic polarity was translated into the overexcitation of nerve fibrils. Even when Müller struck out against the likes of Brown, the foundational notion of a vital stimulant had remained, albeit in a new way. That Müller received the vital substance concept from Romantic sources is clear. But intoxicants also interceded in Müller's bodily existence as well.

Müller had suffered from severe bouts of depression and infirmity throughout his life, notably in 1827, 1840, 1847, and later in 1857 (Otis 2007, 34, 228). During his 1857 episode of illness, Müller began taking massive doses of opium, allegedly to treat insomnia and a pain in his abdomen (Otis 2007, 228). Somewhat inescapably, the opium also perhaps mitigated Müller's depression and melancholia. And it would be opium, again, that led to Müller's death by overdose on April 28, 1858, at the age of 56. But this can hardly be counted as a singular instance of opium use in Müller's life. Sometime in the early 1850s, laudanum had already become something that Müller would recommend to those around him as a curative for a vast assortment of ailments. To this point, du Bois-Reymond sent a letter to Helmholtz in 1853 in order to communicate Müller's recommendation that he treat his colic with laudanum (Otis 2007, 228). While evidence of earlier use of opium products is lacking, Müller's recurrent illness, experimental familiarity with opium, and worsening use toward the end of his life suggests that Müller became well acquainted with the intoxicant at an earlier point than the 1850s. It is very possible that Müller's own experiences with opium, and alcohol, intoxication figured into recognition of these intoxicants as excitatory, rather than merely as poisons. What is clear is that, even as he decried Brunonian ideas, Müller upheld and even advanced the notion of intoxicating vital stimulants, using the methods of physiology.

In his published works and the surviving notes on his given lectures, Johannes Müller remained committed to vitalism, and the vital substance concept, until his death in 1858. Müller mentored many young scientists over the course of his career, including Emil du Bois-Reymond, Hermann Helmholtz, Friedrich Henle, Carl Ludwig, Rudolf Virchow, Ernst Brücke and Theodor Schwann. Du Bois-Reymond, Brücke, and Helmholtz, in particular, would later take it upon themselves to dedicate their life's work to refuting their teacher's assertion of the fact of the vital principle. Of primary importance to their refutation was a concerted effort to erase any enduring argument for not only the vital substance concept but also the empirical methodology that had permitted intoxication as a valid testimony concerning the ultimate nature of the body and mind. Even as they labored to wipe out belief in their mentor's central idea, they remained indebted to Müller's tremendous work as a reformer, whose work directly paved the way for the research of the physiologists who followed.

The Electricity of the Body

Of all of Müller's students, it would be Helmholtz, Brücke, and du Bois-Reymond who would be most crucial in preparing the next generation to oppose Müller's vitalistic ideas, and electrophysiology would be one of their earliest demonstrative tools. Deemed the 1847 group, they conceived of a new approach to physiology that was empirically grounded in physical and chemical principles and divested of vitalism in all its forms (Cranefield 1966). Electrophysiology was still a budding concept when

Helmholtz and du Bois-Reymond began studying medicine. The phenomena of galvanism had been known since the late eighteenth century. The *Naturphilosophen* and the Romantics, among them Alexander von Humboldt, who had written on electric animals encountered in his South American travels, privileged galvanism as a topic of great importance in understanding the nature of organized life. But it was Emil du Bois-Reymond and Hermann von Helmholtz who would establish what might be likened to an early vestige of the modern concept of electrophysiology. An important component of their work, one that fundamentally differentiated their approach from those promoted by Müller, was the notion that the electric currents found in animals were no different from those derived from inorganic, external sources.

They would take issue with their mentor's approach on the topic of vitalism, but they acknowledged the significant philosophical implications of Müller's law of specific nerve energies. Perhaps even more so than Müller, Helmholtz and du Bois-Reymond relied on the empirical psychology of Immanuel Kant to provide a theoretical framework for the new physiology. In Helmholtz's case, his Kantianism led him to study the physiology of perception and the human senses. Deeply inspired by Ernst Weber's declaration that vital forces were in fact physical forces, Helmholtz and du Bois-Reymond shared a vision of developing physiology into an organic physics (Cahan 1993, 25; Cranefield 1966; Königsberger 1906, 25). This meant reducing physiology to the most basic laws of physics and chemistry in order to reconceive of the body on the basis of strict empirical observation. A founding tenet of their new approach was the complete rejection of vitalism. This would present a number of theoretical hurdles. For one, Johannes Müller identified *organization* as the primary difference between organic and inorganic matter. Searching for an empirical explanation for the appearance of organization in life, Helmholtz and du Bois-Reymond became enthusiastic, early adopters of Darwinism (Finkelstein 2013). Their efforts to develop physiology into a fully realized physical science would leave a profound impact on their students, continuing the transition toward greater centralization and institutionalization in the field of physiology.

Yet in elevating physiological neomechanism, or organic physics, as an ontological reality, the neomechanists undermined the epistemological validity of anything but the experimental sciences. Not only that; this approach would come to manifest in an identification of bodily substance as the singular locus of the real with respect to the nature of the body and mind. By the time that Meynert and Wernicke were promoting their theories of cortical localization, this doctrinal principle of the body as the real would lead to an equation of mental states with neural states, the failure of which would pave the way for a radical new science of experiential intoxication. In order to make sense of this, however, it is important to first explore early neomechanism's efforts to repudiate the vitalistic, and vitally intoxicated, ideas that came before them.

Du Bois-Reymond and Helmholtz

It was under Müller's supervision, starting around 1838, that du Bois-Reymond was introduced to Carlo Matteucci's latest work on animal electricity in 1843 and 1844 (Finger et al. 2013, 23; Finkelstein 2013). Müller himself had only become aware of this work thanks to Alexander von Humboldt, who—after observing electric eels encountered during his travels in the Americas from 1799 to 1804—had maintained an enduring interest in the subject (Finger et al. 2013, 23). Humboldt took du Bois-Reymond into his mentorship when he was still a student. Encouraging his work and at times even participating in du Bois-Reymond's experiments, Humboldt's support validated the young pupil's fascination with neuromuscular electrophysiology. Most importantly, Humboldt's immense influence, as much in court as in the academic community, all but assured financial and institutional recognition of the importance of du Bois-Reymond's research (Finger et al. 2013, 26). Under the supervision of Müller and with the support of Alexander von Humboldt, du Bois-Reymond selected electric fishes as the subject of his graduation project and by 1848 published the first section of a multivolume work on animal electricity, a publication that Humboldt is alleged to have read to the king himself (Finger et al. 2013, 28).

From the outset, Emil du Bois-Reymond appears to have been confident in his vision. As early as 1845, he wrote to Humboldt that "[p]hysicists and physiologists will see this dream realized of an electricity operating the movements and perhaps transmitting sensations in animal bodies" (Finger et al. 2013, 27; Finkelstein 2013). One factor that was especially formative of the course of du Bois-Reymond's work was his rivalry with Carlo Matteucci. The Italian scientist had developed what he termed a "rheoscopic frog": an apparatus for detecting current that consisted of the cut nerve of a frog's leg and the attached muscle (Finger et al. 2013, 68; Finkelstein 2013). Functioning under the premise that excitable organic tissues generated direct current when damaged, Matteucci also developed a crude battery of frog thigh muscles, which he called a "frog pile." Seeking to further understand the phenomena, Matteucci charged current from his frog pile through individual muscles and nerves, but he was unable to produce a positive result. In the face of this result, Matteucci concluded that there may be a fundamental difference between currents produced by laboratory preparations and those produced by living animals (Finger et al. 2013, 68). While recognizing the great intellectual debt he owed to his older competitor, du Bois-Reymond was confident that Matteucci had erred in making this assumption, and he set out to prove it (Finkelstein 2013).

At the most basic level, du Bois-Reymond discredited the veracity of this conclusion because he was already convinced of the homogeneity of currents found both within and without organic bodies. Identifying faulty experiment design as the culprit, du Bois-Reymond was able to identify the frog's charge as consisting of component charges found throughout the nerve and muscle tissue. This insight led to the discovery of the law of muscle currents (Finger et al. 2013, 69; Finkelstein 2013). This rivalry with Matteucci also explains du Bois-Reymond's insistence on the use of a galvanometer to measure animal electricity. Whereas Matteucci's rheoscopic frog always left open the possibility of a type of current specific to life, the use of a mechanical device served as testimony that the electricity of organic and inorganic matter was one and the same. After studying the various galvanometers and acquiring suitable expertise, du Bois-Reymond produced the most sensitive galvanometer of its day, the first true nerve galvanometer (Finger et al. 2013, 66).

Like du Bois-Reymond, Hermann von Helmholtz was elevated early on by the patronage of Alexander von Humboldt. While Helmholtz was serving his military duty at Potsdam, Humboldt took notice of his great intellect. Utilizing his influence as royal chamberlain and science adviser in the court of King Friedrich Wilhelm IV, Humboldt was able to shorten the length of Helmholtz's mandatory military service, which he had earlier taken upon himself in exchange for a free medical education (Finger et al. 2013, 24; Finkelstein 2013).

Helmholtz's own first independent foray into physiology also centered on electrophysiology. In 1841, Helmholtz bought a microscope with the little money he could scrape together and decided to write a dissertation under the great Johannes Müller. He undertook a detailed microscopic study of the structure of the nervous system in invertebrates. Nerve fibers, he observed, appeared to originate in the ganglionic cells, a discovery that Müller then had him confirm in other organisms (Cahan 1993, 24). After studying under Müller, Helmholtz moved to Königsberg in 1849 and began to work at ascertaining the speed at which signals traveled along nerve fibers (Cahan 1993; Finger and Wade 2002).

A subject of intense interest since the discovery of the reflex action, the speed of a signal along a nerve fiber was initially thought to be almost instantaneous, perhaps even immeasurably fast (Pantalony 2009, 21; Schmidgen 2002, 142). The central locus of research concerning conduction speeds was then the military, for whom telegraph technology was rapidly growing indispensable (Olesko 1995; Schmidgen 2002, 142). On the basis of an approach first suggested by du Bois-Reymond, Helmholtz wanted to test the speed of nerve conduction for himself, but the technology of the day posed a significant barrier (Schmidgen 2002, 142). In addition to being insensitive, the galvanometers available to him were sluggish and fickle. Helmholtz required precision, an epistemic value at the heart of the new scientific physiology. Kathy Olesko takes this notion a step further. Precision was not merely of crucial facet of the new physiological science. Precision stood in the conceptual foreground of a rapidly industrialized Germany (Olesko 1995). Electricity, telegraph technology, geodesy, and optics—all of which turned on exactingly precise machines—figured centrally within Germany's military and industrial ascendence (Olesko 1995). Questions of microtime, as termed by Jimena Canales, increasingly touched on nearly every aspect of modern life (Canales 2010).

Rather than rely on the galvanometer to measure or detect charges in the muscle or nerve, Helmholtz utilized a technique developed by Claude Pouillet (1790–1868), which used the galvanometer to measure short intervals of time (Schmidgen 2002, 142). This was accomplished by placing the galvanometer in a circuit along with a battery and two switches (Glynn 2013; Schmidgen 2002). The "first switch would be closed at precisely the moment at which a very brief but powerful shock [...] was given to the sciatic nerve of a recently killed frog," while "the second switch, which was closed before the experiment began, would be opened as soon as the muscle began to shorten" (Helmholtz 1850; Glynn 2013). The idea was that current would pass through the galvanometer as soon as Helmholtz stimulated the sciatic nerve and stop as soon as the muscle began to contract (Helmholtz 1850; Glynn 2013; Olesko and Holmes, 1993; Schmidgen 2002). The constant voltage meant that peak needle deflection was a function of the current's duration (Helmholtz 1850; Glynn 2013). Using a light source and a small mirror on the needle, measurements could be made on a scale from the light reflection. Through repeated experiments, by 1850 he was able to ascertain that the speed of the nerve impulse was 24.6–38.4 meters per second (Helmholtz 1850; Glynn 2013; Olesko and Holmes 1993). It was not until 1868 that Julius Bernstein, working in Helmholtz's laboratory in Heidelberg, was able to expand on the insights of Helmholtz and du Bois-Reymond and make convincing measurements of the speed at which action potential moved along the sciatic nerve, clocking it at 28.7 m/s (Helmholtz 1850; Glynn 2013; Olesko and Holmes 1993).

For both du Bois-Reymond and Helmholtz, the establishment of organic activity as reliant on electrical, rather than galvanic or vital, forces was crucial as an experimental demonstration against vitalism. Du Bois-Reymond himself was convinced that the nature of life could be explained through physical laws, yet "he was a hesitant mechanist at best" (Fullinwider 1991, 27). For one, he ascribed to the argument that "force" and "matter" were abstractions. Scientists never have access to force or matter in-themselves. Rather, objects in the world are always already encountered in a reactive state (there is no pure state of force or matter that can be seen in the lab) (Fullinwider 1991, 27). In the 1848 foreword to *Untersuchungen über Thierische Elektricität*, du Bois-Reymond makes this position overwhelmingly clear: "in a word, the so-called life force, in the sense in which it is usually thought in all cases of the living body, is an absurdity [an un-thing]" (du Bois-Reymond 1848, 13).

Du Bois-Reymond's position was emphatically shared with Brücke and Helmholtz, who would write to one another about their certainty that life could be explained on a physical-chemical level (Jones 1953, 45). For Helmholtz, causality and the law of the conservation of energy additionally provide the basis of Helmholtz's rebuke of vitalism. Du Bois-Reymond saw Helmholtz's *Erhaltung* as the killing blow to the idea of a "vital principle" (Fullinwider 1991, 23). At least since the spread of Brunonianism through Germany, the vital force had been conceived of as a measurable energy or force. As the law of the conservation of energy establishes that energy cannot be created or destroyed, the vital principle could not simply appear at birth and disappear at death. This criticism of vitalism would shortly be problematized by cell theory, insofar as it could be said that all living creatures already existed as living cells of their progenitor and that their energy was only lost in death.

Intersubjectivity, a secondary aspect of the question of pre-established harmony, could be supported through measurement (Fullinwider 1991, 26). Helmholtz accepted the conclusions of Müller and Kant that we have no knowledge of the world outside of that which is provided by our nervous systems. For Helmholtz, the "real world" outside of our direct perception is comprised of energy transferences. Many of these changes occur without our awareness until something happens that is available for sensory experience. At this stage, intersubjective awareness of the lawlikeness of perceptions can be explained through unconscious inferences, which depend on a priori judgments. That is to say, our awareness that others perceive the external world according to the same rules that our powers of perception do is rooted in the unconscious application of necessary truths. But they can also be corroborated through measurement. Reminiscent of Alexander von Humboldt's preoccupation with the epistemic significance of measurement, Helmholtz saw measurement as a tool for asserting the veracity of intersubjective perceptions of the "real world" of energy transferences. The Platonic allegory stumbled for wont of a yardstick. The purview of the scientist, as they now understood it, was to understand the physical-chemical relationships that underlay the world—the *real* world—not to speculate about the sources of life or the metaphysical roots of organization in nature.

An illustrative point of emphasis here is found in du Bois-Reymond's comments on the legacy of La Mettrie. La Mettrie was something of a hero to du Bois-Reymond. So much so that in January 1875 du Bois-Reymond dedicated an entire talk to the legacy of the peculiar Frenchman, wreathing his name in honors and esteem. In particular, du Bois-Reymond recognizes La Mettrie as he who "observes the influence of fasting and meat diet, of wine, coffee and opium on the ideas[;] he dissects the conceivable mechanical conditions of memory," among other things (du Bois-Reymond 1875, 193). This excerpt is illustrative of the image of La Mettrie that du Bois-Reymond wanted to impart, namely that of La Mettrie as an undaunted scout in the vanguard of the coming revolution of physiochemical mechanism. Little more needs to be said in asserting that du Bois-Reymond and Helmholtz were psychicalists, if not neomechanists. What this reference to La Mettrie also helps to demonstrate is the kind of role that substances of intoxication had in Helmholtz and du Bois-Reymond's mechanism and how their views departed from those held by their teacher, Müller.

Neomechanism contra Vital Stimulants

In du Bois-Reymond's *Rede* and *Untersuchung über thierische Elekticität*, substances of intoxication are addressed infrequently. When intoxicants are discussed, du Bois-Reymond has compelled them to the witness stand, with du Bois-Reymond assuming the mantle of prosecuting scientists. Standing on trial, as it were, were the principle representatives of galvanism in the first half of the nineteenth century, among them Johann Ritter, Giovanni Aldini, and Carlo Matteucci. Some were confirmed Brunonians. All incorporated vital substances into their experiments. It is to be expected that, in establishing the validity of his own electrophysiological theory, du Bois-Reymond would address the earlier work of galvanism. It is du Bois-Reymond's approach that is of interest here. In discrediting their findings, du Bois-Reymond attempted to undermine not only the significance of their conclusions but also the epistemic validity of their experimental methods, particularly their use of vital substances.

This polemical strategy takes the form of an explicit pattern of critique leveled against the vital substance concept in all its manifestations. Let us begin with Giovanni Aldini, the nephew of none other than Galvani himself. In a single breath, du Bois-Reymond both acknowledges Aldini's importance to the historical study of animal and disparages the Brunonian contamination of Aldini's methodology. Aldini fed "the moist layers themselves with Brunonian stimulants, opium solution and alcoholic chinchona extract," remarkably finding the addition of these vital stimulants to increase the pile's level of excitation (du Bois-Reymond 1848, 98). For du Bois-Reymond, these findings are hardly worthy of acknowledgment, merely remarking, after an additional sentence or two, "thus, have I certainly given the *Essai sur le Galvanisme* more than its due, described the man clearly enough, and do not need to anticipate the readers judgment about the value of both" (du Bois-Reymond 1848, 98). Du Bois-Reymond's dismissal spoke volumes.

Next is Johann Ritter. Du Bois-Reymond's *Untersuchungen* makes frequent references to Ritter, referring to Ritter in more instances than even Müller. Only Matteucci and Volta are afforded greater recognition (du Bois-Reymond 1848). While du Bois-Reymond is quick to praise Ritter's role as a pioneer in the subject, he is swifter still to disparage his findings. Here, du Bois-Reymond is at his most explicit. Ritter's theoretical reconstructions of galvanic activity ought to be disregarded as they proceeded "according to the murky Brunonian categories of depression and excitation, modifications of the no less murky principle of excitability" (du Bois-Reymond 1848, 367). Though far from the singular object of du Bois-Reymond's critique, Ritter's Romantic Brunonianism was identified as a discrete axis along which du Bois-Reymond's rhetorical assault could be advanced (du Bois-Reymond 1848, 367, 317–18).

Skepticism concerning the role of vital stimulants even factored into du Bois-Reymond's famously raucous criticisms of Carlo Matteucci's research on the galvanic frog pile. Du Bois-Reymond's response to Matteucci was particularly crucial. Galvini, Volta, Aldini, Ritter—all were in the ground well before du Bois-Reymond had even enrolled as a student at the University of Berlin. Matteucci, on the other hand, was not merely alive, but a firmly established researcher with an array of institutional resources at his disposal. His work on animal electricity had even won him the Copley Medal in 1844. If du Bois-Reymond was to displace the theory of galvanism, Matteucci was assuredly the person to overcome. In doing so, du

Bois-Reymond would need to exploit every available avenue of critique, and that included Matteucci's employment of Brunonian stimulants.

In one section of du Bois-Reymond's *Untersuchungen* titled "Einfluss verschiedener Todesarten des Thieres auf den Strom" (Influence of Different Causes of Animal Death on Current), du Bois-Reymond compares differences in his and Matteucci's findings concerning the electrical activity of a frog pile comprised of frogs poisoned by opium (du Bois-Reymond 1849, 171). Where Matteucci had discerned—like Aldini—that the opiated frog pile performed more excitedly, du Bois-Reymond was steadfast in his findings that there was no measurable difference (du Bois-Reymond 1849, 171). Du Bois-Reymond described an experiment involving "three frogs which had been poisoned with respect to a acetic acid strychnine solution, a mash of Opium purum and Tinctura Opii simplex, and the Acidum hydrocyanicum Pharm. Bor. and showed at their ischiadicus the current flowed in a lawful direction and was not conspicuously weakened" (du Bois-Reymond 1849, 287). Du Bois-Reymond's experiments—unlike those of the Brunonians—suggested that "vital substances" had an immaterial effect on electricity in the body. Later in that very section, du Bois-Reymond added a further qualification that when nerves were locally bathed in *narkotische Gifte*, specifically hydrocyanic acid, morphine, opium tincture, and strychnine, nervous electrical activity was eventually abolished, without passing through a preceding state of excitation (du Bois-Reymond 1849, 287).

Each of these examples are remarkable departures from Müller's *Handbuch*. Where Müller found intoxicants such as alcohol and opium to have both stimulating and destructive effects on nervous material, du Bois-Reymond either suggested these effects were weaker than otherwise reported or questioned the existence of such a relationship altogether. The scope and persistent nature of du Bois-Reymond's criticisms of Ritter, Aldini, and other galvanists helps explain the marked departure in du Bois-Reymond's conception of the effects of intoxicating substances on nervous material from those espoused by Müller. The vital substance concept, associated as it was with vitalism, needed to be exorcised from the thinking of the new physicalist mechanism. More than merely disagreeing with the concept's theoretical underpinnings, du Bois-Reymond attempts to readdress all of the physiological findings that appear to be even secondarily associated with any notion of "vital substances." Rather than modify his theoretical interpretation to assimilate the trove of findings independently collected by eminent researchers from across Europe that supported some form of excitation associated with intoxicating substances, du Bois-Reymond instead sought to do away with the concept as a whole. This in itself speaks to either the significance that du Bois-Reymond saw in the influence of the vital substance concept or the intensity of its association with vitalism itself.

The case could be made for Helmholtz as well, though Helmholtz employed an entirely different strategy of critique. Famously, "Helmholtz was very reluctant to participate in any scientific debate if he thought it fruitless" (Elkana 1970, 290). This is not merely an observation of Helmholtz's temperament but an overt "rule" spoken on by Helmholtz himself. Helmholtz deemed it "necessary to reply to criticisms of scientific propositions and principles only when new facts were to be brought forward" (Elkana 1970, 291). The expectation was that "when all data have been given, those familiar with the science will ultimately see how to form judgment even without the discursive pleadings and sophistical arts of the contrary part" (Elkana 1970, 291). As a polemic strategy, the effects are twofold. First, there is the immediate result of displacing or writing over the object of the would-be debate—where Helmholtz did not regard the opposing position as bringing forward new facts. Second, and more impactfully, Helmholtz's "rule" entailed a judgment concerning the scientific validity of the position in question. Those ideas that did not merit rebuttal had already been spoken to by the science, so to speak—a rhetorical structure that undermined the rational permissibility of opposing claims on the basis of shifting notions of scientificity. Thus, Helmholtz's silence concerning both Brunonianism and the nature of intoxicating vital substances not only displaced them through his electrophysiological research; it excluded these concepts from consideration on the basis of the epistemic validity.

This segues into what was arguably the most significant element of the neomechanical conflict with the vital substance concept. It has already been established that in the first half of the nineteenth century, self-experimentation with intoxicants was not merely a valid form of empiricism, but constitutive of experimenters' conceptions of both normal and pathological physiology. Self-experimentation had, as Joan Steigerwald suggests, made the human subject a "corporeal part of the experiments," positioned "both as instrument and as subject matter"; researchers could give testimony "from within the phenomena" (Steigerwald 2016, 81).[3] The experiential subject serves "as a tool of mediation between conceptual and sensory apprehension and the material phenomena" (Steigerwald 2016, 81). For Brown and Sertürner, the state of intoxication entailed a tacit attunement to the state of embodiment. This intoxicated way of knowing imparted information about the world, empirically even, and the body's place in it. Gathered in those places at those times, their experiences of intoxicated embodiment gave testimony. What had been so crucial to Romantic science would now become insignificant, even distasteful, to the "new" physiology.

The most potent movement against the last remnants of vital stimulants was the exclusion of the epistemic validity attached to the methodological principles of self-experimentation that had made the vital substance concept possible in the first place. The neomechanical emphasis on an experimentally derived physiochemical study of the body implicitly excluded these forms of testimony. The likes of du Bois-Reymond and Helmholtz parted with the ambiguities of self-experimentation, carrying off the host of perceptional phenomena, from musical tone to vision or muscular reflex, and confining them to the lab. Rigorous experimental induction became the singular valid method of coming to understand the natural world (de Kock 2014). What it meant to see, taste, touch—*to feel*—was realized in the conductivity of the nerve fiber, or the behavior of cells under the microscope (de Kock 2014). They excluded the study *with* the body from the study *of* the body.

Intoxicants could always of course intercede at the level of personal experience; however, it appears that, relative to those who came before them, those involved in the "physicalist turn" made a self-concious move to distance themselves as much as possible from vital substances. In their case, this meant moving away from intoxicants as central objects of scientific research, instead privileging the study of finer physiochemical structures. It would be this shift to privilege certain modes of framing the study of the body, and their corresponding criteria of epistemic validity, that would guide what I have called the "young" neomechanists to an impasse concerning the correspondence between neural states and mental states. It would be unto this very breach that substances of intoxication would, once more, intercede to testify concerning the nature of the relationship between the body and the mind.

✸ 6 ✸
The "Young" Neomechanists and the Problem of the Brain

Localization of Brain Function

Of the state of physics and physiology before the advent of the great physicalist turn, Helmholtz once expressed that they "had, it is true, an almost uncultivated field before [them], in which almost every stroke of the spade might produce remunerative results" (Helmholtz [1877] 1995, 319). This spatial metaphor was not chosen at simple happenstance. Before the European Enlightenment, the "map" was not available as a metaphor for the systematic interrelation of knowledges, nor was "plan" a metonym for strategy (Edney and Pedley 2019). Ceasing to function solely as a representation of geographical relationships, there were now "maps" for all kinds of things. Scientists had become surveyors and cartographers of knowledge, measuring, delineating, and representing the sloping gradients and sprawling plains of epistemic space. For those engaged with the study of physiology, the name bestowed upon *their* murky continental interior—the last truly great, uncharted conceptual vastness whose peripheral features merely gestured at what lay beyond—was "the brain."

By the beginning of the latter half of the nineteenth century, it appears there was a clear sense that physiology was well on its way to understanding the function of much of the body. Yet prior to the emergence of the theory of brain localization in the latter nineteenth century, little was known about how the brain functioned, despite the considerable scientific economy mobilized around the examination of nervous tissue. Of course, the concept of brain localization was not new. Consistent with the Enlightenment "map concept," Franz Gall (1758–1828) developed a doctrine of brain localization at the end of the eighteenth century which led to the creation of phrenology (Hagner 2008; Van Wyhe 2002, 17). His system, which he called the *Schädellehre* (Doctrine of the Skull), divided brain function into different sections, called "organs," which could be discerned through the analysis of the exterior contours of the skull (Van Wyhe 2002, 22). After six years of lecturing, Gall was unexpectedly banned from lecturing in Vienna by decree of Austrian Emperor Franz II in 1801 (Van Wyhe 2002, 25). The harsh ban spectacularly failed to quash interest in Gall's doctrine. The spectacle generated by the emperor's injunction even helped drum up pan-European interest in Gall's work (Van Wyhe, 2002, 25). Gaspar Spurzheim (1776–1832), who worked under Gall as an assistant in Vienna, elaborated on Gall's work, further spreading phrenology to wider audiences in Europe and America (Bilal et al. 2017). The work done by Gall and Spurzheim briefly popularized an early model of brain localization, but by the 1840s their theories had largely fallen into disfavor owing to a lack of clinical substantiation (Simpson 2005).

The nineteenth-century model of cerebral localization would primarily be developed by a new generation of physiologists educated in the school of the organic physicists du Bois-Reymond and Helmholtz. Learning during a period of greater institutional centralization, the researchers behind the modern concept of brain localization also represent the first generation of fully professionalized physiologists, many of whom would go on to further develop different specializations. The discovery of the motor cortex in the 1870s made by Eduard Hitzig and Gustav Fritsch, both of whom studied physiology in Berlin, helped concretize the connection between physiology and psychiatry, paving the way for the brain psychiatry of Theodor Meynert and Karl Wernicke. It would be Wernicke and Meynert who would expand upon the theory of brain localization to establish models of neurophysiological psychiatry founded on scientific principles. Much like their mentors, intoxicants appear to have been far from their minds. Meynert's and Wernicke's desires to see the development of a scientific psychology on the foundations of neurophysiology reflected the values of German physiology toward the latter half of the nineteenth century. They would enlist as intrepid explorers, not of the glacial climes of Chimborazo or the lushness of the Amazon but of the uncharted interiors of the brain. Where others imagined the *sechel*, soul, *Geist*, or "thinking substance," Meynert and Wernicke hoped to discover the physical unity of mental states and brain states, extending the metaphorical mapping of neural cortices into mental experience.

As we will soon see, this would ultimately be their undoing, as their a priori identification of brain matter as the site of valid experimental knowledge concerning the mind would lead to accusations of blurring the lines between psychology and physiology. Most importantly, it precluded them from considering the experimental value that intoxicants might provide, in spite of their active participation in a psychiatric profession increasingly reliant on methods of pharmaceutical intervion. Arguably, it would be the relative failure of brain psychiatry that would help give rise to experimental psychology in the first place. It would be those who eventually introduced substances of intoxication into experimental psychological research who would also prove to be the most steadfast critics of Meynert and Wernicke's approach, which might be dubbed "brain psychiatry." But, before exploring that possibility, it is worth briefly tracing the development of physicalist neurophysiology, beginning with Hitzig and Fritsch's identification of the motor cortex.

The Motor Cortex as an Experimental Demonstration of a New Theory

Eduard Hitzig and Gustav Fritsch created a remarkable disturbance in the physiological community in 1870 when they demonstrated that they were able to cause movement when certain areas of the brain were electrically stimulated (Fritsch and Hitzig 1870; Hagner 2012). While Albrecht Kölliker had previously succeeded in establishing an *anatomical* relationship between motor nerves and the brain, Hitzig and Fritsch were suggesting that they could *physiologically* localize motor function in the human brain.

Gustav Fritsch (1838–1927) studied medicine in Berlin, Breslau, and Heidelberg, learning under none other than Helmholtz, Wilhelm Karl Hartwich Peters, Ludwig Traube, Friedrich Theodor von Frerichs, and Bernhard von Langenbeck (Anonymous 1938). Eduard Hitzig (1838–1907) attended the University of Berlin and the University of Würzburg, where he learned under Rudolf Virchow, du Bois-Reymond, and Moritz Romberg, who had played a similar role as Müller for the field of neurology.[1] In this sense, Hitzig and Fritsch represented the next generation of physiologists—their views, and methodologies, reflecting those propagated by the "organic physicists" of Berlin. At the time, the pair were privatdozenten at the University of Berlin. Barely more than adjunct professors, they were not allowed to conduct their experiments in the university laboratories (Millet 1998, 284). Their conclusions overturned a principle of neurophysiology that had been foundational since Haller: cortical inexcitablity, the belief that the brain could not be artificially stimulated (Millet 1998, 284).

Fritsch and Hitzig "reported that five punctate 'centers' could be distinguished in the anterior cortex of the dog, stimulation of which led to contractions of the contralateral muscles in the neck, legs and neck" (Fritsch and Hitzig 1870, 311; Millet 1998, 284). "Clonic movements and epileptic convulsions" were precipitated by the application of oscillating currents (Millet 1998, 284). Of far greater interest were the localized muscular twitches caused by "constant galvanic stimulation" the moment a circuit was completed (Fritsch and Hitzig 1870, 311; Millet 1998, 284). Reviewing the outcome of their experiments, Hitzig and Fritsch conclude that the "anterior regions of the cerebral cortex is associated with motor function" (Millet 1998, 284). While these conclusions appear simple given the outcome of their experiment, this was a crucial moment in the establishment of the modern understanding of brain localization. From the neomechanistic perspective, Hitzig and Fritsch's work was cause for serious hope concerning an experimental understanding of the brain and mind in the near future. Their study of the motor cortex, in a sense, further demonstrated that the neomechanical research paradigm of organic physics was making good on its initial objectives.

The revolutionary impact of their joint discovery beautifully symbolized the beginning of a new era in German physiology and neurophysiology. The period from 1850 to 1870 had been dominated by the legacy of Johannes Müller. From Albert Kölliker and Rudolf Virchow to Helmholtz and du Bois-Reymond, nearly all the great physiologists of the period had studied under Müller and, whether through modification or principled rejection of his theories, his ideas lived on in their work. Hitzig and Fritsch, as the students of Helmholtz, Virchow, and du Bois-Reymond, represented the next generation of neurophysiologists. With the appearance of a new generation, however, came a transformation in the culture of physiology, a major part of which can be attributed to the formation of the first modern research universities in Germany.

By the 1870s, physiology and biology had retreated even further into the laboratory. Universities across Germany committed to massive expansion projects, investing heavily in laboratory space (Klinge 2004, 128). Historical methods were increasingly displaced by experimentation (Weindling 1993, 44). At this point, the life sciences were almost completely dominated by formal professionals, functioning within established institutional bodies and university research centers (Weindling 1993, 44). Romantic thinkers like Ernst Haeckel were rapidly becoming relics of a bygone era. As Paul Weindling suggests, this increased "professionalism [also] meant a retreat from the overt political activism of Virchow" and others, as researchers became increasingly focused on scientific work (Weindling 1993, 45).[2]

As seen in the case of Hitzig and Fritsch, these changes in the structure of the natural sciences had theoretical ramifications as well, as the new generation assumed the philosophical attitude taken by the "organic physicists." Professionalized physiology was well suited to the stance taken by Helmholtz and du Bois-Reymond, which favored a shared outlook among researchers, a common clinical nomenclature, and methodological agreement. Citing Darwin as the empirical foundation of biology, Hitzig and Fritsch were satisfied to approach the human body and mind as a physiochemical process, rather than as a philosophical subject (Weikart 2016, 115). Another factor was the breakout of the Franco-Prussian War, also in 1870. Like many young physiologists, Hitzig served in the Franco-Prussian War as a physician, where he oversaw the clinical treatment of Joseph Masseau, a twenty-year old French soldier of the Thirtieth Line Infantry Regiment. Published in *Physiologische und klinische Untersuchungen über das Gehirn*, Hitzig was able to connect Masseau's symptoms with damage to the motor cortex in a human subject, reflecting Hitzig and Fritsch's recent experiments with a dog (Hitzig 1874, 119).

The Masseau case was a turning point for Hitzig in more ways than one. It was a demonstration of the clinical significance of Hitzig and Fritsch's work, specifically in the field of psychiatry. Further clinically significant experimentation on the brain was realized through Hitzig's research on induced seizures, once again in dogs. In a rather brutal experiment, Hitzig succeeded in imparting several dogs with seizure disorders, a malady then associated strongly with mental illness, by damaging portions of the brain (Hitzig 1874, 273). Spurred on by newfound supporting evidence, Hitzig's work helped guide neomechanical thinking in the developing field of psychiatry, as the domain of a yet unconquered realm in physiology: the human mind.

Hitzig and Fritsch's study of the motor cortex would change from a critique of the theory of cortical inexcitability to an experimental demonstration of a largely theoretical, conceptual model of neural activity. Already theorized in the work of Theodor Meynert (1833–1892), who suggested there were discernible variations in the histological structure of different regions of the brain in 1867, Hitzig and Fritsch's rebuke of the theory of cortical inexcitability would come to be framed as a crucial experimental demonstration in support of the developing theory of cortical localization (Meynert 1867; Meynert 1872). It would be this model, championed by, above all, Theodor Meynert and Karl Wernicke, that would form the foundation of the most aggressive efforts to see even higher-order mental states equated with neural states, in an effort to settle the final frontier of the neomechanical body.

Mapping an Inner World: Meynert and Cerebral Cytoarchitecture

Meynert had been a student of Carl von Rokitansky (1804–1878) in Vienna, under whom he studied neurological diseases of the brain and spinal cord (Guenther 2015). In Vienna, Meynert was also exposed to Ernst Brücke, who had taken on a professorship at the University of Vienna as early as 1849 (Seebacher 2006). Eventually, Meynert's expertise in neuropathology brought him to psychiatry, assuming a position as head of Vienna's psychiatry clinic in 1870 (Guenther 2015; Hagner 2008). Meynert had no previous psychiatric experience or training, leading some to question his eligibility for the post. Concerns over eligibility arose again in 1873 when Meynert was up for a promotion, but Meynert's former mentor Rokitansky interceded to see that Meynert got the position (Guenther 2015). There is no reason to dismiss this as basic nepotism. Physicalist neomechanism was more than an idea, or even a series of a priori theoretical and methodological commitments (Hagner 2008; Mayer 2013). It was a research community, commited to mobilizing social and institutional manifestations of power in the furtherance of shared ideological commitments. Rokitansky understood that Meynert had a vision: an interdisciplinary approach to brain research, one that would bring the anatomical study of nervous tissue out of the lab and into the clinic (and back again) (Seitelberger 1997; Guenther 2015).

Forecasting later developments, one of the dissenting votes was cast by the alienist Ludwig Schlager (1828–1885), director at Niederösterreichische Irrenenstalt am Brünnlfeld and one of the most celebrated alienists in the Empire (Shrady and Stedman 1885, 211). Unlike Schlager, Meynert was suspicious of the observational and experimental approach to therapy, which would later manifest in an opposition to hypnosis (Mayer 2013). Schlager was so incensed by Meynert's presence that they both eventually refused to work together at the psychiatry clinic (Guenther 2015). Karl Jaspers likened the spat between the two men to a conflict between asylum psychiatry and university psychiatry, or bottom-up versus top-down psychiatry, though it might be better understood as a conflict between alienist psychology and physicalist neurophysiologically-oriented psychiatry (Pichot 2013).[3]

Meynert was not alone in asserting that German psychiatry was at a crossroads. Wilhelm Griesinger (1817–1868), founder and editor of the *Archiv für Psychiatrie und Nervenkrankheiten* (Journal for Psychiatry and Nervous Diseases), prefaced the journal's first 1867 edition with the observation that German psychiatrists were too preoccupied with philosophical attitudes, despite firm evidence of the connection between neuropathology and mental illness (Marx 1970). The moral cause of seeing the mentally ill treated like any other patients was supported by the idea that mental illness was physiological in nature. The only difference was the difficulty of assessing neuropathological conditions (Hagner 2008; Marx 1970). Meynert understood how this functioned in both directions. Neuropathology would remain incomplete as long as it refused to recognize mental illness as physiological in nature, but the study of a given mental illness's neuropathology required external support through clinical observation (Marx 1970). The result is a totalizing project to "map" the diverse functions of the brain through neurophysiological research, in the process subsuming psychiatry and psychology into neurology. Where even du Bois-Reymond had ruled the nature of consciousness beyond the purview of science, Meynert sought to map the physiological landscape of the mind (Pecere 2020).

While Meynert's position was that of the director of a psychiatry clinic, Meynert was first and foremost a neuroanatomist, with limited interest in clinical practice (Hagner 2008; Hakosalo 2006; Mayer 2013). Meynert's visionary approach understood the scope of brain research as an emerging field that required the cooperation of then-conflicting disciplines (Mayer 2013; Seitelberger 1997). Here, brain research was not merely a pursuit of anatomical interest; it meant coming to understand mental illness in scientific terms.

Meynert's theories of cortical localization were built upon his histological research on the brain stem, cerebrum, and fiber systems and were externally corroborated by clinical observations (Seitelberger 1997). An essential component of Meynert's understanding of cortical localization was his discovery of the difference between projection and association fibers. Projection fibers extend from the spinal cord "to the hemispheres and descend from the cerebral cortex through the medulla and white substance to the spinal cord" (Hakosalo 2006, 175; Meynert 1865). Where the projection fibers establish connections between the brain and the rest of the body, association fibers serve the function of forming links between the cortices of the brain (Hakosalo 2006). This distinction between projection and association fibers provides the structural framework for Meynert's map of cortical function. In 1865, Meynert traced the locations of the visual, auditory, and olfactory centers to the occipital and temporal lobes (Meynert 1865). A year later, Meynert was able to study a patient suffering from aphasia, identifying the cause to be growths in the wall of the temporal lobe (Hakosalo 2006).

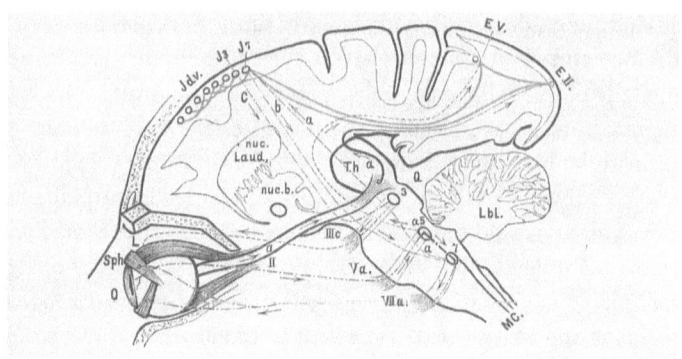

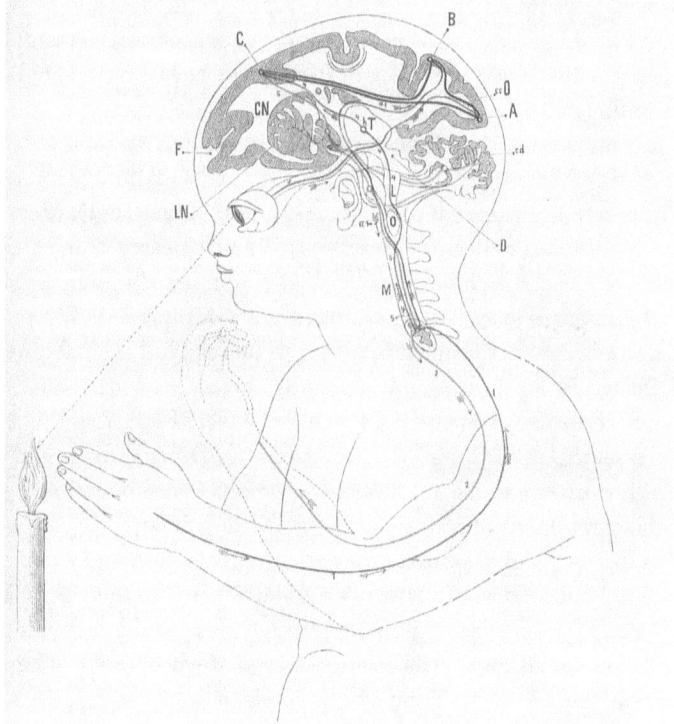

FIGURE 4. Diagrams from Meynert's 1884 *Psychiatrie: Klinik der Erkrankungen des Vorderhirns begründet auf dessen Bau, Leistungen und Ernährung*, depicting his proposed neural pathways for the conscious blink (*top*) and conscious arm movement (*bottom*) (Meynert 1884, 145, 147). Courtesy of the Staatsbibliothek zu Berlin—Preußischer Kulturbesitz, HA 16 Pa 7970.

Though Meynert had been fairly successful anatomically mapping cortices by tracing nerve fibers, Hitzig and Fritsch's 1870 experimental identification of the motor cortex in a dog was a major turning point for Meynert's work. Meynert had actually predicted Hitzig and Fritsch's results, including their conclusions on the excitability of brain tissue, but had not been able to experimentally demonstrate his findings. Hitzig and Fritsch were well aware of this, as they explicitly remarked in their 1870 paper: "The only person who, on the basis of anatomical investigations, took a decisive standpoint that differed from the prevailing opinion [that brain tissue was not excitable], the possibility of which is, of course, doubted by some, was Meynert" (Fristch and Hitzig 1870, 307). Hitzig and Fritsch's work provided experimental confirmation of the clinical significance of Meynert's anatomical research. Two years later, Meynert published "Vom Gehirne der Säugethiere" in the second volume of *Handbuch der Lehre von den Geweben des Menschen und der Thiere* before undertaking a more serious effort to clinically apply his theoretical understanding of brain function (Meynert 1872).

Meynert's idea that the sum of human experience could be explained on a physiological level was not unique or revolutionary. In many ways it was merely the fulfillment of a promise implicit to the theoretical commitments of academic physiology since at least the 1860s. After all, Griesinger had already established himself as an advocate for a shift toward a physiological model in psychiatry, on the shoulders of the physicalist revolution in physiology. What made Meynert a visionary was his eagerness to personally extend the very new concept of cortical localization, as a bodily description of the mind, into the clinic sphere and begin to reconceive of clinical practice from the ground up.

On (the Other) Aphasia: Karl Wernicke's Meynertian Localization of Language

Despite Meynert's optimism and the clear significance of Hitzig and Fritsch's finding, the cytoarchitectural research program that was emerging out of the neomechanical approach in physiology had yet to provide any notable discoveries concerning what were perceived to be higher-order brain functions, such as language. This would change in 1874 with the publication of "Der aphasische Symtomencomplex" (The Aphasic Symptom Complex) by Karl Wernicke, written while studying under Meynert in Vienna (Marx 1970, 364).

The publication's full title is "Der aphasische Symtomencomplex: Eine psychologische Studie auf anatomischer Basis." The subtitle alone leaves no question as to how Wernicke understood the broader theoretical implications of his work. Though if any question remained as to Wernicke's research objectives, the opening paragraphs of "Symtomencomplex" make them perfectly clear:

> In Meynert's theory of brain fibers are contained the beginnings of an exact brain physiology, indeed only in large general features, but in features of such ingenious inner truth that they already now unhesitatingly allow for their application to the individual case.
> The present work is such an attempt to make practical use of Meynert's brain anatomy for a field in which such fundamentals should be most needed, but in fact have been used least so far. (Wernicke 1874, 3)

Wernicke's aphasia research was explicitly intended to not only empirically validate Meynertian neurophysiology but also to further extend it into the practical domain of psychiatry. The combined effect of such an undertaking would, at least theoretically, further establish the potential for neurophysiology to provide physical explanation for a wide variety of mental states.

Thus, it made sense that Wernicke chose aphasia as a starting point. French physician and anatomist Paul Broca (1824–1880) had already provided a firm foundation through his own work on aphasia (Hagner 2008). In France, earlier arguments for cerebral organization had undergone a process of steady decline, in part due to the experimental failings of phrenology (Finger 2000). During the 1860s, Broca forever changed the debate surrounding brain localization when he presented his findings on autopsies done on twelve patients suffering from aphasia (Finger and Wade 2002). All of the autopsies found lesions in the frontal lobe, specifically encompassing a region that is now termed Broca's area.

Reviewing Broca's findings, Wernicke suggested that Broca's area was found adjacent to motor nerves that affected the lips, vocal cords, and parts of the face (Wernicke 1874, 19). This led Wernicke to theorize that this region contained the "rules" for speech production (Wernicke 1874; Geschwind 1970, 941). He linked these findings with his own observations that receptive aphasia was specifically related to damage to a region found in the posterior section of the superior temporal gyrus, later termed Wernicke's area (Wernicke 1874; Geschwind 1970, 940). Wernicke proposed that these two regions of the brain were linked and that this relationship could be used to explain the different ways in which Broca's aphasia and Wernicke's aphasia affected written and spoken language production and reception (Geschwind 1970, 941). One explanation given for the effect of receptive aphasia on written language suggested that written language developed in relation to a spoken language, so damage to Wernicke's area would impair comprehension of both spoken and written language (Geschwind 1970, 941).

Wernicke's Applied Neurophysiology

Following a brief stint of enthusiastic study under Theodor Meynert, Wernicke moved to Berlin in order to work under Karl Westphal at Berlin's Charité, from whom Wernicke would learn about psychiatry. Westphal, in the tradition of Griesinger's advocacy for the recognition of mental illness as brain diseases, saw the future of psychiatry to be in neurological and physiological research (Marx 1970, 361). In this sense, Wernicke came by his interest in synthesizing mechanical and physiological principles in explaining higher brain functions in both his neurophysiological and psychiatric training. Wernicke sought to continue on the path laid out in both Meynert's anatomical theories and Broca's work on language, which set off by asking whether it was even possible to localize higher brain functions (Roth 2002). Wernicke's observations concerning the localization of language function, combined with Hitzig and Fritsch's earlier work on the motor cortex, led Wernicke to theorize about the localization of further brain functions. Utilizing the framework established in his own work, Wernicke soon aspired to use physiology to explain all forms of complex brain function (Marx 1970, 365). He hoped that this new paradigm in cortical localization would provide the foundation for a natural science of psychology in neurophysiology.

This new neurophysiological concept of cortical localization should be distinguished from that put forward by Gall, which ascribed fixed locations for all brain functions (Marx 1970, 365). Although Wernicke conceived of the brain as a network of cortical centers, he theorized that memory and emotion were facets of the structure of the brain itself. Memory was conceived of as "lasting changes" in the structure and organization of nerve cells, whereas emotions were an undetermined "property of living cells" (Marx 1970, 365).

Like many of his predecessors, there is strong evidence to support the inference that Wernicke's neurophysiology was to be understood as strictly physiochemical. In his lectures, Wernicke made a point of qualifying terminology associated with vitalistic concepts in Müller's *Handbuch* with more concrete psychological concepts. For example, as opposed to the vitalistic meaning used by Müller, "organ sensation" (*Organempfindung* or *Organgefühl*) is instead used by Wernicke to refer to sensory input associated with muscle or eye movement in addition to sensory input from the skin (Miller and Dennison 2015, 450).

Publicly, Wernicke took the offensive against those who resisted the neomechanist research program in psychiatry, becoming an outspoken critic of alienist experiential psychology in the treatment of mental illness on the basis of what he perceived to be an absence of physiological support (Engstrom 2003, 101). In spite of this, Wernicke took a holistic approach in the treatment of his own patients as a psychiatrist. This is reflected in Wernicke's lectures on psychiatry, collected in *Grundriss der Psychiatrie* (1894). Here, Wernicke's lectures illustrate his tendency to describe psychiatrically relevant symptoms in a neurophysiological way and yet, in the very same series of lectures, prescribe treatment principles that appear to have no clear physiological foundations.

Wernicke understood the psychiatric institution, or asylum, to be at the absolute center of psychiatric treatment. This was not only practical; the very structure of asylum life was understood to have a curative function. Much of our behavior, Wernicke argued, amounts to processing stimuli in accordance with a given cognitive order, creating a more general order between perceptions, as Wernicke explains:

> As you remember, we called the connection of the ideas associations, the order ruling in the ideas lets us therefore conclude on the possession of certain associations, which are approximately the same for all individuals. Our question may therefore be put thus: In what way are these generally valid associations formed? In discussing the consciousness of the external world, I have already pointed out that the natural order and sequence of things is reflected in our brain to a certain extent, and that thereby a lawful linking of certain phenomena among each other, as they are supplied to us by the external world, is also established in our consciousness. (Wernicke 1900, 69)[4]

Wernicke appeals to an operative theory of mind in which each individual's mind is comprised of an order of sequential associations that reflect a shared reality, and so are relatively similar. In those suffering from mental illnesses, this "proper" order becomes disturbed—dislocated (Wernicke 1900, 69–70). Here, the very structure of the asylum works toward the ordering of perceptions and mental associations. The theoretical foundation of this method is grounded in Wernicke's argument that the behavior of individuals in a group bends to loosely conform with that of the whole (Wernicke 1900, 70). This notion, with the support of a monitored and regulated lifestyle of asylum life, is intended to holistically order a patient's mental associations in accordance with "reality." For Wernicke, "The entire institutional treatment of the mentally ill rests on this foundation" (Wernicke 1900, 70). Of course, in cases where the symptoms are acute, the institution also isolates the patient from broader society, where they can "safely" be sedated with something such as opium or chloral (as required) (Wernicke 1900, 280).

It is in this sense that Wernicke identified articulated speech as the "primary means of acquiring a certain order in the ideas" (Wernicke 1900, 70). This is because the internal structure of language trains the mind to form series of associations in a consistent way, including the most abstract of concepts (Wernicke 1900, 70–71). To this point, it is apparent that Wernicke's focus on language is at least partially a by-product of his own work on aphasia. Citing precisely this research, Wernicke argues that, insofar as those suffering from aphasia still understand concepts and emotions, language is the tool that trains the mind rather than the medium through which associations are maintained (Wernicke 1900, 70–71).

In each of these treatment protocols, Wernicke's approach reflects a robust philosophy of mind, albeit one not explicitly derived from neurophysiological principles. The example of articulated speech reflects instances where Wernicke's approach to understanding mental phenomena appeared to be grounded in brain research; however, this is only in the form of inference from a largely unrelated point of study. There was an apparent disconnect between the perceived significance of recent discoveries surrounding cortical localization and clinical practice.

"Intoxical Inexcitability"

The careful reader is probably wondering what, if anything, this has to do with a story that has been, for the most part, about intoxication, intoxicants, and those who consumed them. But this is precisely the point. The extent to which neither Meynert nor Wernicke appear to have been particularly interested in studying the effects of intoxicants on the mind, despite extensive use of various *Nervina* in their clinical life, *should* be shocking. The psychiatric epoch of Meynert and Wernicke was the heyday of chloral hydrate. From the late 1860s, chloral had quickly become a pillar of asylum life, due to its profound hypnotic effects and the fact that it did not need to be administered by injection (Shorter 1997, 231). Use of chloral at home also became increasingly common, a means for the well-to-do to avoid a hasty and embarrassing visit to the asylum. This led to an explosion in the number of habitual chloral users (real or imagined), with particular societal emphasis on middle-class women (Shorter 1997, 231). The chloral phenomena began in the German world in the 1860s, yet by the 1880s it was already wreaking havoc in America. Harold Kane discussed the growing chloral phenomena in America in his 1881 book *Drugs that Enslave: The Opium, Morphine, Chloral, and Hashisch Habits*, where he remarks at the sheer scope of overlooked chloral use (Kane 1881, 149). Kane himself was convinced of the fact that chloral use was habituating, seemingly on the basis of a Brunonian model of stimulus dependence, though he acknowledges this argument was hotly contested among European physicians (Kane 1881, 149; Perkins-McVey 2023).[5]

Back in Germany, not only were Meynert and Wernicke on the front line of the chloral phenomena, but they were direct participants in the introduction of chloral hydrate to the patient body. Wernicke discusses the use of chloral, morphine, and opium in the treatment of the more frenetic patients under rare circumstances in his *Grundriss* (Wernicke 1900, 292). There is likewise a well-known case of Meynert referring a forty-two-year-old woman to a private clinic on the grounds of an initial diagnosis of a chloral habit, only for her to flip into psychosis upon cessation of chloral use (Eder 1889, 267). As will be seen in a later chapter, they were psychiatrists serving on the front lines of the emerging concept of addiction, a category previously unavailable outside the limited disease concept of alcoholism (Perkins-McVey 2023). Meynert himself was even alleged to be an alcoholic (Shorter 1997). It is astounding that the effects of intoxicants on the mind were practically never discussed by either Meynert or Wernicke, given the prominence of intoxication and intoxicants both as forms of clinical intervention and as possible causes of mental disturbance. The answer to this issue, as will soon be discussed, rests with the way in which any questions surrounding the effects of intoxicants on the mind were theoretically framed and are bound up in the broader issues that ultimately undermined Wernicke and Meynert's shared research program.

Neomechanistic Neurophysiology: Much Ado About Nothing?

By 1891, a young Sigmund Freud would remark in his own *Zur Auffassung der Aphasien* that Wernicke might have conflated psychological concepts with physiological constructs (Freud 1891). Freud would go as far as suggesting that Wernicke's neurophysiological foundations were in fact, as Marx understands it, "the product of psychological assumptions [...] disguised in anatomical and physiological terms" (Marx 1970, 368). Meanwhile, Meynert's grand aspirations to see the development of an interdisciplinary brain science had been dubbed the "brain mythology," an accolade that Karl Jaspers would bestow upon Meynert and Wernicke, jointly, in 1913 (Jaspers 1913). But the prestige of both Meynert and Wernicke had begun to unravel far earlier still, with their downfall arguably well underway by the 1880s and assuredly so by the 1890s. Meynert had been Rokitansky's star pupil; an esteemed friend and colleague to Ernst Brücke, one of the original physicalists; and a respected peer in the eyes of Eduard Hitzig, among many, many others. In sum, Meynert had descended from a place of standing as the presumptive progenitor of a definitive physiochemical reckoning on the nature of the human mind to something of a has-been, bringing Wernicke with him. What is to be made of the sudden collapse in their joint research program?

The crucial issue was the difficulty of practically translating neurophysiological brain research into an empirical understanding of mental illness in particular, and mental states in general. Meynert's conception of brain function was strictly mechanistic. The foundational principles from which Meynert extrapolated psychological functions from cerebral anatomy were little more than the assumption of Bell's law for reflexes and the Müllerian dogma that the role of nerve cells is sensing, while sensations were formed in the forebrain (Meynert 1884, 126; Marx 1970, 362). This was useful for describing certain phenomena, such as blinking, but raised greater complications when marshaled to describe more complicated mental activities (Meynert 1884, 146). As a result, it was all too simple for Meynert to conflate psychological concepts with physiological facts. Meynert's general disinterest in clinical practice may have been his saving grace, given that in Wernicke's case the apparent disconnect between his approach to mental illness and neurophysiology was plain for all to see.

There were other factors that ultimately undermined Meynert's research program. Hakosalo makes the case that Meynert's authoritative footing was eroded by methodological innovations in neuroanatomy, particularly the development of serial sectioning. Coming up in the 1860s and 1870s, Meynert's methodology consisted primarily of the meticulous hand tracing, or "mapping," of nerve fibers. These cortical "maps" were then extended onto mental processes. Meynert particularly described his own *Abfaserung* technique using the verb *einbrechen*, to break in, and describes how the relationships between cortical structures were best observed by pulling apart structures of the brain to view fiber bundles in situ (Meynert 1884, 39–40). The result was something of a roughly hewn severance that, to his credit, preserved the integrity of structures that might otherwise have been occluded by a cross section, although Meynert also hand-sectioned with a razor (Hakosalo 2006, 179). Meynert's great skill at hand-sectioning was admired as tremendously difficult in itself, though it made reproduction of his findings challenging.

While Meynert continued to make use of the difficult *Abfaserung* method, use of the microtome in neuroanatomical research was quickly becoming a standard practice. Though Bernhard von Gudden had been the first to use the microtome on brain tissue no earlier than 1875, use of the microtome in neuroanatomy had become routine by the 1880s (Hakosalo 2006; Monakow 1970; Gudden 1875, 229). This was partially on account of the difficulty of sufficiently hardening brain tissue in order to achieve a clean slice, without altering the appearance of the brain matter itself. This particular issue, that of ascertaining the most effective methods for staining and hardening samples, had plagued German physiology for much of the nineteenth century. The needs of the community had rapidly outstripped the available techniques. Rudolf Virchow had widely used a carmine stain, the use of which can be traced back to late eighteenth-century botany (Alturkistani et al. 2015). Waldeyer developed a hematoxylin stain in 1863, which was flexible and acid-resistant but a relatively weak stain (Alturkistani et al. 2015). Although Italian biologist Camillo Golgi developed the first method for staining nervous tissue using potassium dichromate followed by silver nitrate in 1873, it was not until Kölliker's review of the method, published after visiting Golgi in 1887, that the silver stain technique would be properly introduced to German physiology (Shepherd 1991, 170). Meynert himself used a variation of the carmine staining method, similar to that devised by Rudolf Berlin and used by Virchow, which utilized a fixing agent consisting of potassium dichromate, a carmine-based stain, and a turpentine clearing solution, though some samples were hardened using alcohol and acids (Shorter 1997, 97; Meynert 1867, 198).

Hakosalo argues that while Meynert maintained his fidelity to the older *Abfaserung* method and sectioning by hand, newer work favored serial sectioning with the assistance of a microtome, which enabled consistent transparent brain sections that could be viewed under a microscope (Hakosalo 2006, 185). This can be seen in the almost immediate effect that the advent of serial sectioning had on the anatomy of cerebral pathways, a conceptual focus of Meynert's research program. As Hakosalo points out, Auguste Forel recognized from the moment he first saw sections of the human brain that this spelled trouble for Meynert (Hakosalo 2006, 188; Forel 1935, 74). Things came to a head when Emmanuel Mendel gave a presentation on the structure of the "superior cerebellar peduncle at the Medico-Psychological Society of Berlin in January 1878," questioning "whether the auditory pathway was related to the superior cerebellar peduncle or not" (Hakosalo 2006, 188). Mendel concluded "that the acoustic pathway was related to the superior cerebellar peduncle on the level of the dentate nucleus," a contradiction of Meynert's schematics (Mendel 1878, 402; Hakosalo 2006, 188). Meynert was not present to defend himself. But Wernicke was and, on account of their personal and professional closeness, swiftly moved to undermine the validity of the serial sectioning method on display (Mendel 1878, 403). Wernicke's defense was, ultimately, unsuccessful, and Meynert's anatomical descriptions would fall under continued assault in the coming years, particularly from Constantin von Monakow.

Even before his death in 1891, "malicious tongues claimed that Meynert's only connection with psychiatry was that he had gone through delirium tremens," a slight pointed at Meynert's alleged alcoholism (Shorter 1997, 97). There is a clear case to be made here that the shortfalls of both Meynert and Wernicke's neuropsychiatric research program, though visible, could have been optimistically overlooked in the face of their apparent success in neurophysiology. However, when Meynert's and Wernicke's neuroanatomical work was increasingly threatened by contradictory findings, the broader theoretical argument for late nineteenth-century brain psychiatry was no longer tenable. Though methodological and representational changes in German brain anatomy undermined the Meynert-Wernicke research program of interdisciplinary brain science, brain anatomy evidently continued to flourish, even thrive. The failure of Meynert and Wernicke's approach made way for further developments in the study of brain anatomy, but their competitors, the likes of Emanuel Mendel, Constantin von Monakow, and Heinrich Obersteiner, were deterred from making grand conflations between anatomy and mental life.

The neomechanical physicalist ideal of seeing mental states reduced to neurophysiology had been all but decisively routed. This was, in a sense, the ultimate turning point in the broader project of organic physics. Neomechanism, as a scientific and philosophical concept that reflected the methodological approach of scientific physiology, had been at the heart of remarkable industriousness in nearly every branch of physiology and anatomy. Yet Meynert's and Wernicke's aspirations, the once-great enthusiasm for their approach, also reflected the extent to which, simultaneously, there was an understanding that a neomechanistic description of organic beings would always remain incomplete for as long as it could not also account for mental life. It was on these grounds that the most fervent strains of neomechanism failed.

The difficulty faced by neomechanistic attempts to account for mental states is arguably a by-product of the assumptions underlying the approach itself. There is a consistent conceptual throughline that can be traced from the likes of Helmholtz's relegation of the perceptional world to quasi-illusion to Meynert's cytoarchitectural brain paradigm. In both cases, there is an a priori assumption that the locus of the "real" was not found in questions of perception or representation, but rather in a concealed series of physical interconnections. Helmholtz's world of energy exchanges was simply the ideal description of a neomechanistic, physico-Platonic reality of which Meynert's machine brain was a part. These were underlying theoretical assumptions with methodological and hermeneutical implications: certain possible objects of scientific inquiry were elevated, while others were diminished, and more still all but disappear.

In the case of the neomechanistic encounter with the body, by merit of their approach's underlying theoretical principles, they privileged the anatomical body as the ultimate source of meaningful knowledge about life. It was almost as if, were it possible to debride the concentric meshworks of perception, conceptualization, and interpretation, to cut out the dead tissue once and for all, we would finally seize upon a tangible conception of the truth, of reality. But this focus on the body as scientific object was at the expense of a consideration of the body as something experienced, as *embodied*. This may itself have been a product of the neo-Kantian interpretation of Müller's law of specific nerve energies, from which it followed that when we perceive the world we truly perceive our nervous system. As a result, any scientific inquiry into the experience of embodiment became occluded by the physiology of perception. The issue, of course, was that the summation of conscious experience was irreducible to simple sense perception. Thus, when Meynert and Wernicke made a pioneering effort to extend this approach into mental life, they hoped to describe the mind by way of the body. But they could only *see* the body, and in the end, their efforts fell tragically short. The study of mental states is, after all, the study of the experience of *embodiment*. Prioritization of the anatomical body as the locus of the "real" encounter with life as such was, in this sense, simultaneously an erasure of life as *embodiment*, as the lived, temporal experience of being a body.

It is worth considering how the retraction from any serious discussion of a science of embodiment beyond the physiology of perception or the relationships between neural states also served to legitimize the neomechanistic approach in its nascence. As has been discussed, a great deal of support for the vital substance concept stemmed from the embodied experience of intoxication, from Brown's initial opium insight to Sertürner's identification of the *principia somnifera*. Such insights were empirical, derived as they were from the lived experience of intoxication, and perceived as having real purchase on the ultimate nature of living beings. This association between not merely intoxicants but vitalism and the experience of intoxication may well have extended beyond the study of intoxicants to discourage neomechanists from studying embodiment itself. A hard rejection of vitalism was the initial impetus behind Brücke, du Bois-Reymond, and Helmholtz's physicalistic neomechanical shift. In this sense, an attitude of disinterest toward a scientific study of the experience of embodiment, which did not focus on the physiological foundation of said experience, served to further delegitimize the work of their vitally-minded forebears. The scientific priority of the body in the neomechanical doctrine erased the experience of embodiment as a meaningful object of study, not only on the basis of its underlying theoretical foundations but on a rhetorical basis. But another line of inquiry had risen in the time since the Icarus flight of Meynert's interdisciplinary brain project. Eventually positioning themselves in direct opposition to Meynert's kind of brain psychiatry, Leipzig, first with Fechner and then through Wundt, would become the site of a novel research program dedicated to the development of a new science of embodiment. It would be this new science that would ultimately provide an alternate description of the relationship between bodies and minds, one deeply entangled with the intoxicated experience of embodiment, and which only became possible on a theoretical level by putting down roots in the forgotten dungheap of a retreating "brain psychiatry."

* 7 *
A Tale of Two Cities

BERLIN, LEIPZIG, AND SCIENTIFIC PSYCHOLOGY

The Exception Proves the Rule

You can almost always find an edge case. It is to be expected that there would be some within the physicalist research community for whom experimentation on and with intoxicants could be counted among their research interests. Ludimar Hermann (1838–1914), a student of du Bois-Reymond's, published on the effects of nitrous oxide, ethylene, and other anesthetics in 1864, 1865, and 1866. Though Hermann was initially counted among the "unskilled Jews" who applied for a lecture assistantship in 1856, du Bois-Reymond would come to regard Hermann as a truly talented researcher by 1865 (Finkelstein 2006, 343). Julius Bernstein—another student of du Bois-Reymond's, also Jewish, and arguably his most prominent protégé —likewise published "Ueber die physiologische Wirkung des Chloroforms" in 1870 (Bernstein 1870). Hermann even inhaled nitrous oxide and ethylene in the course of his experiments. What can be made of this seeming contradiction? If intoxicants were so insuperably divested from the physicalist research program, why were Bernstein and Hermann mucking around with anesthetics?

Upon closer examination, however, neither of these cases prove particularly interested in intoxicants themselves and may, to the contrary, reflect a sustained disengagement from earlier research employing substances of intoxication. To this point, Hermann's "Ueber die physiologischen Wirkungen des Stickstoffoxydulgases" was, above all else, a physiochemical critique of Davy's "merkwürdige" (peculiar) Brunonian thesis that nitrous oxide could sustain life, before eventually causing death (Hermann 1864, 521). Davy's conclusion was consistent with Brunonian ideas. If nitrous oxide was a Brunonian stimulant, a vital substance, inhaling such a stimulant had the capacity to sustain life, up to the point of overexcitation.

For Hermann, it was as simple as reducing the problem to a question of the absorption coefficient of nitrogenic oxygen in the blood cell—with Hermann suggesting that the absorbability of nitrous oxide equaled that of the water contents of the blood (Hermann 1864). What of Hermann's self-experimentation with nitrous oxide? Here, too, Hermann's lungs are merely an extension of the apparatus, each breath respired into a gasometer in a demonstration of the limited absorption of free oxygen. Dyspnea takes place, but its immediate realization is obfuscated by the process of anesthesis (Hermann 1864). The effects of nitrous oxide are not only secondary to the research question, but the study explicitly questions the Brunonian bases of Davy's conclusions. Hermann's experiments were, in this sense, in keeping with the broader physicalist efforts to undermine the validity of earlier experiments conducted under the auspices of Brunonian vitalism.

Nor was Hermann ever a straightforward participant in the scientific community of neomechanist physiology. In 1867, Hermann and du Bois-Reymond, his longtime mentor, entered into a bitter, rending debate surrounding the nature of the occurrence of observable electromotive phenomena in nerves and muscles (Finkelstein 2006). Shortly after, Hermann moved to Zürich and the friendship was never reconciled. In a letter to du Bois-Reymond dated May 23, 1853, Brücke had once suggested that for a teacher and a former student to engage in a public row was in violation of their principles (Finkelstein 2006, 345). The case could be made that Hermann had, in this sense, operated outside of the social mores of their representative research community, perhaps even committing a moral transgression against the *bildungsbürgerlich* values of their shared milieu (Finkelstein 2006, 345).[1] Hermann would remain interested in the absorption of gaseous bodies into the body, the chemistry of the blood, and their effects on muscular movement. When he published a toxicology textbook in 1874, intoxicants nevertheless remained a secondary concern. A quarter of the text was committed to methods, half to various caustic compounds, and the last quarter to various medicinal toxins, with even the foremost intoxicants granted scarcely more than one or two pages each (Hermann 1874).

Julius Bernstein, Hermann's friend and colleague from du Bois-Reymond's laboratory, likewise conducted some limited studied on the effects of chloroform, specifically on frog nerves. Bernstein was able to demonstrate that chloroform exerted limited excitation of isolated sensible and motor nerves, instead observing that chloroform exhibited a profound influence on the sensitive ganglion cells of the nerve center (Bernstein 1870). Though limited in the extent to which he could explain how or why this was, Bernstein further noted a visible response reaction to chloroform at the site of the nerve, which he was able to associate with the cholesterol content of the nerve (Bernstein 1870).

Neither of these investigations constituted a substantive departure from the neomechanical physicalist disengagement from intoxicants and intoxication as a central, or even meaningful, object of scientific engagement. Bernstein's chloroform study was infolded by his greater, career-defining project centered on the physiology of nerves and nerve conduction, a veritable synthesis of the du Bois-Reymondian and Helmholtzian research programs. Hermann is trickier, to be sure. But even Ludimar Hermann, "a troublesome student" of the grand social enterprise of neomechanism, entertained only a limited engagement with intoxication, and then strictly as secondary line of inquiry. Chloroform, nitrous oxide—these substances were first and foremost encountered at the site of the blood cell, the nerve ending, or the gasometer's chamber.

Intoxicants would enter the emergent conception of the biological subject not through physiology, but through a novel methodological and theoretical form of encounter, one vested in *embodiment* over and against the body. Getting a handle on what this might possibly mean will require taking leave of the various strongholds of Berlin physicalism and instead stealing into the Saxon climes of Leipzig parallelism. Is it a touch puerile to frame the matter as a tale of two cities? Undoubtedly so. The Berlin and Leipzig schools simply serve as convenient monikers for the two divergent interpretations of the scope and limitations of physiological experimentalism and the corresponding possibilities they engendered both for the body and for being in a body. It would be Leipzig, rather than Berlin, that would countenance the innovation of a novel experimental science of embodiment, with substances of intoxication at its very heart. At that, let us explore the development of this "novel experimental science of embodiment," starting with the likes of Gustav Fechner and his reception in the works of Wilhelm Wundt.

Gustav Fechner, from the Psychospiritual to the Psychological

As legend would have it, the peripeteia in Gustav Fechner's life came in 1839. In a cruel twist of fate, an experiment on vision gone awry saw Fechner partially blinded and resigned to stepping down from his position as professor of physics (Boring 1950). After stewing in pain and poverty for a decade, Fechner imagined, while lying in bed the morning of October 22, 1850, that there might be a connection between geometric relationships in the magnitude of bodily stimulus and logarithmic functions. It was in this sunken state that Fechner is purported to have turned from physics to thoughts on the relationship between the body and the mind. Fechner's would devote the greater part of his remaining life to the development of an experimental understanding of the relationship between consciousness and external stimuli: *Psychophysik*.

Drawing heavily on the legacy of his former teacher and colleague, Ernst Weber, Fechner's psychophysical research program set the foundation for Leipzig as the epicenter of a new science of psychology, sweeping aside Kant's dismissal. Notably, while Weber had been an influential proponent of physiological reductionism, Fechner infused his own antimaterialist perspective, drawing heavily on his budding, adolsecent endeavors in the study of the psychical. The psychological distinction between the mental and the physical would gain further expression in the work of Wilhelm Wundt, a student of Fechner's who established the first laboratory for experimental psychology in Leipzig. Here, Wundt would erect an influential, scientific approach to studying the psychical, one that is best described as a template for the study of embodiment rather than the body. If Berlin was to be the capital city of the "body," Leipzig would become the capital of the embodied mind—and it is here, in the embodied mind, that intoxicants will once again be brought into conversation with the body.

For the purposes of this story of the intoxicated body, making sense of the conditions underlying the emergence of Leipzigian experimental psychology as a science of embodiment, its theoretical alterity relative to the neomechanist's bodily description of the mind, and its underlying conceptual tensions is critical to understanding how intoxicants were marshaled as testifying witnesses in the trial over the nature of the biological subject. The Fechnerian project of psychophysics will provide an alternative model for not only the nature of the relationship between the mind and body, but the study of mental processes. It would be Wundt's controversial interpretation of both Fechner's principle of parallelism and his radical empiricism that would give rise to Wundt's distinct notion of the psychical, taking root in the theoretical opening created by the seeming failure of the "brain psychiatrists." Possibly, the rational validity of the entire research program of experimental psychology was a product of this failure, and Wundt, along with his students, would soon be counted among the most outspoken critics of Meynert's approach to the study of the mind. It is this conceptual context—one raised on a foundation developed in opposition to the neomechanical conception of the body—that will make possible the experimental encounter with the state of intoxication as the manipulable expression of dynamic psychological processes.

Fechner as Philosopher, from the Eternity of the Soul to Psychophysical Parallelism

The stated goal of taking any part of consciousness as an object of scientific inquiry already meant clearing the hurdle of Kant's scientific epistemology, if Fechner's undertaking was to be considered authentically scientific. At first glance, Fechner's conception of consciousness relative to the outside world falls squarely within a post-Kantian understanding of embodiment: perceptions, impressions, and sensations are correlated with an external reality but they are not correlative—one's experience of the world is always an experience first of one's senses. But this is not on account of Fechner's Kantianism (Heidelberger 2004). Rather, Fechner takes up the bodily description of the physiological Kantianism of the likes of Müller, his contemporary, to go to war with not only philosophical Kantianism but neomechanistic materialism.

The key was Weber's study of just discernible differences. Ernst Weber, a teacher of Fechner, had attempted to experimentally measure bodily stimulus, specifically studying people's ability to distinguish objects of similar appearance but different weight (Fancher and Rutherford 2016, 157). This amounted to a test of human sensory limit thresholds, a concept first introduced in 1824 by Johann Herbart (Gescheider 2015, 1). Weber was not only able to establish that relative rather than absolute weight was the determining factor, but he was able to identify that the smallest discernible difference between weights was consistently about 3 percent (Fancher and Rutherford 2016, 158). Weber repeated this approach with other sensory tasks. When it came to discerning the difference in length between two lines, Weber found the smallest noticeable difference was by 1 percent, while for musical pitches the just-noticeable difference between frequencies was approximately 0.6 percent (Fancher and Rutherford 2016, 158).

Ernst Weber's experimental study of barely discernible differences in sensation would be elevated as a great leap forward for experimental psychology by the late nineteenth century, a legacy that has endured to the modern day. Yet Weber himself, as well as many of his contemporaries, evidently understood his research as primarily concerned with physiology. Weber's eulogy in *Nature*, for example, plainly paints Weber as a talented and decorated physiologist (Anonymous 1878, 286). Further still, Weber was an adamant, early proponent of the idea that vital forces were entirely fictitious and that the body could only be understood in physical terms (Königsberger 1906, 25). It was Ernst Weber whose philosophical attitude toward the science of physiology had most inspired the neomechanical revolution headed by du Bois-Reymond, Brücke, and Helmholtz (Königsberger 1906). It is clear that Weber's experiments on the just-noticeable difference was intended to be descriptive of the holistic physiological function of the sensory organs in tandem with a physical brain, as opposed to psychological phenomena.

Weber's experiments were, without a doubt, profoundly influential for Fechner. When Fechner sought to mathematically describe the relationship between psychological and physical stimulus across multiple senses, Fechner called it Weber's law (Fechner [1860] 1888, 17). The law in question was a formula wherein the intensity of the perceived stimulus (S) of a sensation is the logarithm of the physical input (P) multiplied by a constant (k), or $S = k\log P$ (Fechner 1888; Fancher and Rutherford 2016). These initial insights, building off the foundation set by Weber, formed the justification for the pursuit of psychophysical research as a science, as it satisfied the Kantian stipulations of mathematicity, apodictic certainty, and systematicity. Further still, Fechner absorbed Weber's method to develop a template for the measurement of all kinds of psychical events (Gescheider 2015, 1).

However, Fechner differed markedly from Weber as to the underlying theoretical implications of these discoveries and, thus, they had radically divergent research objectives. Early on in his medical education, Fechner lost the Christian faith that had shaped his upbringing and had begun to approach the natural world mechanistically, a course that was suddenly changed by Fechner's exposure to the naturphilosophical work of Lorenz Oken (Heidelberger 2004, 22). For Fechner, his discovery of Lorenz Oken "suddenly shed new light on the whole world," invoking language reminiscent of a religious epiphany (Heidelberger 2004, 22). So profound was the impact of Schelling-Oken for Fechner that he even sought a career as a *Naturphilosoph* all the way through his qualifying degrees. Struggling to find solid ground in *Naturphilosophie*, Fechner's naturphilosophic aspirations ended with frustration soon after Fechner finished his degrees, which precipitated a further commitment to physics and chemistry (though he soon found inspiration in the philosophy of Johann Herbart and the Romanticism of his friend Martin Gottlieb Schulze) (Heidelberger 2004, 25–26, 35). Yet as late as November 1862 (well after the publication of *Psychophysik*), Fechner identifies himself as a *Naturphilosoph* in his personal diary, though it's not clear in what sense this was intended (Meischner-Metge 2010, 416).

Fechner is not one of the many scientists who set down their work to become a philosopher (or vice versa). On the contrary, Fechner's extraordinary productivity saw him publishing influential works on psychological measurement alongside his philosophical works. A selected list of Fechner's philosophical writings includes *Das Büchlein vom Leben nach dem Tode* (1836), *Ueber das höchste Gut* ("Concerning the Highest Good") (1846), *Zend-Avesta oder über die Dinge des Himmels und des Jenseits* (1851), *Ueber die physikalische und philosophische Atomenlehre* (1855), *Ueber die Seelenfrage* ("Concerning the Soul") (1861), *Die drei Motive und Gründe des Glaubens* (1863), and *Die Tagesansicht gegenüber der Nachtansicht* (1879). From the titles alone, it is clear that Fechner's interests far exceeded merely measuring sensation. Yet topics as diverse as the eternal soul, the atom doctrine, and the basis for faith do find a certain degree of systematicity in the form of Fechner's doctrine of parallelism.

Fechner may personally trace his conception of psychophysical parallelism to Leibniz and Spinoza, with Fechner, or at least Fechner scholarship, possibly originating the attribution of the label of parallelism to Spinoza (Fechner 1888, 5; Yakira 2010, 106). Fechner's parallelism argued that the psychical, mental, or spiritual existed independently of the body, but alongside and in close conjunction with the physical. Unlike Spinoza, Fechner's mind-spirit and body are not different expressions of an intermediary substance. For Fechner, the body is the seat of the soul; they are one and the same. To this point, Fechner did recognize Leibniz first in his *Psychophysik*, but he also did not subscribe to the Leibnizian doctrine of pre-established harmony (Fechner 1888, 1). Rather, Leibniz and Fechner share a basic underlying principle of functional congruity between the body and the mental/spiritual, with Fechner arguing that the psychical and physical share a common basis in the individual organism. In later works including *Zend-Avesta*, Fechner raised psychophysical parallelism to its most panpsychistic formulation, with Fechner proposing that the entirety of the physical world has a psychospiritual parallel (Fechner 1901, vi; Heidelberger 2004, 172–73).

This principle of the psychospiritual served as the foundation for Fechner's argument for the existence of the eternal soul and the atom doctrine. Fechner's 1855 *Atomlehre* is an interesting topic in light of the divisiveness surrounding atomism at the time, with the thought of Schelling and Hegel coming up against physicists such as Augustin-Jean Fresnel and Augustin-Louis Cauchy. Some German materialists struck out in support of atomism, as Karl Vogt had in 1855's *Koehlerglaube und Wissenschaft*—but Fechner's atomism is decisively antimaterialistic. Fechner's approach to atomism reflected his own psychophysical ontology. Fechner was a point atomist, meaning that he subscribed to the notion that an atom was a center of force or a monadological "point" and that such points of force made up the entirety of the physical world (Fechner 1855, 33–36;

Heidelberger 2004, 138, 147). Of course, atoms only comprise the most rudimentary unities in Fechner's ontology. Atoms themselves are not souls, nor do they have souls. Fechner saw souls in the lawlikeness of the behavior of what he perceived to be organic wholes: humans, animals, plants, as well as planets, leading up to the essential unity of all things in God. This was not intended in the sense of an unconscious reflexivity; Fechner saw living consciousness reflected in the behavior of all of these dynamic wholes. In Fechner's psychophysical ontology, this meant that the sum of atomic reality is just the bodily accompaniment of the universal psychospiritual. Individual souls, in this sense, live on after death insofar as they were always only a specific unfolding of the universal soul and thus can never be destroyed, just as the atoms perpetually cycle through the physical world.

Although Fechner's philosophical, or theoretical, works and Fechner's scientific research were published separately, they are inextricably intertwined. Experimentally demonstrating a systematic relationship between the bodily and the psychical had been the grand motivation behind Fechner's description of Weber's law. Theoretically, Weber's law not only demonstrated that the psychical was studiable; it made real the connection between the physical and the mental-spiritual as an object of scientific rather than philosophical inquiry, an objective relayed in the opening lines of *Elemente der Psychophysik* (Fechner 1888, 1). One of the most striking expressions of Fechner's approach to mechanistic thinking in science was the opposition he developed between what he called the *Tagesansicht* (day view) and the *Nachtansicht* (night view) (Adler 1991, 10). Mechanism and physiological reductionism were what he termed the *Nachtansicht*, a term seemingly alluding to Goethe's own evaluation of Baron d'Holbachs's mechanistic *Système de la nature* as depicting a hollow, lifeless, and twilit world (Goethe [1811] 1989, 490). The contrast between *Tagesansicht* and *Nachtansicht* has roots in the very beginnings of Fechner's psychophysical project, with the *Tages-* and *Nachtansichte* reflecting the double aspect of the psychical and physical first introduced in *Zend-Avesta*. Insofar as Fechner argued for the parallel, nonreductive functioning of the mind and body, the vacuity of the *Nachtansicht* lay in its fundamental rejection of precisely this underlying duality, at the expense of the possibility of any truly meaningful scientific inquiry. The science of the *Tagesansicht* needed to reflect the unity that underlies the psychophysical reality, a position that securely identifies Fechner's scientific research program as a radical empiricist manifestation of the *Naturphilosophen* that inspired Fechner early in his career.

Materialism and Antimaterialism

The materialist and antimaterialist debates that surrounded the legacy of Gustav Fechner were as real to his contemporaries as they are today. There are some scholars who are hesitant to identify Fechner as an antimaterialist. Michael Heidelberger, who has written several impressive books on the life and work of Gustav Fechner, argues that Fechner is a nonreductive materialist:

> The position [Fechner's] is nonreductive, because it describes life and consciousness as having an independent, original nature that cannot be further reduced to physical phenomena, or—as the nineteenth century would have it—reduced to "the mechanics of atoms." Yet, at the same time, Fechner's position is materialistic. He sees every change in the physical world as wholly explicable by laws of nature, and he sees for every mental change some change in the physical world that precedes it. (Heidelberger 2004, 73)

In many senses, this is an excellent argument. It washes away the position held by some modern scholars and some of Fechner's contemporaries that Fechner was something of an empirically minded scientist at work and a philosopher on the side. The connection between his philosophical and methodological approaches formed the basis of Fechner's entire psychophysical research program. Furthermore, Heidelberger's evaluation of Fechner as a nonreductive materialist helps bring Fechner's day view approach to science and radical empiricism to the forefront of his portrait of the thinker. But to ascribe the materialist label to Fechner in any capacity is already to undermine the psychophysical parallelism that forms the foundations of his work. Material, atomic reality is, in every sense, at one with the psychospiritual parallel; they are one and the same, neither existing without the other.

Gustav Fechner and the Cure for Opium Poisoning

Between the years 1834 and 1838, Gustav Fechner was in a period of financial desperation. As he had before, Fechner sought to earn a living by writing, this time securing a job as the editor and primary author behind the *Hauslexikon: Vollständiges Handbuch praktischer Lebenskentnisse für alle Stände*. Given the breadth involved in what Fechner described as a guide to life as relayed by the "neuste und beste Quellen," the *Hauslexikon* is a unique source of insight into Fechner's approach to a vast and varied array of subjects (Fechner 1834, III). Fechner personally wrote over a third of the entire eight-volume encyclopedia. For those portions that he did not personally write, Fechner did edit their content and approve them for publishing, providing at least some semblance of tacit endorsement for the ideas present within (Fechner 1841, 103). Thus, when references to vitalistic theories of medicine appear throughout the *Hauslexikon*, they merit some degree of serious consideration.

Not only do a number of the medical entries in the *Hauslexikon* reflect vitalistic theories of bodily function, but there are overt references to Brunonian concepts, and the dictates of the *Hauslexikon* contained examples of the vital substance concept. The following passage addressed the appropriate medicine to provide in the context of a difficult childbirth, among other circumstances:

> Therefore, coffee is an indispensable remedy or rather palliative in all cases of overstimulated vitality, which it reduces in its aftereffects. It is preferable to opium in all pains where the sick scream and howl loudly—in overstimulation of women in childbirth in the case of too-violent births. (Fechner 1841, 272)

Here, the curative efficacy of both coffee and opium are aligned with their capacity to affect the vital force, even sporting a characteristic word of caution arising from the extent of opium's vital effects. Elsewhere, steam baths are described as being medicinal because "the whole body is so hot, & all vital forces are so stimulated, that the impression of external cold is immediately defeated, & has its other consequences, as an increased reaction of the whole organism" (Fechner 1841, 411). Both of these examples testify to the suggestion that Fechner at least tacitly supported concepts imported from a Brunonian or Röschlaubian model of vital activity that entailed some conception of vital substances.

What is known is that Fechner had more than a passing familiarity with the literature surrounding the emerging category of alkaloids. Fechner's translation of Thénard, titled *Repertorium der organischen Chemie*, contained a review of the literature on meconic acid as well as the adjacent discovery of morphine, discussing publications by Sertürner, Hermann Vogel, Robiquet, Thénard, Johann Ludwig Choulant, and Seguin (Fechner 1826, 238). The *Repertorium* also contains passing references to *Lebenskraft*. Fechner added the commentary that Nicholas de Saussure found that Jacques Étienne Bérard erred in his finding that fruits off-gas more during the day because the experimental design exposed the fruit to daytime conditions that depleted their "Lebenskraft" (Fechner 1826, 45). Fechner himself makes passing, perhaps rhetorical, use of the term *Lebenskraft* in *Vorschule der Aesthetik* (Fechner 1876, 124, 215). None of the examples provided above could be described as definitive evidence of mainstream vitalistic in the thought of Gustav Fechner. What these references demonstrate is that Fechner had a familiarity and association with not only theories of vitalism but also Brunonian etiology and therapeutics with vital stimulants.

All of these points provide a backdrop against which to understand the discussions of intoxicating substances in Fechner's major works. Here, Fechner's references to intoxication are infrequent, albeit reflective of an underlying framework. Speaking to research on the ability to perceive the distance between two points on the skin, Fechner remarks that "after chloroforming or ingestion of narcotic substances (morphine, atropine, daturin), the tips of the circles must be placed much wider than usual in order for them to be perceived as distant" (Fechner 1889, 323). The research Fechner cites here comes from Lichtenfels and Fröhlich, who "have shown through extensive and repeatedly modified experiments [the effects of intoxicants]" (Fechner 1889, 323).[2] Interest in such research is to be expected: it specifically concerns the measurement of a bodily sensation. But not all of Fechner's references are so clinical.

In fact, the majority of the references to substances of intoxication found throughout Fechner's *Psychophysik* do not directly concern the measurement of the senses whatsoever, instead appealing to the phenomenological, lived experience of intoxication.

Thus Granier de Cassagnac writes of the effect of etherization: "It was to me, as if everything outside disappeared; I felt no longer the flacon in my hand, hardly noticed that I had clothes on the body, and the ground, on which I stood, seemed to me to have lost its original reality. . . . the outer and material world is no longer present[. . .]." Another observer indicates: "I felt from the outside world at all, even from my own body nothing more. The soul was, as it were, completely isolated and separated from the body." Madden (Fror . Not . XXVI , p. 14) writes of the effect of opium intoxication: "While walking, I hardly noticed that my feet touched the ground; I felt as if I were gliding along the street, driven by an invisible force, and as if my blood consisted of some ethereal fluid that made my body lighter than the air."- Another observer says of the effect of hashish +): "The sensations produced were such an extraordinary lightness, so to speak airiness," . . . and further: "the limiting sensation (the feeling of confinement within the boundaries of flesh and blood) fell away instantly. The walls of the organic body burst and fell into ruins, and without knowing what form I was wearing, sincse I lost the face, even any idea of form, I only felt that I had expanded to an immeasurable circumference of space." (Fechner 1889, 326)

Fechner likens these sensory effects to a form of paralysis of the function of sensory nerves in the skin, even to a kind of death (death being the great paralysis) (Fechner 1889, 327). Notably, none of these scraps of testimony were the product of Fechner's firsthand research, but were rather pulled together from accounts of various sorts. The hashish narrative, for example, came from an 1854 account found in *Magazin für Literatur des Auslands* (Fechner 1889, 326). This suggests that, although the study of the effects of intoxicants fell outside of his own research program of measuring sensation, Fechner regardless found the information gleaned from such accounts to be scientifically meaningful.

It was precisely this sort of outlook, an openness to the embodied perceptional world couched in the language of psychospiritual parallelism, that made Fechner an object of ridicule from the younger generation of neomechanically-minded physiologists. Unsurprisingly, the reception of Fechner's ideas among the neomechanists could hardly have been worse. As early as 1849, letters between Carl Ludwig and du Bois-Reymond on Fechner's work contains a tone of what can only be described as mockery, du Bois-Reymond going as far as suggesting that Fechner must suffer from an organic brain tumor (du Bois-Reymond 1847/1927; Meischner-Metge 2010, 417). Fechner, two months and twenty-six days Johannes Müller's senior, would have seemed to the younger generation of Berlin physiologists a lasting vestige of the era swiftly fading into the distance. Even when Carl Ludwig spoke of him warmly in the letter, Ludwig wrote of Fechner as one might an amiable, albeit kooky, old man (du Bois-Reymond 1847/1927; Meischner-Metge 2010, 417).

With an eye to the course taken by neomechanists regarding the nature of the connection between the body and the mind, Fechner's radical empiricism stands out for its willingness to take on the testimony of experiences that might otherwise be described as being plucked from the murky dream world of perception. Like John Brown or Friedrich Sertürner, Fechner appears to have recognized that the holistic, embodied experience of intoxication imparted information that was irreducible to any number of physiological processes. Looking to the foundations of Fechner's psychophysical thought, it is clear that Fechner's radical empiricism is drawn from his naturphilosophical background and perhaps even that his openness to experiences such as intoxication was shaped by a latent influence of the vital substance concept. The result was that Fechner's psychophysics functioned as a restrictive science of embodiment. Still principally concerned with measurements of sensation, Fechner's work was nevertheless theoretically open to information gleaned from the sum of perceptional experience. This is only made clearer when contrasted with the neomechanists, for whom embodiment was a superficial distraction from the more fundamental physiology of sensation. The issue for Fechner was that he had a framework but lacked a distinct methodology: Fechner's psychophysics was still firmly planted in the methods of physiology. The innovation of Fechner's approach was to project the findings of Weberian sensory physiology onto a psychospiritual entity. The shift would occur when Wundt took up Fechner's framework and imbued it with a radical new methodological approach, one that led to the emergence of a new *independent* science of embodiment.

Wundt's Science of Embodiment and the Shift from Psychophysics to Psychology

On its face, many aspects of a pre-1875 biographical sketch of Wilhelm Wundt's life would portray Wundt as an unlikely candidate to carry Fechner's legacy into the future. Pages could be filled musing about the kind of research Wundt might have led had his early electromuscular work been well received by du Bois-Reymond, or even had he not been passed over to take the position of his former mentor, Hermann von Helmholtz, in 1870. But these things did happen, and Wundt's legacy as arguably the most influential figure in the history of scientific psychology is today all but assured. However, the development of Wundtian psychology was far from straightforward. Beginning as a physiologist under the mentorships of Helmholtz, Müller, and du Bois-Reymond, the emergence of Wundt's earlier psychological theories was as much an unfolding of neomechanistic physiology as it was an opposing position. It would be the influence of Lotze, Herbart, Weber, and, above all, Fechner's *Psychophysik* that would collide with Wundt's personal struggle in academic physiology to ultimately give rise to Wundtian parallelism. This partial parallelism, grounded in a secularization of Fechner psychospiritual/material divide, was the foundation of the argument for the requirement that psychology be formulated as its own experimental science. Before this could be done though, Wundt needed to develop a foundational research program through which to ground the validity of the broader project of scientific psychology, and for that Wundt would choose reaction time, a choice that would put Wundt in direct conflict with his former peers in physiology.

This new experimental approach to psychology would establish Wundtian psychology as a science of embodiment, concerned above all else with the lifeworld of the subject over and against an unseen physiological substrate.

Wundt began his medical degree at Tübingen in 1851, before ultimately moving to Heidelberg in 1852 and continuing his studies there until 1855 (Jones 1994, 49). In 1856, Wundt had the opportunity to spend a semester in Berlin studying physiology with du Bois-Reymond and Johannes Müller, before returning to Heidelberg to complete a medical doctorate and habilitate (Boring 1950, 318). With Müller, Wundt had been tasked with "extirpating nerve centers in invertebrates" while du Bois-Reymond had Wundt assessing the merits of Ernst Weber's 1846 findings concerning the effects of varied loads on muscular extension and contraction (Diamond 2001, 20). Upon returning to Heidelberg, Wundt lectured on physiology and, from 1858 to 1865, worked as an assistant to the newly appointed head of the physiological institute, Hermann von Helmholtz (Boring 1950, 318). It seemed that Wundt had completed the set of German organic physicists, and this period saw Wundt publish a number of physiological texts, among them "Ueber die Elasticität der thierischen Gewebe" (1857), *Die Lehre von der Muskelbewegung* (1858), and "Ueber das Gesetz der Zuckungen und die Veränderungen der Erregbarkeit durch geschlossene Ketten" (1858).

Up to this point, Wundt's biography reads like a perfect template for a successful career in academic physiology in mid-nineteenth-century Germany. Wundt could almost be counted among Johannes Müller's students, all of whom by then occupied the top positions in medical and physiological faculties across the German-speaking world. Yet it was in the years that followed that Wundt made his debut as a psychologist with the publication of the first of the articles that comprised *Beiträge zur Theorie der Sinneswahrnehmung* (1858–1863), followed by a series of lectures later published as *Vorlesungen über die Menschen- und Tier-Seele* (1863). While Wundt was still associated with the physiological institute in Heidelberg, Wundt's *Beiträge* marks the beginning of the trajectory that would see him founding the first laboratory for experimental psychology roughly a decade later in Leipzig.

The intellectual origins and personal reasons for this shift in Wundt's interest remain, to this day, a subject of considerable debate. The tendency of many histories of Wundt has been to emphasize the philosophical foundations of his psychological approach. A complicating factor is that Wundt received no formal philosophical training. Wassmann (2009) makes a compelling case for an appraisal of Wundt's early psychological work in light of his established physiological training. Rather than speculate as to the depth of the philosophical influences on early Wundt, Wassmann points to Wundt's concepts of brain function and emotion as examples of how the foundations of Wundt's work can be found in the physiological research that surrounded him (Wassmann 2009, 215). The tremendous merit of such an approach is that it resists the scholarly temptation to read someone's earlier work in light of their later work, an enduring example of which is the scholarly treatment of Freud's early anatomical and neurophysiological research with Carl Claus, Brücke, and Meynert.

But Wundt's short career in physiology was hardly the idyll it might seem at first glance, and the years that surrounded the publication of Wundt's early psychological work were colored by professional disappointment and feelings of betrayal. In 1858, Wundt published his first book as a physiologist, *Die Lehre von der Muskelbewegung*. Drawing on some of the research done under his mentorship in Berlin, Wundt's *Lehre* was dedicated to Emil du Bois-Reymond. Much to his supreme disappointment, Wundt's debut book received little to no recognition, even from the man to whom Wundt dedicated the work (Wundt 2013; Diamond 2001, 23). Wundt attributed this failure to a conspiracy led by none other than his former mentor himself (Wundt 2013; Diamond 2001, 23). This falling out with du Bois-Reymond and feelings of betrayal from those at the top of the field, as well as a professional conflict with Hermann Munk in 1861, provide more than sufficient grounds to ask what kind of future Wundt saw for himself in physiology.

What then of the foundations of Wundt's shifts to psychology? One particularly persistent attribution of intellectual indebtedness accords much of Wundt's early psychological thinking to the influence of Gottfried Leibniz (Fahrenberg 2016). Klempe (2021) makes the argument that Wundt's debt to Leibniz draws primarily from Leibniz's mathematical thought, in particular the Leibnizian critique of mechanical causality and physicalist dynamics (Klempe 2021). It is worth considering if any recognition of intellectual indebtedness on the part of Wundt himself is at least partially retrospective. Wundt's later publications do identify Leibniz as an important influence in his research program, though this increased emphasis on the significance of Leibniz is seen only after the year 1900 (Fahrenberg 2016, 13). In the 1850s and 1860s, Wundt was clearly interested in Leibniz, but as Araujo suggests, the extent of this early influence is fragmentary. For example, Wundt's discussion of the Leibniz quotation that opens the *Beiträge* was intended to show the (ambiguous) middle ground Wundt was assuming between Leibniz's innatism and Locke's sensualism (Araujo 2016, 60). Alternatively, efforts to paint the young Wundt as essentially a Herbartian are themselves mired by profound contradictions between

Wundt and Herbart with regard to the scope and nature of psychology. This association is understandable, as Wundt himself cited Kant and Herbart as being among his greatest influences in *Grundzüge der physiologischen Psychologie* (Wundt 1874a, 3). Yet Herbart, quite famously, opposed the possibility of experimentation in human psychology, the very program Wundt sought to realize (Herbart 1850, 4).

There is also the matter of the theoretical influence of Helmholtz. Much has been made of Helmholtz's legacy in the history of psychology, both in regard to his work on the physiology of perception as well as his theory of unconscious inference. As has already been discussed, Helmholtz's approach to the physiology of the senses, as well as the nervous system, was bound up in the neomechanical physiology of the body, rather than any explicit psychological research program. Though the historical impact of Helmholtz's work on the physiology of perception and nervous activity is important for modern neuropsychology, there is no reason to suspect Helmholtz's influence was of primary significance to Wundt's early psychological project. On the contrary, though Helmholtz and Wundt refer to each other respectfully in their work, their relationship has been described as "uncongenial" (Titchener 1921, 162). If not uncongenial, the relationship itself did not bear much fruit. Ivan Sechenov, a student working in the laboratory in which Wundt was an assistant, reports that Wundt often sat in the company of his books in silence (Diamond 2001, 28). Wundt even seemed hesitant to make use of the relationship he had with as eminent a figure as Helmholtz for his own professional development (Diamond 2001, 28).

Wundt's Beiträge

It is clear is that Wundt was evidently influenced by the philosophies of Kant, Leibniz, and Herbart, but that their inspiration for Wundt's psychological turn was likely fragmentary and complimentary to the more immediate influences of Ernst Weber and Gustav Fechner. Both the *Beiträge* and the *Vorlesungen*, the defining works of Wundt's turn to psychology, came into their final published forms over several years, in the midst of which Fechner published his *Psychophysik*. This makes it relatively simple to trace the impact of *Psychophysik* by examining those components of the *Beiträge* or the *Vorlesungen* published from 1860 onward. Wundt's *Beiträge* might thus be read less as science than as bildungsroman. It becomes the story of a man's departure from sure-footed physiology into the crumbling, craggy heights of consciousness, sealed by the transformative declaration that the Fechner-Weber law was applicable across the entire arc of psychic life (Wundt 1862).

The "Erste Abhandlung" of Wundt's *Beiträge*, published in 1858, focused on the sense of touch and space perception. There can be no denying the Weberian influence on the treatise. Wundt himself identifies Weber's experimental approach as a primary inspiration for his own project (Wundt 1862, 1). Though some of the ideas present in the article resemble Helmholtz's concept of unconscious inference, the "Erster Abhandlung" is unlikely to bear any Helmholtzian influences. As Araujo points out, Helmholtz himself did not assume his post in Heidelberg until October 1858, meaning the first portion of the *Beiträge* could not have been written under Helmholtz's supervision (Araujo 2014, 53). Although Wundt freely takes on Weber's approach, the initial purpose of Wundt's ambitious project was actually to demonstrate that Weber's theory of tactile localization was incorrect, building on the criticisms of Kölliker and Lotze (Wundt 1862, 5, 8). But Wundt doesn't stop there. The "Erster Abhandlung" turns a critique of Weberian tactile perception into a broader framework for approaching further questions in the physiology of perception as such.

On this point, Wundt quotes Theodor Waitz, specifically that "the essence of the soul contradicts the simultaneous conception of a manifold, and it is precisely the inability to do so that compels it to place next to one another the manifold that is given to it at the same time" (Wundt 1862, 10). "Herein," argues Waitz, "lies the origin of special conception" (Wundt 1862, 10). Wundt pairs this idea with Lotze's observation "that the perception of space is an original, a priori property belonging to the nature of the soul," which Lotze understands physiologically through his concept of *Lokalzeichens* (local signs) (Wundt 1862, 12–13). Wundt takes these observations as instructive concerning the entire host of available sense perceptions to argue that some unconscious process mediates the passage from sensations into perception, a notion familiar to Kantians everywhere (Wundt 1862, 65). Though tempting to interpret this model as a forerunner to Wundt's psychological parallelism, there is no clear evidence that Wundt's 1858 "unconscious process" is any less physiological than du Bois-Reymond's, Helmholtz's, or Müller's neo-Kantianism on similar questions. For this reason, Wundt's "Erster Abhandlung" can hardly be called psychology in the sense of his later work—there is no necessity of a psychological parallel. It is a physiology of sensation in the methodological tradition of Ernst Weber, articulated by way of Waitzian philosophy. Both published in 1859, the second essay of the *Beiträge* was a historical review of different theories of vision and the third was a discussion of monocular vision, relying on the framework introduced in the "Erster Abhandlung."

The first three sections of Wundt's *Beiträge* seems to have garnered limited attention. A notable exception to this cool reception was the glowing praise that Wundt's "Erster Abhandlung" received from Fechner in *Psychophysik* (Fechner 1888, 296; Fechner 1889, 315, 317, 323). Given the painful anxieties of rejection and perceptions of conspiracy that had plagued the immediately preceding years, Wundt very likely would have been receptive to such recognition from a senior figure such as Fechner. Wundt may have also already been somewhat aware of Fechner's work pre-*Psychophysik*. Much of Helmholtz's writing of the *Handbuch der physiologischen Optik* (1856–1867) overlapped with Wundt's assistantship, and Helmholtz relied heavily on Fechner's research on contrasting colors and afterimages, citing Fechner throughout the text (Helmholtz 1867, 313, 387, 403, 418, 542, 793, 836, 868). The fourth and fifth essays of Wundt's *Beiträge* continued the theme of the preceding sections and now discussed the physiology of binocular vision, so that Wundt and Helmholtz would have been consulting many of the same sources while working in the same lab. In any case, both Wundt's sixth and final essay and the introduction added in 1862 once the essays were published together as a book leave no question as to the profound impact of Fechner's *Psychophysik*, replete as they were with references to the old man of Leipzig.

The fourth and fifth articles had relied on Fechner's psychophysical studies to assess the role of attention and other factors in the function of visual perception. It is here that one observes a remarkable shift in the tenor of Wundt's approach. The "Sechster Abhandlung," in contrast to the preceding sections, is overtly psychological, even philosophical, in its objectives, though it seeks to ground its conclusions in the principles derived from the preceding visual studies. Here, Wundt overturns Herbart's thesis that competing ideas exist in the conscious mind simultaneously to argue that conscious thoughts first undergo an unconscious unification process (Wundt 1862, 382). By Wundt's reasoning, Wundt had demonstrated this phenomenon in spatial perception, where he argued that the conflict between the sum of different, competing stimuli and the unity of perception entailed an unconscious law-like synthesis process. This was Wundt's attempt to empirically demonstrate the psychical fact of an unconscious law-like, logical process, aided by psychophysical research pioneered by Weber and then Fechner (Wundt 1862, 416–417).

It was with the publication of the completed collection that Wundt effectively inaugurated the principle underlying thesis of what would become his psychological research program. The introduction to the completed collection heaps praise upon Fechner as the one who above all pioneered an experimental and theoretical method of approaching the barrier between the physical and the psychical (Wundt 1862, xxx). Most important was the recognition "that [Fechner's law] is in fact not a physical law," which, Wundt argues, means "that the same law retains its validity also in the field of higher psychic activities" (Wundt 1862, xxx). This extension of Fechner's thesis, which Wundt suggests is apparent upon common-sense reflection, established Fechner's law as not merely psychophysical but a universal psychical law, a theoretical innovation of tremendous consequence. Wundt translated the Weber-Fechner law of dependence between stimulus and sensation into a psychological law, which stated "that where two mental functions are directly dependent on each other, the dependent function always grows in proportion to the logarithm of the original variable" (Wundt 1862, xxxi). Wundt had modified Fechner's logarithmic relationship between sensation and stimulus in order to define a connection between perception and sensation. All higher mental life—as in spatial perception—now functioned in accordance with an underlying logic, a law-like process that could be empirically encountered.

Perhaps most importantly, this represents the clear emergence of psychophysical parallelism in Wundt's thinking. Wundt had, somewhat vaguely, alluded to the weakness of strict neomechanism in describing organic beings, in favor of some yet unknown future framework, as early as his introduction in *Die Lehre von der Muskelbewegung* (Wundt 1858a, 2). In the completed *Beiträge*, Wundt openly demonstrates the early development of his own doctrine of psychological parallelism, which disinvested the psychological from any physiological dependence. This idea would see significant further development and formalization in Wundt's famous *Grundzüge der physiologischen Psychologie* in 1874, though Wundt's 1862 *Beiträge* demonstrates that Wundt had the foundation of his own conception of a novel approach to psychology as an empirical science of *embodiment* as early as the 1860s.

The Janus moment in Wundt's career seems to be a series of events in 1870–1871. In the span of less than two years, Helmholtz left Heidelberg for Berlin, Wundt was passed over as Helmholtz's replacement as the head of the physiological institute, and much of Wundt's work was subjected to intense physiological scrutiny by Julius Bernstein. By all accounts, Wundt still saw himself as a professional physiologist as late as 1870, although he had increasingly attended to psychological, philosophical, and anthropological themes in his lectures in the preceding years. If anything, Wundt's interest in Helmholtz's position in the first place is demonstrative of Wundt's continued identification with physiology, even if only aspirationally. In fact, the Wundt of the early 1870s outwardly seems more determined to continue his career as a physiologist than he had in the preceding years, publishing an ambitious project titled *Untersuchungen zur Mechanik der Nerven und Nervencentren* (1871). Diamond credits this renewed effort to make a mark in neurophysiology to the opportunities afforded by Heidelberg's rising prestige, owing to their accomplished faculty members (Diamond 2001, 51), though it is worth recalling that Wundt's *Mechanik der Nerven* was being placed on a rising wave in neurophysiology. Hitzig and Fritsch's findings on the location of the motor cortex in dogs had been published in 1870, a project that identified itself as a confirmation of Meynert's own theorizing about the nature of brain function. This was Wundt's opportunity to stake a bold claim in the field of physiological research that was seeing the beginning of a massive conflagration of development. Unfortunately for Wundt, *Mechanik der Nerven* was received with much of the same silence that had accompanied his previous endeavors in physiology.

That very same year, Wundt became aware of a book that threatened much of what he had attempted to achieve in his challenging career. The book in question was written by Julius Bernstein, Wundt's replacement in Helmholtz's lab. Titled *Untersuchungen über den Erregungsvorgang im Nerven- und Muskelsysteme,* Bernstein's book was also published in 1871 and, much like Wundt's work, it discussed the physiology of spatial perception, among other things. The primary thesis of Bernstein's *Untersuchungen* advanced his notion of cortical irradiation to provide a physiological hypothesis for Weber's two-point touch threshold. (Bernstein 1871, 167–169; Diamond 2001, 54). This expanded upon Bernstein's 1868 "Zur Theorie des Fechner'schen Gesetzes der Empfindung," a boldly named text that established that cortical irradiation operated according to Fechner's law, assuming relatively consistent localized inhibition and that the intensity of the irradiation of stimulus was proportional to the intensity of the stimulus (Bernstein 1868). Bernstein had not only proven to be a far stronger physiologist than Wundt, but his hypotheses had encroached on the psychological thesis of Wundt's *Beiträge* and subordinated Fechner's psychophysical law to physiological reductionism.

As a physiologist, Bernstein was everything that Wundt was not: recognized for his pioneering work, a personal favorite of du Bois-Reymond and Helmholtz—in every sense the fresh face of physicalism (Araujo 2014, 53; Seyfarth 2006, 2). This is not to say that Wundt was by any stretch an incapable or incompetent physiologist. Rather, Bernstein was far more literate in the styles of reasoning favored by neomechanistic physiology, particularly mathematics. The radically different impulses behind the two figures are expressed through their different conclusions on how Fechner's law should be interpreted. Wundt emphasized the psychical dimension of Fechner's law and sought to extend it into higher-order mental processes. Bernstein saw Fechner's conclusions as empirically merited but sought to understand them physiologically (even du Bois-Reymond recognized Fechner's knack for measurement) (du Bois-Reymond 1847/1927). By the time of the events of 1870–1871, Wundt may have begun to realize that his aspirations would not be realized as a professional physiologist. All of this is important for understanding the context out of which Wundt's famous *Grundzüge der physiologischen Psychologie* emerged in 1873 and Wundt came to be in Leipzig by 1875. Though Wundt had been hesitant to break off from the only stable field in which he was established, circumstances leading up to the year 1872 had proven that Wundt's future as a physiologist was as at least as uncertain as any other path. It is no surprise, then, that by 1873 Wundt struck out against the neomechanistic physiologists in order to expand upon the foundations set out in the *Beiträge* and develop it into a full-fledged research program.

The Grundzüge

In 1874, Wundt would be called to Zürich to assume a professorship in inductive philosophy, only to be called to Leipzig in 1875. The contents of both inaugural addresses reflect the transformation that Wundt had undergone in the course of writing and publishing the *Grundzuge*. Their language carried the sentiment that the materialists of the natural sciences had forgotten their origins—that the foundations of their exclusive reductionism was derived from the realm of philosophy, rather than nature itself (Wundt 1876). The man who gave these inaugural addresses was no longer the Wundt who pandered for the recognition of his colleagues in physiology. This was a Wundt who had broken off, who was trying to forge his own path, who had picked a side. This was the ultimate function of Wundt's *Grundzüge*: it was the formation of a decisive stance, born through the culmination of over a decade of labor in psychology, psychophysics, and physiology. Most importantly for this discussion, it is a window through which to view everything that was to come. It would be through the lens of experimental psychological parallelism that the state of intoxication itself would come to be examined via a new science of embodiment, and ultimately come to bear on the arrival of the biological subject at the end of the nineteenth century.

At times deemed the most important publication in the history of psychology, the objective behind Wundt's 1874 *Grundzüge der physiologischen*

Psychologie was "to delineate a new field of science" (Boring 1950, 322; Wundt 1874a, III). The title in itself more than sufficiently described the book's contents, as Wundt himself remarked in the introduction (Wundt 1874a, 1). True to its title, the *Grundzüge* was intended to be the foundational text for a new science that reached beyond the purview of physiology or psychophysics. The reversal of *"psycho-physik"* into *"physiologischen Psychologie"* is telling in this regard. Just as the ordering of the words *psycho-* and *physical* reflected Fechner's hope that physical measurement could uncover the relationship to the psychical, Wundt's *physical-psychology* was concerned with the direct, experimental study of the conscious process, via the methodological interface of physiological science (Wundt 1874a, 2). Thus, physiological psychology meant using the methods of physiology to study the psychical, as opposed to studying physiological phenomena in order to study the connection to the psychical.

Central to Wundt's argument for a science of psychology were the failures of physiology and existing psychology's speculative character. Psychology, Wundt argued, had always had a role in physiology (Wundt 1874a, 2). Particularly where the function of nervous tissue was concerned, the physiologist's observations relied upon the recognition of psychological symptoms (Wundt 1874a, 2). Yet the dominant trend in neurophysiology had already begun to favor the reduction of even higher-order mental functions to neural states, in a sense subordinating the study of psychological phenomena to neurophysiology. Conversely, many theories that laid claim to the title of psychology lacked any interest in the physical, bordering on the speculative (Wundt 1874a, 2). In this sense, the primary subjects of Wundt's criticism were psychologies grounded in metaphysical presuppositions or simple observation (Wundt 1874a, 2). Physiological psychology was to be an experimental science, in every sense as empirically valid as physiology, except its object was to be the conscious process (Wundt 1874a, 2–3).

The question was how to demarcate the psychical from the physical in such a way that an independent science is not only possible but necessary. For Wundt, the answer was a secularization of Fechnerian parallelism, combined with a Herbartian critique of faculty psychology. Fechner's concept of the soul was inseparable from his understanding of the psychical. The physical and mental ran in parallel, the soul being eternal where a specific body is not. Wundt had readily taken up Fechner's stance concerning psychological parallelism; however, Wundt stopped at supporting Fechner's stance on spiritism (Meischner-Metge 2010, 420–21). This is preserved in a letter from Fechner to his younger colleague dated June 25,

1879, with Fechner conceding that, though the pair had long debated the question of spiritism, he was finally willing to concede that Wundt's mind could not be changed (Fechner 1879; Meischner-Metge 2010, 420–21). An intermediary influence here was Friedrich Lange, who had advocated for a psychology without a soul and held the chair Wundt assumed in Zürich prior to his departure for Leipzig (Engstrom 2015, 153). Wundt upheld the parallelistic principle while reducing the soul (*Seele*) to a term that denoted the logical subject of inner experience, rather than a distinct metaphysical substance (Wundt 1874a, 9). Instead, Wundt defined terms like *Seele, Geist* (spirit/mind), *Leib* (old term for body associated with life), and *Körper* (body) in phenomenological terms. *Seele* would be to *Geist* what *Leib* was to *Körper* (Wundt 1874a, 9–10). *Geist* and *Körper* were conceptual representations of their subjects as objects, the mind and body, respectively. *Leib*, meanwhile, stood in for the lived experience of having a body, and *Seele* for the experience of having an inner world. Thus, Wundt conceived of *Seele* and *Geist* as terms for the same thing at different levels of self-reflection.

This phenomenological shift that emphasized the actuality of the conscious process over speculation on the structure of mind reflected the influence of Herbartian antifaculty psychology. Wundt enthusiastically embraced Herbart's critique of the faculty model found in Kant and Wolff, arguing that the notion of faculties remained a mere possibility that contradicted experience (Wundt 1874a, 18). As Wundt read Herbart, conscious perceptions were received as total unities—there are no synthesizing faculties, only direct ideas, feelings, perceptions (Wundt 1874a, 18). Wundt ultimately saw Herbart as errant in his approach, but he drew inspiration from Herbart in emphasizing the actuality of the conscious process as the object of psychological inquiry (Wundt 1874a, vi, 18–19). The structure of the conscious process would then be investigated through rigorously controlled experimentation.

The physiological referent for rudimentary psychical events such as sensation was evidently something that Wundt considered knowable in the *Grundzüge*. The entire first volume of the *Grundzüge* was dedicated to outlining the physiology of nervous tissues and their function. He further discussed the structure of various organs involved in sensation (Wundt 1874a). In these cases, physiological and psychical events had a clear relationship. This ceases to be the case for higher-order psychical events, perhaps even for all psychical processes beyond the realm of sensation. It was toward this Byzantine realm of psychical processes, only remotely relatable to physiological events, that Wundt's psychology would guide its inquiry. What Wundt had proposed was not spiritual parallelism; it was a science of *embodiment*, the dynamic lifeworld of the subject, within the framework of partial parallelism.

The Lab and the Question Concerning Reaction Time

However grand, Wundt's self-conscious efforts to found a new branch of scientific inquiry would remain unrealized as long as they failed to generate a concrete research program. Wundt got an opportunity to realize his project once he arrived in Leipzig. As for the subject of his foundational psychological study, Wundt chose reaction time—a decision that was not without controversy.

The origins of reaction time as an object of experimental inquiry are essentially twofold. There was, of course, Helmholtz's work on the propagation speed of an action potential across a nerve fiber, which had demonstrated that the speed of an action potential was much slower than previously assumed (Canales 2010). Of at least comparable significance was Friedrich Bessel's impactful realization that independent observers of the same astronomical events consistently had minute but mathematically significant variations in their recorded values, which Bessel termed the "persönliche Gleichung" (personal equation) (Canales 2010; Hoffmann 2006, 172; Exner 1873, 606). Bessel had begun to recognize this phenomenon as early as 1818, but only fully recognized and sought to investigate the universality of the effect in the early 1820s (Hoffmann 2006, 147, 166, 172–79). In 1861–1863, Swiss-born astronomer Adolph Hirsch first used Matthäus Hipp's chronoscope device to measure the delay in what would come to be known as simple reaction time (Canales 2010; Robinson 2001, 164; Schmidgen 2002).

Bessel's personal equation had demonstrated the scope of the phenomena, although it was Helmholtz's work on propagation speed that validated further investigation into personal equation as an object of physiological inquiry. Though the emergence of the modern concept of reaction time is rightly located in the 1870s, Franz Donders's work in the 1860s was arguably the first physiological study of the matter (Schmidgen 2002; Schmidgen 2005). In 1868, Donders demonstrated that simple response times were shorter than response times that involved more complex mental tasks, in this case recognition of a vowel sound (Donders 1868, 423). In order to come to these results, Donders pioneered the subtraction method as a means of determining what he called "physiological time" (Donders 1868, 417, 428; Robinson 2001, 164). This was a basic but effective methodological approach that entailed the experimenters initially taking a simple response time, followed by a second reading involving a more difficult mental process. The first time was then subtracted from the second, allowing the experimenter to determine separate values (Robinson 2001, 164). Rather than use Hipp's chronoscope as Hirsch had, Donders used a different chronograph. Importantly, Donders understood the phenomena in physiological terms, and perceived his experiment as a continuation of the work pioneered by Müller, du Bois-Reymond, and Helmholtz (Donders 1868, 415; Schmidgen 2005).

The modern language of reaction time is derived from Sigmund Exner's 1873 "Experimentelle Untersuchung der einfachsten psychischen Processe, Erster Abhandlung." Exner had studied with Helmholtz in Heidelberg from 1867 to 1868, and Exner's research on the neurology of perception was a self-conscious continuation of Helmholtz's work on the physiology of the nerve impulse. Exner's research would become a series on the neurophysiology of simple psychological phenomena, namely reaction time (Exner 1875, 404). Exner's project was inspired first by Helmholtz and Nikolay Baxt's 1870 discovery that variations in temperature affect the propagation speed of an action potential across a motor nerve fiber (Exner 1873, 601, 606). The second component was Exner's awareness of the limited nature of the work done on the physiological nature of Bessel's personal equation (Exner 1873, 606). Understanding this phenomenon would theoretically have helped determine "how to reduce personal error to a minimum" (Exner 1873, 608). As Exner understood it, the physiological question posed by the problem of the personal equation could be reduced to the minimum time elapsed in generating a bodily reaction to a direct stimulus (Exner 1873, 609). Thus, Exner dubbed this metric "Reactionzeit" (reaction time), thereby coining the term (Exner 1873, 609).

Given the dominance of the physiological conception of reaction time, how was it that the early reaction time studies conducted in Wundt's lab would come to, in the words of the historian Kurt Danziger, "constitute the first historical example of a coherent research program, explicitly directed toward psychological issues and involving a number of interlocking studies" (Danziger 1980, 106)? The answer is that Wundt took up physiological reaction time as an object of experimental inquiry and reconceived of reaction time as a *psychological event*. As early as the *Grundzüge*, Wundt had begun to consider reaction time to largely be a psychological, rather than a physiological, phenomenon. Wundt was well aware of the findings of Donders, Exner, Baxt, and Helmholtz, all of whom are addressed in the first edition of the *Grundzüge* (Wundt 1874a, 498, 728, 740–41, 752). It came down to how reaction time was broken down into separate subprocesses, and where those subprocesses figured within Wundtian parallelism.[3] The initial sense perception and the muscular movements necessary to signal a reaction were naturally physiological events. However, between these physiological events there were at least three psychological events: perception, the focusing of apperception, and the act of willing (Wundt 1874a, 728, 734–36). This made reaction time, as something that could easily be measured and placed in a table alongside other measurements, an ideal candidate for experimental psychology's founding research program.

The critical innovations that made Wundt's argument possible were his definitions of differentiation, choice, and will. As Robinson suggests, Wundt was critical of the distinctions Donders made between choice and differentiation in his experimental design (Robinson 2001, 166). Donders had proposed models for simple, differentiation, and choice reactions. The choice reaction entailed hearing a vowel sound and responding with the same sound, while the discrimination reaction required that the subject respond when they heard a predetermined syllable (Robinson 2001, 165–66; Wundt 1874a 744–45). Wundt's critique of Donders's experimental design was that Donders's discrimination reaction still entailed motor selection, and therefore choice (Robinson 2001, 165–66; Wundt 1874a, 744–45). This would set Wundt on a course to develop an experimental design for a pure discrimination reaction, which would debut in his Leipzig laboratory and be included in subsequent editions of the *Grundzüge*. Further still, Wundt took up Donders's subtraction method, though Wundt followed through with his preference for Hipp's chronoscope that Wundt had expressed in the initial *Grundzüge* (Wundt 1874a, 732–33). Donders's research thus provided a backdrop against which Wundt developed his own psychological conception of reaction time.

The psychological definition of reaction time would form the initial justification for the new experimental science. By the time Wundt's psychological laboratory was founded in Leipzig in 1879, psychological reaction time had crystallized into a robust and experimentally dynamic concept. A telling detail to this effect was the creation of a new kind of expert, professional scientists whose research was centered on the study of psychological reaction time in keeping with the methodologies of experimental psychology. Almost immediately after the formation of Wundt's psychology laboratory, a young doctoral student of Wundt's named Max Friedrich began conducting research on reaction time in response to visual stimuli in the lab (Domanski 2004, 311). Friedrich wrote the first dissertation in the field of experimental psychology; titled "Über die Apperzeptionsdauer bei einfachen und zusammengesetzten Vorstellungen," it was published in 1881 after being officially awarded in 1880 (Behrens 1980, 19; Domanski 2004, 311–312; Friedrich 1883). Friedrich would the first, but Wundt's lab quickly filled with eager students, would-be professionals in this new science of embodiment. More than Wundt himself, it would be his students who would ultimately go about institutionalizing the science of experimental psychology and disseminating Wundtian ideas across the world.

Wundt had faced down an uncertain future in academic physiology and come out as the founder of a new branch of the sciences. This was in large part owed to the influence of Weber and Fechner, whose parallelistic doctrine provided an avenue of critique through which Wundt could approach what he had seen over the course of his education in physiology. While not a critique of the science of physiology, the emergence and eventual success of Wundt's new science of experimental psychology was in contrast to the totalizing equivalence between mental and neural states championed in neurophysiology. In this sense, it was implicitly critical of neomechanism, while still nevertheless upholding the value of physiology. This was made possible by Wundt's shift from a position characterized by an a priori recognition of the concealed world of the body as the locus of the real to the study of the ready-to-hand, perceptional lifeworld, to the experience of embodiment. Given the enduring attention that Wundt allotted to neuroanatomy and the experimental focus he took from physiology, this new science was not predicated on a rupture in criteria of validity. Rather, an adequate description of higher-order mental processes had demonstrably eluded neurophysiological reductionism. This ambiguity concerning the physiological definition of mental processes left sufficient space for Fechnerian parallelism to have a conceptual influence.

Though Wundt would eventually attain world renown, in the early years of the laboratory experimental psychology was hardly an established science and was not without detractors. It fell on Wundt's students to carry the new science forward, and there was one student in particular whose impact on not only psychology, but psychiatry and biology more broadly, has only begun to be accounted for. That student was Emil Kraepelin, whose place in this story is about to be told. But before diving into the Kraepelinian legacy, it is important to contextualize his work with an accounting of their age. After all, their age, the epoch of Wundt, Meynert, and Kraepelin, was a golden one, albeit not only for the life sciences. It was also a golden age of intoxication. It is perhaps this detail that lies at the foundation of the modern biological subject.

III
The Intoxicated Subject

※ 8 ※
A Postalkaloidal "Golden Age" of Intoxication

Novel Intoxicants

The years running into the 1870s and 1880s introduced us to bold conceptions of a mechanical brain and a novel way forward in psychology, a new science of embodiment. Yet much of this sound and fury largely ignored an ever-growing substrate of nineteenth-century medical, biosocial, and material existence: the rapid explosion in the range of publicly available intoxicants. The alkaloidal revolution of the early nineteenth century, sparked by Sertürner's identification of morphine in 1817, had been the dawn of a new age, but it was in the latter half of the German nineteenth century that the full implications of this event stepped into the light of day.

The number of intoxicants available to physicians and the public alike in Europe at the beginning of the nineteenth century had been a short list consisting of opium, cannabis, ether, and alcohol—on top of lighter stimulants like tobacco, coffee, and tea. By the 1880s, these proven fixtures were supplemented by a growing array of alkaloidal isolates available in thousands of different formulations. There had never been so many different kinds of intoxicating substances or ways to consume them, most of which were pushed into the hands of customers by way of questionable advertisements. Of course, the lion's share of the intoxicants consumed still fell to alcohol in its varied preparations. But even the trade in alcohol was captured by a spirit of proliferation and diversification. International trade in alcohol increased like never before and with it came radical shifts in how alcohol was consumed. Working-class Britons of the 1860s began to meaningfully participate in the consumption of traditionally more class-restricted alcohols such as wine, port, and sherry, the adulteration of which likely gave rise to the modern alcohol branding (Duguid 2003). It was a "golden age" of intoxication. Why does it warrant such a lofty moniker? Because of the unprecedented public access to different powerful intoxicants, because of the range of novel experiences, because of the rise of industrial pharmacy, and because of the sheer scale of drug and alcohol consumption in the era. But, most of all, it was a "golden age" because it was the twisted and distorted realization of a dream of a brighter future, one shared by organic and alkaloidal chemists of the mid nineteenth century alike. It was the consummate expression of a utopian dream at the core of nineteenth-century Liebigian organic chemistry, the self-same dream that called forth the advent of modern pharmacological science. To make sense of this, let us look at the early career of Justus Liebig and the first synthesis of chloral hydrate, before beginning to unpack what this "golden age" can tell us about birth of the biological subject.

New Synthetic Substances

The emergence of the alkaloid as a new kind of thing to be in the world had already forever changed pharmacy and medicine. The discovery of morphine had almost immediately altered how therapeutics were deployed in a clinical context by all but confirming the active principle theory of physiological action. Opium was one of nature's curious manufactures, like honeycomb, shellac, or even alcohol. Morphine remained an organic plant principle, something that was already present in the opium latex produced by the poppy. The discovery of morphine itself had no direct bearing on the possibility of synthesizing a new drug. However, the very idea of using the methodologies of the chemical sciences to isolate an active plant principle, thereby making it subject to a battery of chemical analyses, very well might have paved the way for modern synthetic pharmacy.

The first synthetically derived intoxicant, or drug of any kind, was chloral hydrate—the same chloral hydrate that drove asylum life in the latter half of the nineteenth century. It was created by Justus von Liebig in 1832, who had fled from the beginnings of a career in practical pharmacy to pursue chemistry (Hofmann 1876, 102; Schmitz 1985, 63). Liebig's legacy as a primary founder of organic chemistry was already secured in the 1830s when he established the "first large-scale teaching, research laboratory in any science" (Werner and Holmes 2002, 422). The Liebigian "Giessen Model" would rapidly spread across central Europe, owing as much to its pioneering use of "paper tools" as to Liebig's force of personality (Rocke 2023). But he pushed further still. In large part owing to the foundations set by Liebig and his friend and colleague Friedrich Wöhler in the 1830s, the subfield of organic chemistry had begun to develop more quickly than inorganic chemistry by the 1840s (Werner and Holmes 2002, 422). Much like his contemporaries Fechner and Müller, though, many of his theoretical positions seem at odds with the scientific heritage fostered by his work. For one, Liebig was, and remained, a vitalist. His 1843 *Die Thier-Chemie oder die organische Chemie in ihrer Anwendung auf Physiologie und Pathologie* boldly asserted, in just one of many similar statements, that "all processes in the organism are under the influence of an immaterial activity [life force], which the chemist cannot dispose of arbitrarily" (Liebig 1843, 148). Proclamations such as these lead some historians to ponder the question of whether such an eminent scientific figure as Justus von Liebig can really be considered a vitalist (Lipman 1967, 167). A better question is: In what way was Liebig a vitalist and how did his vitalism factor into the research program of organic chemistry?[1]

Liebig's vitalism was traditional insofar as it upheld the notion that a vital *Lebenskraft* was determinative of the physiological character of organisms and the primary quality distinguishing organic and inorganic matter (Liebig 1843, 9, 11, 86, 143, 176). In this sense, Liebig's vitalism was more reminiscent of eighteenth-century vitalists like Reil and fundamentally at odds with the vital monism of the *Naturphilosophen*, whose theories Liebig vehemently opposed. But on the chemical level, Liebig famously argued that there was no essential chemical difference between organic and inorganic material, as stated in an 1838 paper with Wöhler: "The philosophy of chemistry will draw from this work the conclusion that the production of all organic matter, in so far as it no longer belongs to the organism, must not only be probable in our laboratories, but must be regarded as certain" (Liebig and Wöhler 1838, 242). At first glance, this idea appears to reflect the one-in-many principle of *Naturphilosophie*: a vital monism that erodes the distinctions between organic and inorganic by realizing all that exists as a single organic whole.

There were many more ways in which Liebig's ideas superficially resembled those of *Naturphilosophen*. Like the *Naturphilosophen*, Liebig was a teleological thinker (Brock 2002, 311). Like the *Naturphilosophen*, Liebig rejected material reductionism. Liebig himself had even been a student of Schelling's for two years in Erlangen (Snelders 1970, 193). And yet, Liebig would go on to be one of *Naturphilosophie*'s strongest opponents, looking back on his years with Schelling as lost time (Snelders 1970, 193). Liebig later summarized *Naturphilosophie* as "so rich in words and ideas, yet so poor in truthful knowledge and genuine research" (Liebig 1874, 34). These were kinder words than Liebig had used in 1840, when he described *Naturphilosophie* as the "Black Death" (Brock 2002, 67). Liebig allegedly never found Schelling's *Naturphilosophie* compelling, specifically because Liebig was keenly aware that Schelling himself had a limited understanding of the natural sciences (Brock 2002, 27). This does not necessarily mean that

Liebig was always hostile toward *Naturphilosophen*. Liebig's mentor in physics and chemistry, Karl Kastner (1783–1857), was also a *Naturphilosoph*, and yet Liebig followed Kastner from Bonn to Erlangen, cleaving to his mentor through his doctorate, all the way up until Kastner secured a grant for Liebig to study in Paris with Gay-Lussac.

However, the Liebig of the 1820s was decidedly at odds with the naturphilosophic tradition. In 1824, Liebig wrote to Platen about his commitment to materialism (Brock 2002, 309). Though compelling, it is not entirely clear what Liebig meant by this. Was he identifying with materialism in contrast with *Naturphilosophie*? What does this say about Liebig's vitalism in his early years? Unfortunately, Liebig's passing reference to materialism does little to suggest he was not a vitalist in his student days and early career. Vitalism can be a central part of a materialist doctrine. Brock and Hall speculate that Liebig's belief in the vitalism of organized bodies was informed by Johannes Müller's 1835 to 1837 publication of the *Handbuch*, although there is no direct evidence to suggest this connection (Brock 2002, 310). More importantly, Liebig's vitalism sharply contrasts with that of Müller's, a postnaturphilosophic vitalism grounded in his notion of organizing principles. Liebig's *Thierische-Chemie* instead implicitly argued that a vital force was needed to account for the capacities for growth, tissue repair, and complexity characteristic of all living beings (Lipman 1967, 176). Regarding the chemical composition of organic and inorganic bodies, it is true that Liebig saw no difference, except that the chemical products of living beings are of sufficient complexity that they could only naturally originate from a vital entity. What likely started as a matter-of-fact vitalism became, by the 1850s, a principled effort to carve a middle way between what Liebig saw as the extreme reductionism of the neomechanists and the wanton speculation of *Naturphilosophie* (Brock 2002, 311).

This was the perspective that Liebig carried into his professorship in Giessen, where he would rise to fame. Alexander von Humboldt recommended Liebig for a professorship at the University of Giessen in 1824 (Brock 2002, 35). Initially restricted from accessing laboratory time, Liebig proposed the formation of a pharmaceutical institute in 1825, but when his proposal was rejected by the university he was forced to organize it as a private venture until the university agreed to include Liebig's lab in 1833 (Brock 2002, 43, 47). Given that Liebig's introduction to chemistry was a pharmaceutical apprenticeship in Heppenheim, it was only fitting that Liebig would place pharmacy at the center of his chemical research program (Schmitz 1985, 63).

It was at this point, in the early 1830s, that Liebig began to research the structure of various "strychnines, morphines, narcotines, atropines and almost all alkaloids known at that time," a project that would later see Liebig declare that everything from sugar to morphine would soon be manufacturable not only in nature but in the lab (Volhard 1898, 39; Liebig and Wöhler 1838, 242). The first step was to establish a methodology for determining the elemental constitution of organic compounds. Sertürner had been the first to fully demonstrate that a pharmaceutically active organic substance could be extracted and crystallized; now Liebig sought to create the same substances from their basic chemical components, an astounding theoretical innovation. To this purpose, Liebig developed a combustion method for determining the carbon and hydrogen content of organic substances (Liebig 1831, 1–3). Morphine was among the first substances put through Liebig's combustion analysis, followed shortly after by strychnine and atropine (Liebig 1831, 9). The inaugural focus of the world's first large-scale pharmaceutical research laboratory had been the chemical analysis of morphine and other intoxicants, just as alkaloidal chemistry had begun with morphine decades earlier.

How fitting then that the following year Justus von Liebig synthesized the first drug when he created chloral hydrate alongside chloroform, though Liebig was unaware of chloral's effects at the time. His process "led chlorine gas through alcohol," which had previously been dried with calcium chloride (Liebig 1832b, 252). This process was continued until hydrochloric acid ceased to precipitate at the top of the apparatus (Liebig 1832b, 253). The colorless liquid remaining afterward formed an oil-like droplet when put in water, dissipating in warm water (Liebig 1832b, 256). The oily substance behaved very differently when only a small amount of water was added to it. Here, the substance "immediately combines with it when shaken, heating up strongly; a few moments later the mixture solidifies into a transparent white crystal mass" (Liebig 1832b, 256). Liebig named the oily liquid "Chloral" in reference to the word "Aethal," recognizing "that in a complete decomposition of the alcohol, the chlorine separates out [the alcohol's] hydrogen and takes [the hydrogen's] place" (Liebig 1832b, 252). Liebig further noted how an alkali, soda, decomposed chloral to produce *Chlorkohlenstoff* (chloroform) as well as formic acid (Liebig 1832b, 256). However remarkable a development in the history of pharmacy and intoxication he had just made, Liebig was unaware of chloral's hypnotic properties, which would only be recognized in 1861 by Rudolf Buchheim and written on in 1868/1869 by Oskar Liebreich (Liebreich 1869, 15).

In association with the Liebigian professionalization of pharmaceutical chemistry, the *Annalen der Pharmacie* was formed in 1832 by Justus

Liebig, Rudolph Brandes, and the alkaloidal chemist Philip Lorenz Geiger. It was around this time that Friedrich Wöhler also began working in Liebig's lab, for lack of resources in Kassel (Liebig 1832b; Hoppe 2007, 195).[2] Liebig and Wöhler quickly formed a highly productive professional friendship, supported by a new journal in which to share their findings. They jointly published a vast number of papers on foundational topics in organic chemistry, pioneering the "radical theory" in response to their shared work on benzoyl (Rocke 1993, 51–52).

What does this have to do with claims of a "golden age of intoxication"? It has already been made clear that the Liebigian research program of organic chemistry would come to form the central locus of the emerging field of scientific pharmacy. Novel intoxicants did not simply come "out of the ether," but out of the etheric solvents prepared by Liebigian-trained pharmaceutical chemists. This would coincide with the rise of industrial pharmacy in the Germanies, where Liebigian scientific pharmacy granted an aegis of scientific validity to mass pharmaceutical enterprise. Germany would in no way be the first to mass produce alkaloids, but the German industry would rapidly rise to a position of foremost prominence and be among the first to produce new intoxicants such as chloral and make them available for mass consumption. As new technologies of intoxication were developed over the course of the nineteenth century, it would inescapably be in relation to the emerging study of scientific pharmacy. One important intoxicant directly related to the laboratories of Liebig and Wöhler was cocaine.

Wöhler's Stimulating Discovery

The effect of coca leaves had been known to Europeans for quite some time, but the remarkable difficulty of transporting a large amount of coca leaves from the South American colonies to Europe had significantly hampered the efforts of interested researchers. The earliest attempt to render an extraction from coca leaves was carried out by Heinrich Wackenroder in 1853, using an isinglass solution to yield a precipitate from ethanol and coca leaves (Wackenroder 1853, 24). But Wackenroder was only able to produce a crude extract.

Further investigations were stalled, first by a shortage of coca leaves and then by Wackenroder's death the following year. Efforts were renewed two years later when, in 1855, Friedrich Gaedcke published an article titled "Ueber das Erythroxylin, dargestellt aus den Blättern des in Südamerika cultivirten Strauches Erythroxylon Coca Lam" in *Archiv der Pharmazie* (Gaedcke 1855, 141). Gaedcke detailed the process by which he extracted the active alkaloid found in the South American coca leaf, adorning the new substance with the catchy name *Erythroxylin* after the coca leaf's genus, *Erythroxylon* (Gaedcke 1855, 141). The brief paper ran fewer than ten pages, and much of its content was committed to the discussion of the social and cultural significance of coca leaf in its South American homeland (Gaedcke 1855, 141–44). Nor was Gaedcke able to establish meaningful accomplishments in excess of what had already been put forward by Wackenroder. By Gaedcke's own report, "[t]his attempt is not sufficient to prove the identity of this substance I have named erythroxylin," citing an insufficient supply of coca leaves (Gaedcke 1855, 150). The quantities available were inadequate for Gaedcke to even say for certain whether the extract simply contained caffeine or a different, related, substance (Gaedcke 1855, 148, 150). An additional complication, one that likely precluded Gaedcke from reaching any decision about his precipitate, was the quality of Gaedcke's isolate. Gaedcke was able to produce a bitter crystalline substance, though it had emerged alongside a slick, oil-like substance (Gaedcke 1855, 147–48). By his own admission, Gaedcke had rushed to establish priority in naming the substance, without even identifying the novelty of the substance itself. Others in Italy and Great Britain had also tried, with worse luck still (Niemann 1860, 149–50).

Shortly after Gaedcke's publication in *Archiv der Pharmazie*, Friedrich Wöhler took it upon himself to acquire a stock of coca leaves in order to study their composition. This was eventually accomplished by enlisting the assistance of Austrian explorer Karl von Scherzer, who acquired thirty pounds of coca leaves in Lima while circumnavigating the world with the *Novara* expedition (1857–1859) (Niemann 1860, 133). Upon receiving the parcel in 1859, Wöhler gave the project to one of his graduate students, Albert Niemann, reflecting his pedagogical alignment with the Liebigian method of establishing students as independent chemical researchers (Niemann 1860, 132–133). Niemann took it on as the topic of his dissertation, with the resulting paper being published in 1860 under the title "Ueber eine neue organische Base in den Cocablättern." Niemann would dub his alkaloidal extract *Cocaïn*, the name it is known by to this day.

As Niemann himself admits, the presence of an alkaloid in coca leaves was already expected, insofar as pharmaceutical chemists had come to suspect that yet undiscovered alkaloids were behind the physiological effects of all the remedies of the materia medica (Niemann 1860, 130–31). This blanket assumption, Niemann attested, had already been proven false, while many well-supported inquiries into presumed alkaloids were wholly insufficient to make any definitive claims (Niemann 1860, 130–31). In these opening pages, Niemann drew a clear demarcation between what he understood to be the overeager findings of ambitious alkaloidal chemists and those claims whose alkaloidal objects merited recognition as distinct substances, such as Sertürner's work in 1816/1817. Sertürner's name even appears in the first line of Niemann's text, in a clear stint to position his own alkaloidal discovery within the authority of a Sertürnerian tradition.

Niemann's method involved leaving 200 grams of chopped coca leaves in a bath of alcohol, previously mixed with diluted sulfuric acid, at a warm temperature for four days (Niemann 1860, 151). After straining out the diced coca leaves, Niemann was left with a green, light brown liquid, which he noted was still somewhat acidic (Niemann 1860, 151). After drying the leaves and finding them so void of any bitterness that a second extraction would be unnecessary (though he nevertheless did), Niemann mixed the fluid extract with a watery paste consisting of hydrated lime and allowed this blend to sit at room temperature, with regular shaking (Niemann 1860, 151). The precipitate was filtered out and set aside, while the tincture was distilled in a warm water bath until the remaining solvent had dissipated, at which point distilled water was added to the syrupy residue (Niemann 1860, 151–52). The syrupy residue had become an undissolved, resin-like mass, which Niemann filtered out and washed several more times (Niemann 1860, 152). Rather than pour out the last body of distilled water, Niemann added carbonate of soda, which turned the liquid a deep, dark red (Niemann 1860, 153). The crucial final step was the introduction of a generous amount of solvent, after which the container was sealed and shaken, with the ether left to evaporate on a shallow tray (Niemann 1860, 153). What remained after twenty-four hours was a strongly alkaline crystallized mass, flecked with yellow-brown particulates. He persisted in purifying his isolate further and was eventually successful at rendering a purely white final product (Niemann 1860, 154–55). Niemann's cocaine isolate was still chemically impure, but he had nonetheless succeeded at making real a novel alkaloidal entity.

The broader consequences for organic chemistry in general, and alkaloidal chemistry in particular, did not go unconsidered by Albert Niemann. In the decades between discovering chloral and Niemann's extraction, Justus von Liebig steered organic chemistry toward ever-broader horizons. Organic chemistry, Liebig suggested, had far-reaching social consequences, and he began to write on the chemistry of agriculture, living beings, and food, publishing *Die organische Chemie in ihrer Anwendung auf Agricultur und Physiologie* in 1840, *Die Thier-Chemie, oder die organische Chemie in ihrer Anwendung auf Physiologie und Pathologie* in 1843, and *Chemische Untersuchung über das Fleisch* in 1847. Though at times marred by Liebig's own lack of familiarity with the actual practice of agriculture, Liebig's agricultural chemistry was deeply impactful, with Liebig inventing chemical fertilizer and emerging as the pope of agricultural chemistry in his time (Brock 2002, 128, 149, 176). Organic chemistry, Liebig imagined, would finally stay the hungry masses and bring healing to the sick and weary, unlocking medical and nutritive secrets that had eluded humankind since time immemorial, ushering in a newer, better condition for humanity. Liebig's vision was of nothing less than a brave new world cooked up in the lab. It was precisely this dream that Niemann saw refracted in his shimmering crystalline precipitate.

Like Gaedcke's paper, Niemann's analysis was preceded by a meandering discussion of the South American practices and cultural understanding surrounding coca leaves.[3] It is difficult not to understand these remarks through the lens provided by Niemann himself in the opening pages: that of the methodological triumph of chemical analysis over not only the vagaries of prealkaloidal medicine but also the rushed conflations of his colleagues. Niemann thus explicitly associated the South American, and older European, consumption of whole plant remedies with imprecision, even backwardness, while simultaneously aligning alkaloidal extracts with precision and futurality. Niemann's remarks on the significance of alkaloidal chemistry for the medical profession further illustrate this point. "No physician would understand" that plant remedies "contain other effective bodies which substantially modify the effect of the pure alkaloid" (Niemann 1860, 131). The physician determines which remedy to employ or how it should be delivered; however, it is the alkaloidal chemist, Niemann argued, who bears responsibility for ultimately building the bridge between the murky uncertainties of medieval medicine and a future of medicochemical precision. Niemann's work was but a pharmaceutical expression of the dream shared with Liebig, Wöhler, and other organic chemists of the mid-nineteenth century. The discovery of cocaine, like morphine, was a step into a new world order. Albert Niemann would die in the year 1861 at the age of twenty-six years old, just one year after publishing his research on cocaine. He never lived to see the radical future his work would help build.[4]

Industrial Pharmacy: A Brave New World Cooked Up in the Lab

Theoretical and experimental developments fueled the diversification of intoxicants throughout the nineteenth century, though the magnitude of these events could hardly have been as noteworthy were it not for the simultaneous emergence of industrial pharmacy, further bolstered by the development of new technologies of intoxication. Just as mass consumption of opium entailed a sprawling field of poppies, public access to morphine of any significance required chemicoindustrial infrastructure. Merck, Pfizer, Bayer AG—the titans of modern pharmacy trace their humble beginnings to the nineteenth-century industrial production of alkaloids.

Merck was among the first, if not the first, modern industrial pharmaceutical drug companies in history. The company traces its origins to Friedrich Jacob Merck's acquisition of Darmstadt's Engel-Apotheke in 1668, though it wasn't until Heinrich Emanuel Merck assumed control in 1816 that the Merck name would be associated with anything more than a local pharmacy business. Heinrich Merck had studied in Johann Trommsdorff's famous pharmaceutical institute from 1810 to 1812 (Friedrich 1998, 508). In a world where pharmacists still typically learned through practical apprenticeships, Trommsdorff's institute was one of only a few that deigned to offer prospective pharmacists with both practical and theoretical training in general chemistry (Rocke 2023). When Liebig set about founding his own research institute, he initially intended to establish a laboratory-based pharmacy school much like Trommsdorff's (Rocke 2023). After working in state and court pharmacies in Eisenach, Frankfurt, and Straßburg, Merck wrote the Prussian *Provisorexamen* in Berlin in 1816 and assumed control of the family pharmacy in Darmstadt (Friedrich 1998, 509). In light of his studies with Trommsdorff on pharmaceutical chemistry, Merck was very quick to develop an interest in alkaloidal research after Sertürner identified and named morphine in 1816–1817 (Friedrich 1998, 509). The enterprising young pharmacist began to focus his energies on perfecting the extraction processes for the new alkaloids that were now regularly being discovered, in order to realize pure alkaloidal isolates (Friedrich 1998, 509). By the year 1827, Merck's new alkaloid production company started selling morphine, wholesale.

Standing on the shoulders of chemicopharmaceutical research science, German industrial pharmacy swiftly ascended to a position of global eminence (Bartmann 2003). As other pharmacies, among them J. D. Riedel and E. Schering, flocked to reproduce the success of Merck and chemical dye manufacturers, such as Hoechst, expanded into pharmacy, industrial pharmacy flourished across the German Empire (Cramer 2015).[5] The majority of the emerging industrial pharmaceutical manufacturers operated as wholesalers, typically to local pharmacies in the surrounding areas and—later—to patent medicine compounders (Bartmann 2003). If this approach failed, recourse was found in specialization (Cramer 2015). Specialization in the manufacture of one or more drugs, bolstered by the regional localization of pharmacy science, enabled a select group of German firms to assume control of the global sale in particular pharmaceuticals. In the case of Riedel, the opposite prevailed. Johann Riedel's pharmacy business attempted to establish itself as a specialty producer of quinine in 1827, only for the business to falter (Huhle-Kreutzer 1989). Saving the enterprise, Riedel pivoted, becoming a pharmaceutical wholesale business serving local Berlin pharmacies (Huhle-Kreutzer 1989). Whatever the circumstances faced by particular manufacturers, the net effect was that of Germany's meteoric ascent to a position of global centrality in the industrial manufacture of pharmaceutical drugs. By the Wilhelmine period, Germany would be known as the "world's pharmacy" (Bartmann 2003, 20).

The English morphine business followed a similar trajectory, albeit with more modest successes. Thomas Morson, after studying medicine in Paris, began producing morphine in the back of a retail business on Farringdon St., London, and selling it between 1821 and 1822 (Berridge 1999, 136). Around this time, the Society for the Encouragement of Arts, Manufactures, and Commerce also began offering premiums "for promoting and improving the cultivation of the *papaver somniferum*" (Jeston and Dick 1823, 17). Morphine's "English variety sold at eighteen shilling per drachm, with the acetate and sulphate selling at the same rate" (Berridge 1999, 136). Edinburgh's Macfarlan and Company began morphine production in the early 1830s, buying opium wholesale from London and selling it back as muriate of morphine (Berridge 1999, 136).

A major purchaser of morphine were patent medicine manufacturers. Patent medicines were remedies that were produced by entrepreneurs and sold directly to the consumer, generally without any obligation to divulge the content of the "medicine" (Corley 1987, 112; Woycke 1992, 42). Though the patent medicine industry had existed since the eighteenth century, patent medicines only achieved significant large-scale commercial success in the mid-nineteenth century. By the 1870s, it is estimated that a quarter of all posted advertising was for patent medicines (Petty 2019, 289). Contrasted with the 1820s, when De Quincey famously reported "that the number of amateur opium-eaters (as I may term them) was at this time immense," the public source for intoxicants such as opium and morphine was increasingly patent medicines (De Quincey 1885, 4).[6] The annual sale of patent remedies in Britain increased from 500,000 pounds sterling in the 1850s to 5,000,000 pounds by 1914 (Digby 1999).

British economists of the 1880s described the patent medicine industry as filling the gaps in medical care for those who sought to treat their ailments without seeking the care of a doctor they could not afford (Corley 1987, 112). In this respect, there were droves of poor and working-class people in industrial England, France, and America—the dispossessed, *les misérables*—who, despite working long hours in dangerous conditions, were largely overlooked by the medical system (Corley 1987, 112). The receptiveness of the American market functioned on a similar principle: in light of the rural character of the American population, self-medication using commercial remedies was often the most familiar method of treating pain and disease, and in other cases perhaps the only way (Petty 2019, 288). Many of the remedies themselves promised to not only put an end to pain and infirmity, but also to actually improve the customer's overall health and well-being. Ultimately, the medicines generally consisted of little more than opium, morphine, and other alkaloids, which could quickly and cheaply allay the customer's physical or mental discomfort without furnishing any of the advertised benefits. But, beyond merely treating ailments, patent medicines were represented as a means of chemically enhancing life (Black 2022). Avenues of pleasure and respite just as readily as they were sleep aids, painkillers, and stimulants, patent medicines were represented as modern solutions to age-old problems (Black 2022). Much as Liebigian chemists dared to conceive of organic chemistry as the solution to the challenges of health, hunger, nutrition, and happiness, patent medicines promised a future of convenience, one void of pain or tire (Black 2022). In short, patent medicines were a part of what it meant to be modern.

Following Niemann's publication on the isolation of cocaine in 1860, pharmaceutical manufacturers were swift to develop a means of manufacturing this alkaloidal extract too. Merck began selling cocaine as early as 1862, albeit in limited quantities (Courtwright 2002, 47). When Freud wrote *Über Coca* in 1885, he exclusively used Merck-brand cocaine and credited Merck with ultimately justifying the belief that the entirety of coca's power stemmed from alkaloidal cocaine (Freud 1885c, 9, 11). From 1862 to 1879, there was considerable interest in the alkaloid among chemists from England to Russia, with some physiologists conducting variably successful experiments on the effects of cocaine on various animals (Freud 1885c, 8).

Cocaine typically reached the general public through a different means: the cocaine tonic. In 1863, a Corsican chemist and businessman by the name of Angelo Mariani developed a tonic consisting of cocaine and Bordeaux wine for sale on the international market (Karch 2006, 40).[7] Dubbed "Vin Mariani," Mariani's cocaine tonic was available everywhere in France by 1870, soon reaching London and New York (Karch 2006, 40). Though competitors sprung up as the tonic's popularity grew, Vin Mariani went on to have tremendous brand success in America and across the European continent, with the label itself becoming synonymous with cocaine (Smith 2008, 42; Karch 2006, 41). This was largely made possible by Mariani's aggressive advertisement strategy, which would not have been out of place in the twenty-first century. Mariani sent free samples of his *vin* to prominent physicians and celebrities, seeking endorsements to feature prominently in the pamphlets, books, and flyers found wherever Vin Mariani was sold (Karch 2006, 43–44). Pope Leo XIII, Pope Saint Pius X, Thomas Edison, and Ulysses S. Grant would all become disciples of Vin Mariani (Kennedy 1985, 86; Karsch 2006, 44). Pope Leo even offered his photo in a celebrity endorsement of Vin Mariani, and Edison put his fame behind a published endorsement as well (Karch 2006, 43). What was remarkable about cocaine was how it ushered in a radically novel experience of being in the body. Stimulants of cocaine's sort were in short supply on the European continent. When coffee reached Europe, it fundamentally changed European society; coffeehouses became known as sites of free-flowing ideas and excited chatter (Pendergrast 2010, 9). Jürgen Habermas famously identifies the coffee house as crucial to the symbolic and literal emergence of the bourgeois public sphere (Habermas 1990). Cocaine, much like coffee, lifted the spirits, but it carried those spirits to dizzying, euphoric, and heretofore unseen heights.

The German patent medicine and tonic boom did not take place until after 1869. Earlier in the nineteenth century, patent medicines were not available in Germany in the same way they had been in England and America. The majority of the medicines containing opium, morphine, and so on that had been purchased from German pharmacies had been by prescription, or at least with the customer's relative knowledge of their content (Woycke 1992, 42). The German moratorium on patent medications ended in 1869 when the liberal Berlin Medical Society argued that the medical profession would be better served by a free-market spirit— meaning that physicians would be free to choose who to treat and at what rates (Woycke 1992, 42). One particularly catalytic facet of the Berlin Medical Society's recommendations allowed anyone to manufacture medications on grounds of operational freedom (Woycke 1992, 50). By the end of 1871, the year that their recommendations were included in the commercial code by the imperial government, over 1,000 different patent remedies were available on the German market (Woycke 1992, 43). One of the most ubiquitous "brands" in German patent medicine were those manufactured by F. A. Richter, who by the 1880s was making five million marks a year on his "Pain Expeller" and "Swiss Pills" alone (Woycke 1992, 44). Like their French, English, and American counterparts, German patent medicines generally consisted of some blend of morphine, opium, chloral, hashish, and cocaine, depending on their advertised effects. Although one might expect the institution of obligatory health insurance in 1883 to shift profits away from patent medicines and toward traditional pharmacies, the profits of both patent medicine firms and pharmacies would continue to swell through the remainder of the nineteenth century (Cramer 2015; Woycke 1992). By the year 1900, pharmacists founded roughly a quarter of all manufacturing companies (Cramer 2015).

It was around this time that the first technological apparatus was developed for the explicit purpose of taking morphine into the body.[8] Even though morphine was increasingly available at an industrial scale, morphine still generally entered the body by way of the age-old methods used for opium. Ettmüller, a German physician, experimented with the medical use of opium vapors as early as 1809, yet by the mid-nineteenth century, German pharmaceutical chemistry had all but concluded that morphine could not be vaporized (Ettmüller 1809; Waldenburg 1864). Morphine was briefly administered through percutaneous cones, a method requiring a wound to be left open, so that hot morphine poultices could be administered directly, as used by Heinrich Heine (Auf der Horst and Labisch 1999). The hypodermic method would represent everything these methods were not. Precise, rapid, technological—the hypodermic method simultaneously reflected both the chemico-utopian aspirations of pharmaceutical chemistry and the instrumentalist commitments of experimental physiology. The idea of delivering a narcotic solution into the body by way of injection beneath the skin had been around since the seventeenth century. Christopher Wren and Robert Boyle had somewhat successfully injected a dog with a wine and opium solution in 1656, though the dog did die (Gibson 1970, 334). In the German world, Hamburg's Daniel Major published *Chirurgia Infusoria* in 1664 and Berlin's Johann Sigismund Elsholtz wrote *Clysmatica Nova* in 1665 on intravenous injection (Reinbacher 1998, 32; Gladstone 1933, 190–191). Interest in injection waned until the nineteenth century, when Irish physician Francis Rynd published the results of what is believed to the first ever successful subcutaneous injection (Rynd 1845, 167–68). Rynd produced a solution of "fifteen grains of acetate of morphia, dissolved in one drachm of creosote," which was "introduced to the supraorbital nerve, and along the course of the temporal, malar, and buccal nerves, by four punctures of an instrument made for the purpose" (Rynd 1845, 167–168).

In spite of Rynd's successful procedure, the invention of the modern syringe is generally accredited to the Scot Alexander Wood, who in 1853 developed a glass syringe attached to a hollow needle, though Charles Hunter suggested that Wood first employed the method in 1843 (Brunton 2000, 349; Hunter 1865, 10). The subcutaneous syringe was designed to treat pain at the nerve site, with Wood citing Johannes Müller's discussion of the effects of morphine on the excitability of exposed frog nerves (Brunton 2000, 349).[9] By 1858, Wood declared that his method was being applied almost universally to treat pain in Edinburgh (Brunton 2000, 350). The term "hypodermic" comes from Charles Hunter, an English surgeon, whose 1865 "On the Speedy Relief of Pain and Other Nervous Activities by Means of the Hypodermic Method" made the case that hypodermic injection of an alkaloidal solution had generalized rather than merely local effects (Hunter 1865, 11–12; Brunton 2000, 350). Hunter's methodology very quickly prevailed, though by the time of publication the use of the syringe had already spread onto the European continent.

Spread of the hypodermic method directly mirrored the growing trend of morphine use in European hospitals. On the European continent, Paris hospitals in 1855 doled out a measly 272 grams of morphine from pharmacies to their patients, but by 1875 that number increased by 3,576 percent to 10,000 grams (Courtwright 2002, 37). In Germany, too, hypodermic morphine use had grown to such a level that the Berliner Eduard Levinstein identified morphinism as a distinct physiological disease for the first time in his 1877 publication *Morphiumsucht nach eigenen Beobachtungen* (Levinstein 1877). Levinstein recognized that Lähr's "Ueber Missbrauch mit Morphium-Injectionen" (1872) and Fiedler's "Ueber den Missbrauch der Morphium-Injectionen" (1876) were the first to earnestly identify hypodermic morphinism as a medical concern, though both attribute morphinism to "Psychose" (psychoses) rather than a distinct physiological phenomenon (Levinstein 1877, 8). Levinstein attributed the spread of hypodermic morphinism to morphine's generous deployment in the 1866 Austro-Prussian War, much as it had been in the American Civil War (1861–1865), though the latter association has since fallen into doubt (Levinstein 1877, 3; Courtwright 1978, 101).[10] Levinstein further reported that, from then on, not only did physicians increasingly begin to prescribe morphine injections for "every abnormal sensation" but the general public also developed their own enthusiasm for the remedy and soon "[even] psychic pain was annihilated by morphine injections" (Levinstein 1877, 3–4).

Chloral hydrate entered medical parlance in 1869 when Oscar Liebreich published *Das Chloralhydrat ein neues Hypnoticum und Anaestheticum und dessen Anwendung in der Medicin* on his discovery of the effects of Liebig's chloral hydrate. Practically overnight, the first synthetic drug came into widespread use, both within and without the medical space. In the months following Liebreich's paper, the cost to produce chloral decreased by nearly 90 percent, and chloral quickly became more popular than other sedatives, such as bromide and opium (Snelders et al. 2006, 110). From late

1869 to 1871, sufficient chloral had been imported to England to provide half a million doses, while by the summer of 1870 German factories were already producing 70,000 doses of chloral *per day* (Snelders et al. 2006, 111). When Heinrich Byk founded his pharmaceutical factory in 1873, it was initially founded with the express purpose of manufacturing chloral (Fischer 1998, 13, 39).[11] Chloral possibly found its way into European and American asylums, hospitals, pharmacies, and patent medicines faster than any prior intoxicant in history.

A child could be born in a time where alkaloids were beyond consideration as a type of thing to be in the world and yet, by their sixtieth birthday, treat an old wound with morphine-based patent medicines, toast their health over whiskey, use chloral as a sleep aid, and treat the hangover with a cocaine tonic—should they so desire. The latter nineteenth century had been revolutionized not only by the ever-broadening range of never-before-seen intoxicants in different preparations but also by the general public's practically unfettered access to inebriates. Patent medicine companies and pharmaceutical manufacturers were able to seize the opportunity before their eyes, capitalizing on the emergent mass media environment in seeking to reach a new consumer base.

The purpose, here, has not been to pen the definitive chronicle of nineteenth-century drug normalization. There are others who do this perfectly well. Writing on France in the long nineteenth century, Sara Black likewise aligns the mass normalization of various intoxicants with a broader cultural enterprise, epitomized as "the chemical enhancement of modern life" (Black 2022, 3). Even this is a smidge modest. Chemical enhancement was but one facet of a far grander, almost utopian, dream, a New Atlantis, a Xanadu of crystalline spires, regally plotted aface a sheening plain of glassware. Yes, powerful intoxicants had swiftly become a critical facet of the changing social order. But—so much more than that—the rapid proliferation of potent inebriates that characterized the nineteenth century was inextricable from the developing research sciences of pharmacology and pharmaceutical chemistry. Here, intoxicants-as-scientific-objects were eminently participatory in the ontological and epistemological transformations at hand in the research program of Liebigian chemistry.

Yet as the din of pharmaceutical production grew ever louder, questions about the significance of intoxicants in the study of living bodies generally remained far afield. If anything, this discussion has only further stressed the rift between experimental physiology, neurophysiological psychiatry, and the developing science of pharmacy. As we will now see, it was the emerging science of experimental psychology, then principally a science concerned specifically with the experience of embodiment, that opened the door to a novel experimental encounter with intoxicants. Here, the immediate experiential interface between mind, body, and world made possible in intoxication could at last garner rational validity as the object of an experimental science, radically shaping the emergence of the biological subject in the twilight years of the nineteenth century. At the forefront of this encounter with intoxicants, testifying with his own body and mind, was the psychiatrist and acolyte of Wundtian psychology, Emil Kraepelin.

✳ 9 ✳
The Life and Times of Emil Kraepelin

DRUGS, BODIES, AND MINDS

I Dream of Morphine

Years before he moved to Leipzig and joined Wundt's lab, Emil Kraepelin had already developed a sensitive understanding of intoxicants. It happened late one evening while Kraepelin was a student researcher of Franz von Rinecker's in Würzburg. The night in question was sometime in the winter of 1877/1878. At the time, Kraepelin, then twenty-two, had been anxiously preparing for his qualifying *Staatsexamen*, to be completed that summer (Kraepelin 1983, 9). Busy practicing sphygmographic pulse measurements, Kraepelin often found himself working "deep into the night, and thus lost sleep at times" (Kraepelin 1983, 9). It is easy to imagine the collective impact of youthful insecurity, academic pressure, and sleepless nights. This was also taking place in the 1870s—a bold new world of intoxicating remedies. Thus, it felt unproblematic, even reasonable, when, one night, Kraepelin gave himself "a morphine injection of 0.02 gr," around 20 miligrams, "in order to be fresh the next morning" (Kraepelin 1983, 9). Having prepped a syringe with what he believed to be the proper dose, Kraepelin gave himself the morphine injection, presumably subcutaneously. What followed was not a night of much-needed rest, of respite—but a long sleepless night of clammy nausea, retching, and vomiting (Kraepelin 1983, 9). The pièce de résistance was the very first of the many migraines that would plague Kraepelin's life (Krapelin 1983, 9).

The silver lining, as Kraepelin acknowledged in hindsight, was that this visceral experience shielded him from coming to use morphine as a sleep aid, a judgment already passed on some of his colleagues (Kraepelin 1983, 9). This was Kraepelin's personal introduction to the new world of intoxication, a relationship that would in many ways shape the course of his professional life. Kraepelin would carry this experience into Wundt's lab, where he, in the early 1880s, would conduct the first experiments on the psychological effects of various intoxicants. It was here that Kraepelin planted the seeds of a parallelistic nosological psychiatry premised on experimental psychology, the roots of which would tightly coil around intoxicants and intoxicated ways of knowing.

Before Wundt's Lab

So, who was Emil Kraepelin and how did he come to study the effects of intoxicants on the conscious process in Wundt's lab? We can begin easily enough. Born in 1856 in Neustrelitz, Emil Kraepelin was the youngest child of Karl Kraepelin, an actor who filled Kraepelin's childhood with literature, music, and theater (Kraepelin 1983, 1). Kraepelin relays in his memoirs that it was through his older brother Karl, eight years Emil's senior, that he initially became familiar with the work of Wilhelm Wundt, whom his brother reportedly held in high esteem (Kraepelin 1983, 3).[1] A second influence on the course of Kraepelin's early life, and at least comparable to his brother's, was that of his father's friend, the doctor Louis Krueger, who allowed the young Emil to visit on hospital rounds and from whose library Kraepelin borrowed Wundt's *Vorlesungen über die Menschen- und Thier-Seele* (Kraepelin 1983, 3). On the guidance of Krueger, Kraepelin began studying medicine at Leipzig in 1874 with the hope of pursuing psychiatry.

At Leipzig, Kraepelin attended zoology lectures with Rudolf Leuckart, as well as courses on chemistry and dissection with Gustav Wiedemann (Kraepelin 1983, 3-4). During the summer of 1875 Kraepelin continued his studies in Würzburg. Here, Kraepelin studied chemistry with Johannes Wislicenus and anatomy with Kölliker, and sat in on the psychiatry training courses led by Franz von Rinecker, the pharmacologist (Kraepelin 1983, 4). Kraepelin also became well acquainted with the anatomist Hans Gierke, who allegedly informed Kraepelin of the recent publication of Wundt's *Physiologische Psychologie* as well as Wundt's plans to leave Zürich for Leipzig. Kraepelin claims to have then decided to return to Leipzig and seize on the chance to study under one of his teenage heroes, the wellspring of his initial interest in psychology (Kraepeli 1983, 5). Though he was able to meet Wundt around Easter of 1877, Kraepelin returned to Würzburg to accept the position of a psychiatry assistant under Rinecker—where Kraepelin had his ill-fated experience with morphine injection (Kraepelin 1983, 6). Here, his role as a student doctor had him treating all manner of cases, in addition to updating Rinecker on the status of psychiatric patients and measuring the contours of their skulls (Kraepelin 1983, 7). Kraepelin recalls this being a particularly challenging period in his young life. The psychiatry hospital was poorly staffed and ill-equipped, and the inability to monitor patients at risk of suicide led to liberal use of chloral hydrate (Kraepelin 1983, 7).

According to Kraepelin's memoirs, it was while he was working in the psychiatry hospital in Würzburg that he wrote "Ueber den Einfluss acuter Krankheiten auf die Entstehung von Geisteskrankheiten," though some secondary literature suggests that this was written while Kraepelin was in Leipzig (Kraepelin 1983, 9; Allik and Tammiksaar 2016, 319). This paper is often referred to as "prize-winning," although Kraepelin recalls that technically no one else competed (Kraepelin 1983, 9). This publication nevertheless provides insight into Kraepelin's identity as a student psychiatrist/psychologist, prior to completing his doctorate in München and working in Wundt's lab. Kraepelin himself described "Ueber den Einfluss" as a review of the extant literature on the subject, unified through the lens of Wundt's "Mechanik der Nerven und Nervenzentren" (Kraepelin 1983, 9). What "Ueber den Einfluss" demonstrates is the consistency of Kraepelin's early admiration and awareness of Wundt's work, here even extending it into new territories of study.

After completing his *Staatsexamen* in July 1878, Kraepelin took on an assistant position at an asylum in München under the supervision of Bernhard von Gudden, where he completed his doctorate. The period between 1878 and 1881 was formative for Kraepelin, in large part because it helped define Kraepelin's perspective as a psychiatrist. Gudden was a brain psychiatrist, an empirical neomechanist paradigmatically associated with Meynert—though Gudden's careful approach to brain anatomy led him to criticize Meynert's audacity (Kraepelin 1983, 11–17). As discussed, it had been Gudden who pioneered the method for using a microtome in brain sectioning. By that time, Kraepelin had already become a personal disciple of Wundt—very likely privately adhering to a doctrine of psychological parallelism rather than material reductionism. Kraepelin saw neomechanistic physicalism as a reaction to the shortcomings of *Naturphilosophie*, with one extreme being exchanged for the other (Engstrom 2016). Thus, Kraepelin was weary of Gudden's understanding of mental illness, which saw the advancement of the medical understanding of mental illness as reliant on further developments in neurophysiology and cortical anatomy (Kraepelin 1983, 11–17; Steinberg and Himmerich 2013, 249). If he wanted to witness a real shift in the understanding of the nature of mental illness and help patients rather than merely sedate them, Kraepelin understood that he would need to delve into the psychomental phenomena that comprised the experience of mental illness itself (Steinberg and Himmerich 2013, 249). The pursuit of this goal finally led Kraepelin back to where he had long intended to be: in Leipzig, working alongside Wilhelm Wundt.

Leipzig, Flechsig, and Wundt

Kraepelin made his way to Leipzig in February 1882, where he made short work of pursuing employment with Wundt (Steinberg and Himmerich 2013, 249). Unfortunately, Wundt was unable to provide Kraepelin with a paying job. Instead, Wundt suggested that Kraepelin apply to be an assistant psychiatric physician in the new university hospital with Paul Flechsig. What followed would mark the beginning of a lifelong personal and professional dispute, with Flechsig on one side and Kraepelin/Wundt on the other (Hlade 2021, 5). Flechsig accepted Kraepelin's application, but their relationship seems to have almost immediately run awry. By Kraepelin's account, their relationship grew strained when Kraepelin refused Flechsig's offer to assist in his habilitation—however, this story is not supported by any external evidence (Steinberg and Himmerich 2013, 249). Whether such an interaction ever took place, it is clear that Kraepelin had little interest in Flechsig beyond the financial support the position provided. From their letters, we learn that Kraepelin submitted an informal habilitation project proposal to Wundt directly after being accepted for the position in the university hospital (Steinberg and Himmerich 2013, 249). While certainly far from honorable, Kraepelin's conduct here is at the very least understandable. Flechsig was yet another brain psychiatrist, a neurophysiologist who saw psychological phenomena reflected in the sheen of a neural cross section. In short, Flechsig was precisely what Kraepelin hoped to escape in finally joining Wundt in Leipzig. It is likely that Flechsig could have come to appreciate that Kraepelin ultimately had a greater interest in experimental psychology, as long as he fulfilled his duties in the hospital effectively. Sadly, that never happened. It appears that, as Kraepelin increasingly focused on his research in Wundt's lab, tensions between Kraepelin's and Flechsig's competing interests resulted in Kraepelin's dishonorable dismissal in mid-June 1882 (Steinberg and Himmerich 2013, 249–50; Hlade 2021, 5). Flechsig naturally felt taken advantage of, a factor that complicated the grounds of Kraepelin's dismissal, but it seems to be the case that Kraepelin did neglect his duties and instead focused his time on his psychological research with Wundt (Steinberg and Himmerich 2013, 249–50; Hlade 2021, 5). This all provides context for the purposes of our story, which is concerned above all with the nature of precisely the psychological research Kraepelin was doing instead of fulfilling his employment obligations.

Intoxicated Testimonies I: Kraepelin's "Proposal," Exner, Dietl, and Vintschgau

In a series of enthusiastic letters to Wundt leading up to Kraepelin's move to Leipzig, Kraepelin wrote to a hero of his teenage years about his hopes and aspiration for the future. Perhaps most crucially, Kraepelin wrote to Wundt about his plans to habilitate in Leipzig, putting forward the following informal proposal:

> Alone—a kingdom for a subject! In order to have a topic which would proceed at least with a certain probability to any results and which one could limit arbitrarily, I would think, for example, to examine in the manner of Dietl and Vintschgau some [2/3] of the better known Nervina (chloral hydrate, bromine potassium, hashish, amyl nitrite, strychnine, etc.) and their influence on the duration of the reaction time. This would not be exactly original, but it would be appropriate for the purpose. My further plans to investigate fever patients, nervous persons, asthenics, alcoholics, etc., in order to come closer to the essence of the "psychopathic disposition," would, of course, take years or decades and thus be much too far-reaching for a mere habilitation thesis ad hoc. (Kraepelin 1881a, 1–2)[2]

With his typical reticence, Kraepelin describes this proposal as "not exactly original" (Kraepelin 1881a, 2). His claim to unoriginality likely refers to Kraepelin's awareness of Sigmund Exner's 1873 "Experimentelle Untersuchung der einfachsten psychischen Processe, Erster Abhandlung" and Michael Dietl and Maximilian von Vintschgau's quite extensive 1877 "Das Verhalten der physiologischen Reactionszeit unter dem Einfluss von Morphium, Caffee und Wein."[3] There was also his focus on psychological reaction time, already Wundt's foundational research program. The question then becomes: How were these previous studies conducted, and how did Kraepelin's proposed project differ?

As previously discussed, Exner's 1873 "Experimentelle Untersuchung" combined Bessel's personal equation with Helmholtz and Baxt's observations on the influence of temperature on propagation speed to introduce the physiological concept of reaction time. Helmholtz and Baxt's findings further led Exner to propose "that the nerve conduction velocity is a very variable quantity, and that its value may depend on many circumstances other than temperature" (Exner 1873, 601). A factor that Exner found to consistently affect reaction time was tiredness or fatigue (Exner 1873, 627). Via this association between tiredness and reaction time, Exner arrived at the idea of using narcotics and stimulants in some of his experiments, on the premise that they might function as modifiers of the subject's state of exhaustion (Exner 1873, 627). Much to Exner's astonishment, though, neither three cups of strong tea nor a subcutaneous injection of morphine had any meaningful effect on the subject's physiological reaction time (Exner 1873, 627-28). Two bottles of Rhine wine did have a measurable effect, making the subject's scores more erratic and inconsistent (Exner 1873, 628). As far as an investigation into the effects of intoxicants was concerned, this was the extent of Exner's consideration: the inclusion of tea, morphine, and wine were only secondary, founded on an association that Exner's own finding appeared to dispute. The readings from the morphine and tea trials were not even included in the data tables. Further still, intoxicants were only employed in a limited battery of eye-to-hand reaction time tests, and not applied in tests on different stimulus points, such as reaction time from hand to hand, right hand to right foot, and so forth.

In 1877, Michael Dietl and the Austrian physiologist Maximilian von Vintschgau took it upon themselves to further Exner's limited drug experiments at the physiological lab in Innsbruck. The project was an entailment of their own interest as physiologists in the temporal relationship between nervous activity and muscle function. Vintschgau had been a protégé of Ernst Brücke, a foundational neomechanist, and had served as his assistant in Vienna's physiological institute from 1856 to 1857, where Exner had also studied in 1865. Dietl and Vintschgau were very upfront in identifying their research as an effort to pursue an otherwise incomplete component of Exner's research program, though they made a series of noteworthy methodological changes. They substituted tea for black coffee, a somewhat personal choice (Dietl and Vintschgau 1877, 316). Dietl and Vintschgau further decided to limit their experiments to the reaction time of a tactile sensation on the inside of the middle finger, as opposed to the broad range of stimulus points examined by Exner (Dietl and Vintschgau 1877, 318). The rationale for this was greater experimental control, and because a diverse set of tests was not needed to establish if there was a quantitative difference when morphine, wine, or coffee were introduced (Dietl and Vintschgau 1877, 318-19). Dietl and Vintschgau also differed from Exner in the design of their measurement apparatus. The apparatus used by Dietl and Vintschgau was one that Vintschgau had used in an earlier 1875 study on physiological reaction time with Hönigschmied. Here, when the subject's hand was touched with a brush device, their reaction closed a circuit that caused a fixed pen to leave a mark on a paper covered cylinder, rotated at a set speed by a Helmholtz electromotor (Vintschgau and Hönigschmied 1875, 2-8; Dietl and Vintschgau 1877, 319-20).

For the wine trials, Dietl and Vintschgau determined that champagne was the most agreeable wine, with the most excitatory effects; therefore they procured a stash of Roederer champagne, with one attempt made with Tyrol wine (Dietl and Vintschgau 1877, 368). Much like Exner, Dietl and Vintschgau found that wine greatly affected the consistency of their results (Dietl and Vintschgau 1877, 376–77), though they themselves admitted that "if we take a closer look at this wine experiment, we cannot really find a certain result" (Dietl and Vintschgau 1877, 381). In this sense, their wine trials were generally consistent with Exner's results. They had slightly better results with morphine and coffee. By doing series of tests back-to-back, Dietl and Vintschgau observed that reaction time consistently reached peak lengths twenty minutes following a morphine injection, only for the reaction time to shorten again after forty to sixty minutes (Dietl and Vintschgau 1877, 357). As for the coffee trials, they found "that the reaction time can be remarkably shortened by coffee" (Dietl and Vintschgau 1877, 368). Dietl and Vintschgau concluded with the suggestion that the divergence between their findings and Exner's could be explained by methodological differences.

Exner, as well as Dietl and Vintschgau, conducted studies on the relationship between reaction time and intoxication. But the nature of reaction time itself had been the landscape across which the battle for the legitimacy of Wundt's experimental psychology had been fought and, in the eyes of some, won. Exner, Dietl, and Vintschgau comparatively understood reaction time as a strictly physiological phenomenon. We might look to Henning Schmidgen's framing of the history of reaction time as a history of machines. As Schmidgen argues, the "Donders machine" emerges as a "temporally limited" installation composed of "partial objects derived from the experimenter and the experimental subject," "a dynamic meshing of flows of materiality and semiotic stuff" (Schmidgen 2005, 211, 217). These temporalized chimeras become points of spatialized convalescence, pouring out all manner of heretofore unseen measurements, modes, and ways of being a self. Coffee, wine, and tea were components of the apparatus, intended to modify a chimera subject's state of tiredness or alertness. This put their research at odds with Wundtian psychology. In his research on reaction time, Wundt had uncovered a founding research program, one that justified the realization of experimental psychology as a distinct branch of science. Yet Wundt's successes raised further theoretical difficulties. At the very core of Wundt's conception of psychology was the principle that a certain share of mental life was psychical and could not be reduced to the physiology of the brain. Wundt's colleagues and competitors, among them Flechsig, Meynert, and Wernicke, had in many ways failed to make good on the promises of neural reductionism in their approach to understanding mental states.

Wundt hadn't sought to overturn the initial premises of the neomechanists, but rather to broaden the scope of scientific research in order to account for the shortcomings of their conclusions. This raises a burning question. What experimental evidence was there to account for these two seemingly contradictory statements, the body is a physiochemical entity, while mental states are not reducible to neural states? Was it merely a speculative assertion, a secularization of Fechner's psychospiritual? It is very possible this was not at all an issue for Wundt himself. Nevertheless, Kraepelin's proposal implicitly, and perhaps explicitly, provided a bridge, an experimental subject through which to realize the nature of this connection and, through it, understand the nature of the body and mind. As he expressed in his letter, Kraepelin hoped that psychological research on states of intoxication would glean valuable insights about psychopathology, and thereby the ultimate nature of the psychical (Kraepelin 1881a, 1–2). Exner, Dietl, and Vintschgau's research meanwhile had been a natural entailment of Helmholtz's research program centered on the nerve impulse and the physiological nature of reaction time, an elucidation of the neomechanistic conception of the body. Kraepelin's proposed project shifted the emphasis from the physiology of intoxicated reaction time to the psychomental, *embodied*, phenomena of intoxication.

Intoxicated Testimonies II: Kraepelin's Early Reaction Time Trials

Published in Wundt's *Philosophische Studien* in 1883, the first section of Kraepelin's "Ueber die Einwirkung einiger medicamentöser Stoffe auf die Dauer einfacher psychicher Vorgänge" discussed Kraepelin's reaction time experiments with alcohol, morphine, chloral, tea, amyl nitrate, ether, and chloroform, though only the trials on amyl nitrate, ether, and chloroform were sufficiently conclusive to merit inclusion at the time. "Ueber die Einwirkung" was preceded in 1881/1882 by a cursory outline of some of Kraepelin's earliest results, published under the titles "Ueber psychische Zeitmessung" and "Ueber die Dauer einfacher psychischer Vorgänge." Though comparably light on details, "Ueber psychische Zeitmessung" serves as something of an introduction to the sprawling, multidecades-long project that Kraepelin had just begun. Here, for the first time, Kraepelin introduces the principle that would guide all of his work with intoxicants: intoxication was artificial mental illness, or—in twentieth-century parlance—a model psychosis (Kraepelin 1881c; Ban 2006; Müller et al. 2006, 135). It is this concept of intoxication as model psychosis that would ultimately make possible not only the unity of experimental psychology with psychiatry, but also a novel conception of the relationship between the body and mind.

The characterization of intoxication as some form of madness has a long history. In all likelihood, Kraepelin's reception of this concept can be attributed to Paul Julius Möbius, Kraepelin's colleague in Leipzig. Though it appears that he spoke of it sooner, it was no later than the 1882 publication of *Die Nervosität* that Möbius advanced the argument that "acute alcohol intoxication or intoxication is a temporary insanity," and that this was true of other intoxicants as well (Möbius 1882, 103). Möbius would further influence Kraepelin's dual categories of endogenous and exogenous psychoses (such as intoxication), with Möbius drawing on Valentin Magnan and the French tradition of degeneration in his identification of endogenous psychoses in 1875, tentatively asserting a duality of exogenous and endogenous psychoses in 1886, and formalizing the distinction in 1892 (Beer 1996; Bürgy 2008; Steinberg and Müller 2005). Methodologically, Kraepelin's reaction time trials built on the experimental design more or less standardized in Wundt's psychology laboratory, and thus the apparatus fundamentally differed from those used by Exner or Vintschgau and Dietl. The experiments were conducted with the assistance of the newest version of Hipp's chronoscope, a favorite of Wundt's (Kraepelin 1883b,

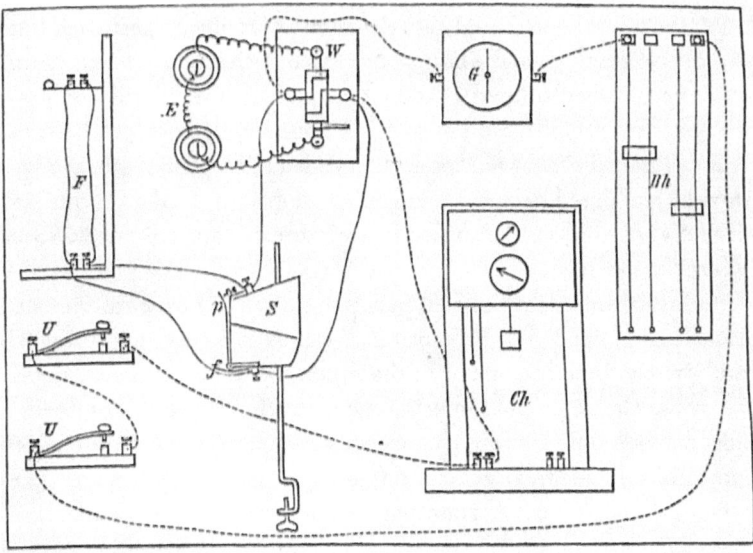

FIGURE 5. Diagram of Kraepelin's apparatus (Kraepelin 1883b, 421). Courtesy of the National Library of Israel.

419). The newer design distinguished itself not only by its more robust construction but "by applying the principle of shunt closure, it is also possible to produce those combinations of experiments with a single galvanic current which previously required two separate currents" (Kraepelin 1883b, 420). This made the following experiment design possible:

> In our experiments, the setup was such that a current was opened both by the stimulus and by the reaction. For this purpose, a commutator W was connected to the current coming from the battery E, from which two paths were open, namely through the [. . .] bell S or the drop apparatus F, on the one hand, and through the galvanoscope G, the rheochord Rh, the interrupters U, and finally the chronoscope Ch, on the other hand. (Kraepelin 1883b, 420)

Using the apparatus, Kraepelin would test not only the simple reaction time of subjects, but also their discrimination reaction and choice reaction times. As the stimulus, Kraepelin elected for vocal vowel sounds, largely because of the relative ease and flexibility of this method (Kraepelin 1883b, 421). While not particularly groundbreaking, the ability to control for the timing of the stimulus within the experimental apparatus itself added a great deal of credence to Kraepelin's experiments, as well as reflecting the strength of apparatus-based experimentation in Wundt's laboratory in the years following its formation.

For the trials themselves, Kraepelin relays that subjects abstained from tea, coffee, or alcohol in the hours prior to a test, though how many hours is not specifically identified. In order to administer the chloroform, Kraepelin attached an inhalation mask to an iron frame, into which he could attach a chloroform-soaked sponge as needed (Kraepelin 1883b, 430–31). Amyl nitrate could have left a lingering odor, so instead a glass funnel containing a handkerchief soaked in a few drops of amyl nitrate was used (Kraepelin 1883b, 431).

Both versions of this inhalation apparatus raised novel experimental obstacles that needed to be controlled against in Kraepelin's study. Beyond the confounding effects of the experimental setting itself, merely the awkwardness entailed in using the inhalation rig could foreseeably affect reaction time, on top of concerns surrounding the extent to which a powerful odor—intoxicating or not—might further alter reaction times (Kraepelin 1883b, 431). Accounting for the former concern was accomplished by running the pre-intoxication control tests with the mask in position. Out of an abundance of caution for the latter objection, Kraepelin ran a set of tests "with deep inhalation of rose petal water," which

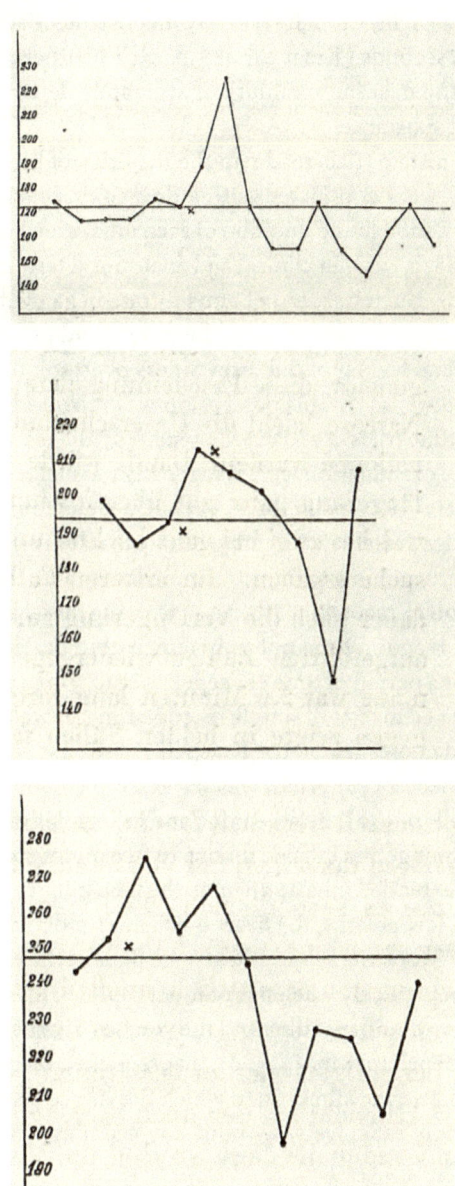

FIGURE 6. Amyl nitrate's effect on simple reaction (*top left*), discrimination reaction (*top right*), and choice reaction (*bottom*) (Kraepelin 1883b, 435, 437, 439). Courtesy of the National Library of Israel.

did result in a slight, though relatively inconsequential, delay in reaction time of 0.015 seconds (Kraepelin 1883b, 431). Neither adjustments could account for the near impossibility of accurately determining the dosage of an inhaled substance.

The amyl nitrate trials made up the majority of the inhalation experiments. They were conducted with a dosage that ranged between four and ten drops, though the duration of each inhalation varied considerably (Kraepelin 1883b, 433). In clinical and anatomical terminology, Kraepelin described how the inhalation of amyl nitrate began with an irritation in the nose, followed by an increased heart rate and pulsation in the head, before a general feeling of dull drowsiness overtook the subject's senses (Kraepelin 1883b, 433). This perception was mirrored in the findings of the reaction time trials. As Kraepelin remarks, "the results of the objective psychometric examination were remarkably parallel to this change of the subjective state" (Kraepelin 1883b, 433). Kraepelin found that the experience of intoxication elicited by amyl nitrate markedly slowed the subject's reaction time, beginning to improve immediately after the cessation of further inhalation (Kraepelin 1883b, 435). After the initial lengthening phase, the subject's reaction time not only returned to baseline but very often shortened for a period of time (Kraepelin 1883b, 437–439). These findings were reproduced in tests of simple reaction time, as well as discrimination reactions and choice reactions (Kraepelin 1883b, 437–439).

Ether, Kraepelin reported, was far more pleasant than amyl nitrate (Kraepelin 1883b, 442). Between one and two grams of diethyl ether was applied to a sponge and inhaled for one to two minutes on average, though one attempt extended inhalation up to thirteen minutes (Kraepelin 1883b, 442). After a few breaths, the heart quickened and a mounting feeling of sleepiness overtook the subject. Kraepelin likened the experience to when one falls asleep quickly: the environment rapidly fades away, sounds become dampened and metallic, and the course of mental activities comes to a halt (Kraepelin 1883b, 442). All that remained of any serious content was the perception of stimulus, to which the subject, Kraepelin reported, responded only reflexively, without any seeming act of will or intent (Kraepelin 1883b, 442).

With such an account of the effects, it is little surprise that Kraepelin consistently found that ether lengthened reaction times, though, much like with amyl nitrate, there was a consistent shortening of reaction times in the second phase that followed the initial lengthening phase. Here, Kraepelin further differentiated between *leichte* and *tiefere Narcose* (light and deep narcosis), with deep narcosis having a markedly greater effect both in lengthening and in shortening (Kraepelin 1883b, 445).

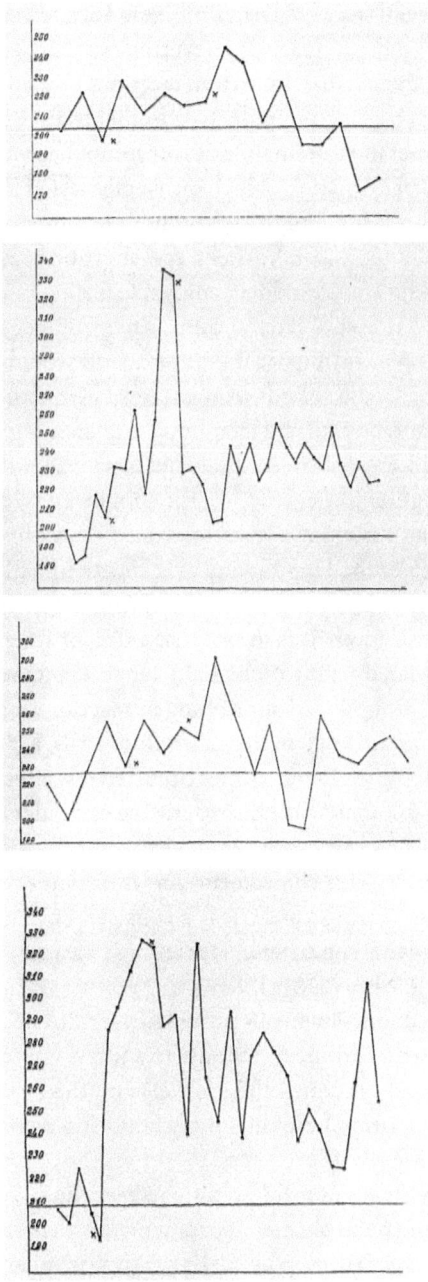

FIGURE 7. Effects of ether on simple reaction (*top*), discrimination reaction (*middle left*), choice reaction (deep narcosis, *middle right*), and choice reaction (light narcosis, *bottom*) (Kraepelin 1883b, 443, 447, 450, 449). Courtesy of the National Library of Israel.

Discrimination reactions were tested, albeit in such few numbers so as to be inconclusive—though Kraepelin nevertheless shared his findings that ether further lengthened discrimination reactions (Kraepelin 1883b, 446).

The chloroform trials were similar to the ether trials. Regarding dosage, Kraepelin found that the amount of chloroform applied was less important than the duration of inhalation, which on average lasted two to three minutes with some only lasting for thirty seconds and others going up to nine minutes (Kraepelin 1883b, 452). After a few short breaths, the nose grew irritated, and shortly after, a sudden, enjoyable feeling of fatigue overtook the subject, coupled with a general calm. Kraepelin goes on to describe how the subject's perceptions of the outer world would then grow ever more indistinct, with the sounds of apparatus beginning to sound tinny, muffled, and distant, punctuated later but a sudden return to wakefulness (Kraepelin 1883b, 452). Much like the amyl nitrate and ether trials, there was an initial lengthening in reaction time followed by a second phase, characterized by a shortening of reaction time (Kraepelin 1883b, 453–54).

The magnitude of the lengthening and shortening phases relative to baseline was found to be strongly associated with the intensity of the subject's state of intoxication. This meant that a deeper state of narcosis was consistently correlated with a more acute lengthening of reaction times, as well as a more pronounced shortening in reaction time in the second phase (Kraepelin 1883b, 454). In the chloroform trials, Kraepelin observed that peak lengthening in reaction times occurred just after the cessation of inhalation, corresponding with the experience of suddenly returning to a state of alertness (Kraepelin 1883b, 455). Very few differentiation or choice reactions were tested with chloroform. As with simple reactions, the intensity of the state of intoxication was a consistent predictor of reaction time, both in the lengthening and shortening phases of the chloroform trials (Kraepelin 1883b, 456–457).

For the first section, Kraepelin provided very little by way of closing remarks, though what little he did choose to address was telling. Naturally, Kraepelin addressed his compelling findings on the two distinct phases formed in the association between intoxication and reaction time (Kraepelin 1883b, 461). He also felt the need to address the dynamic phenomena of individual differences in reaction time (Kraepelin 1883b, 462). But his most crucial point spoke to the importance of further research into this subject matter. First of all, even between relatively similar intoxicants like chloroform, amyl nitrate, and ether, Kraepelin found that there were measurable differences in their effects, validating further questions about the ever-growing host of intoxicants in nineteenth-century society (Kraepelin

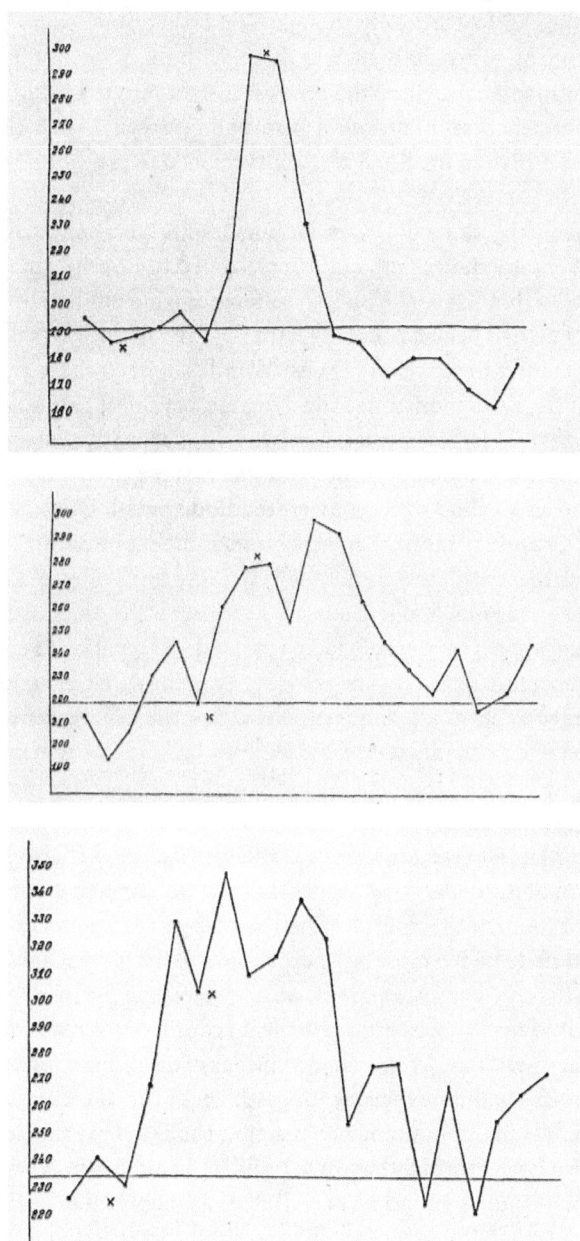

FIGURE 8. Chloroform's effect on simple reaction (*left*), discrimination reaction (*top*), and choice reaction (*right*) (Kraepelin 1883b, 454, 456, 459). Courtesy of the National Library of Israel.

1883b, 461–62). Most crucially, though, Kraepelin expressed his belief that experiments on intoxicants such as these, with the advent of further isolated experiments, provided the greatest insight into the ultimate nature of the inner structures of the mind (Kraepelin 1883b, 461–62). This insight would prove pivotal for the next twenty years of Kraepelin's research, if not his entire career.

The second section of "Ueber die Einwirkung" focused exclusively on alcohol. While Exner, Dietl, and Vintschgau had not researched any of the inhalants that Kraepelin had, the effects of wine on the physiology of reaction time had been a primary feature of their trials. This was Kraepelin's opportunity to strike out against Wundt's detractors by producing a study that clearly demonstrated the dynamic, psychical nature of reaction time. In light of this, Kraepelin is quick to critique the value of these earlier studies. One need look no further than the first page.

Some of Kraepelin's criticisms were methodological: Dietl and Vintschgau didn't conduct more than seven experiments between them, and the arbitrariness with which they selected one form of alcohol over another raised reasonable questions about whether the alcohol content of the chosen wines was the sole factor at hand (Kraepelin 1883c, 573–74). Given Kraepelin's strictures surrounding even the smell of various inhalants, there were grounds to question whether the effervescence of Dietl and Vintschgau's champagne factored into their measurements, not to mention the variable chemical contents of different wines. The principle ground of Kraepelin's rebuke was that Dietl and Vintschgau had relied exclusively on simple reaction time (Kraepelin 1883c, 574). This was arguably a reflection of the competing concepts of reaction time that pervaded the different studies. Both Exner and Dietl-Vintschgau had understood simple reaction time to be the most basic expression of the physiological process common to more complex actions, such as recording the time at which an object passed over a telescope's reticule. In coordinating his experimental methodology with the Wundtian definitions of discrimination and choice reactions, Kraepelin's experiments established the veracity of psychological, rather than physiological, reaction time. Kraepelin, like Wundt, would have characterized the initial problem in astronomy that gave rise to Bessel's personal equation as a choice reaction rather than a simple reaction—exemplifying Kraepelin's assertion that reaction time needed to be treated as a psychically dynamic phenomenon. Though Kraepelin's trials on amyl nitrate, chloroform, and ether themselves contained only a limited number of more complex reactions, the sum of Kraepelin's tests involving choice or discrimination reactions well exceeded the total number of drug trials of any kind conducted by Exner or Dietl and Vintschgau.

Such remarks were far from sectarian polemics. Kraepelin took his own criticisms of Exner's and Dietl-Vintschgau's studies to heart and incorporated these criticisms into the design of his own experiment. To account for the ambiguous nature of drinks like wine and beer, subjects were given a solution of pure alcohol and water, with nothing but a small amount of raspberry syrup to cut the flavor (Kraepelin 1883c, 574-74). Rather than limit his study to simple reaction time, Kraepelin incorporated more complex reactions, as he had in the first section, and ran these trials in greater numbers.

Dosages of pure alcohol ranged between 7.5 and 60 grams, with trials involving smaller doses lasting approximately forty to fifty minutes and larger doses up to an hour and a half (Kraepelin 1883c, 575).[4] Subjects who took smaller doses, under thirty grams, presented as only slightly excited (Kraepelin 1883c, 576). When larger doses of up to sixty grams were taken, Kraepelin described a pleasant feeling of intoxication that arose after six to eight minutes, a slight dizziness that kept the sense of personality intact without succumbing to serious drunkenness (Kraepelin 1883c, 576-77). He goes on to describe how ideas flew by with great clarity and a colorful liveliness, while reactions appeared to occur faster and with a pronounced decisiveness (Kraepelin 1883c, 577). To control against the possibility that a full stomach might affect "psychical times," Kraepelin ran control trials using carbonated water instead of alcohol (Kraepelin 1883c, 575-76). These trials led to a slight lengthening of psychical times followed later by a slight shortening, leaving Kraepelin to speculate about whether the carbonated water had some excitatory effect or whether it was coincidental (Kraepelin 1883c, 576). Whatever the case, the effect was so slight that any sufficiently strong results uncovered in the alcohol trials were irreducible to a full stomach (Kraepelin 1883c, 576).

In simple reaction trials, Kraepelin found that low alcohol doses were immediately followed by a shortening in reaction, after which there was a marked, albeit erratic, lengthening in reaction time (Kraepelin 1883c, 579). The higher doses had a quite different effect: not only was the fluctuating lengthening phase that followed the initial shortening phase at lower doses more erratic and variable, but the initial shortening in reaction time at times disappeared altogether (Kraepelin 1883c, 581-82). In all cases, "the absolute magnitude of the lengthening is generally consistently smaller than that of the shortening" (Kraepelin 1883c, 582). Though the results were fairly consistent across subjects, Kraepelin recognized that the magnitude of the shifts varied significantly between individuals, which he attributed both to each individual's natural constitution as well as to personal familiarity with alcohol (Kraepelin 1883b, 579, 582).

In trials on discrimination reaction, low-dose alcohol produced a relatively consistent shortening in psychical time overall (Kraepelin 1883c, 584–85). It was at the higher doses that the pattern most closely mirrored those seen in simple reaction time, where an initial shortening in reaction time was followed by an erratic lengthening phase (Kraepelin 1883c, 588). In the choice reaction trials, "the paradigmatic picture of the alcohol effect was more unclear due to multiple fluctuations of the observed values," although the familiar trends were still visible (Kraepelin 1883c, 590). Kraepelin attributed this, at least in part, to the characteristic of all choice reactions to be slower and more variable (Kraepelin 1883c, 590). There remained, nevertheless, an average shortening phase, followed by a lengthening phase (Kraepelin 1883c, 590–592).

It is clear that Kraepelin led the first serious research project into the effects of intoxicants on the mind or the body. But how does Kraepelin himself conclude his initial project? How did he understand it? In the conclusion of the second section, Kraepelin offers insight into how he conceives of the nature of the effects of these different intoxicants on reaction time. Regarding differentiation and choice reactions, Kraepelin argued that the common element "is, as already indicated above, apperceptive activity, which in the latter case represents itself as the apprehension of the external impression by the attention, in the former case as the comparison of the same with the ready held memory image" (Kraepelin 1883c, 599). It is this "activity that undergoes a moderate acceleration in the first stage of alcohol action, but a pronounced deceleration in the second" (Kraepelin 1883c, 599). Alcohol's effect was an initial quickening, followed by a slowing, of the power of apperception, which ultimately suppressed the subject's abilities through an "impediment to the perception of external impressions that regularly becomes noticeable with the use of larger doses" (Kraepelin 1883c, 600). This manifests itself as a difficulty in apprehending sense impressions, as well as discerning between them.

This slowing of apperception accounts for the sensory effects of acute alcohol intoxication, but it does not account for the full array of symptoms. For example, Kraepelin found that alcohol led his subjects to anticipate the stimulus, as well as respond to the apparatus more forcefully. To this latter point, Kraepelin identified a disassociation of the power of will from the apperceptional faculty (Kraepelin 1883c, 600–601).[5] Apperception, Kraepelin pointed out, is the sole avenue by which the subjects had an awareness of their setting, and so these anticipatory responses seen under the effect of alcohol are described not as willful acts but actually as demonstrative of a loss of free will (Kraepelin 1883c, 600–601). This slight lapse in free will emerges out of the perceptional ambiguity that

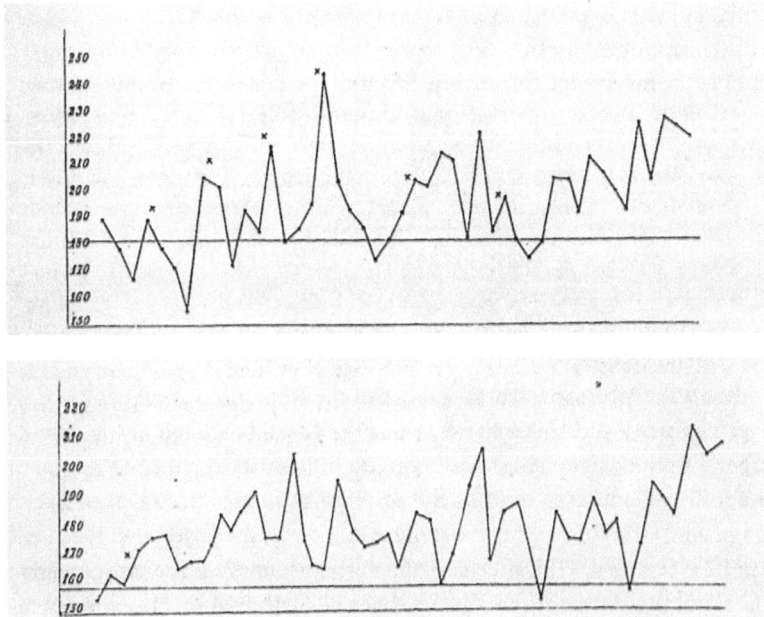

FIGURE 9. Alcohol's effect on simple reaction low dose (*top*) and high dose (*bottom*) (Kraepelin 1883c, 579, 581). Courtesy of the National Library of Israel.

arises from the "abolition of the ability to discriminate" (Kraepelin 1883c, 600–601).

The Experimental Frame and a Nosology of Intoxication

Employing psychological language to describe the effects of intoxicants in his study, Kraepelin clearly delineates the objectives of his research: this is an experimental rendering of the experience of embodiment, of the subject's existence in the milieu of the lifeworld. The scientificity of such a pursuit, a notion that underscored Wundt's entire research paradigm, relied on appeals to robust experimentalism. Kraepelin's earliest drug trials were in many cases exceptionally thorough, striving to outstrip the levels of experimental control seen in early efforts to measure physiological reaction. Where necessary, Kraepelin made every effort to control against confounding factors, from the possible influences of foul odors to the role of a full stomach. This was the experimental crucible that psychic phenomena would pass through in becoming constituted as scientifically valid.

Yet, in essence, Kraepelin's experiments were consistent with the radical empiricism that Wundt inherited from Fechner, and thereby the naturphilosophical, even Brunonian, empiricism of Fechner's time and place. This is supported by the identities of the subjects enrolled in the experiments themselves. Collectively, fifty-three trials were conducted using amyl nitrate, chloroform, and ether, with twenty-six trials being conducted with amyl nitrate, twelve with ether, and fifteen with chloroform. Well over half of those experiments, thirty-three attempts, were conducted on Kraepelin himself, while six and fourteen attempts were carried out on two willing subjects (Kraepelin 1883b, 430). For the alcohol experiments, nineteen of the forty-seven trials were conducted on Kraepelin himself (Kraepelin 1883b, 575). Kraepelin's experiments might more accurately be described as an experiential science of embodiment.

But unlike the experiential science of the Romantics, Kraepelin makes every effort to put the experiential in the distal language of the experimental frame. Kraepelin is forthcoming in identifying himself as one of the four test subjects, even as the primary test subject; however, in the language of the study, what most often remains is the abstracted *Reagirenden* (respondents). This is clear in the descriptions given for the subjective effects of the different intoxicants that are attributed to the experiences of all the test subjects, and yet upon review almost require that they be written by someone who had experienced these varied states of intoxication firsthand. The experiential components of his research are freely admitted, only to be obscured by the rhetorical expectations that underlie the scientific publication as a vehicle of epistemic permissibility. Where the findings of researchers working with vital substances had been eschewed by early neomechanistics on a methodological level, Kraepelin was working within an experimental methodology that could approach the phenomena of intoxication as an expression of embodiment, while still aspiring to scientificity.

The role of self-experimentation and experiential science has interesting implications with respect to Kraepelin's solution to the underlying problem of the uncertain dosage of an inhaled intoxicant, as well as the analogy Kraepelin draws between psychopathology and intoxication. As mentioned, an uncontrollable element in the initial experiments had been the ambiguity surrounding the dosage of an inhalant, and Kraepelin's solution had been to develop a method of categorizing *leichte* and *tiefere Narcose* (Kraepelin 1883b, 432). The categorization process necessarily relied on identification of a differentiating criteria of symptoms, which were then applied both to Kraepelin's own state of intoxication as well as those of the other participants. What Kraepelin had developed was a functional psychiatric nosology of intoxication, for the purposes of his experiment.

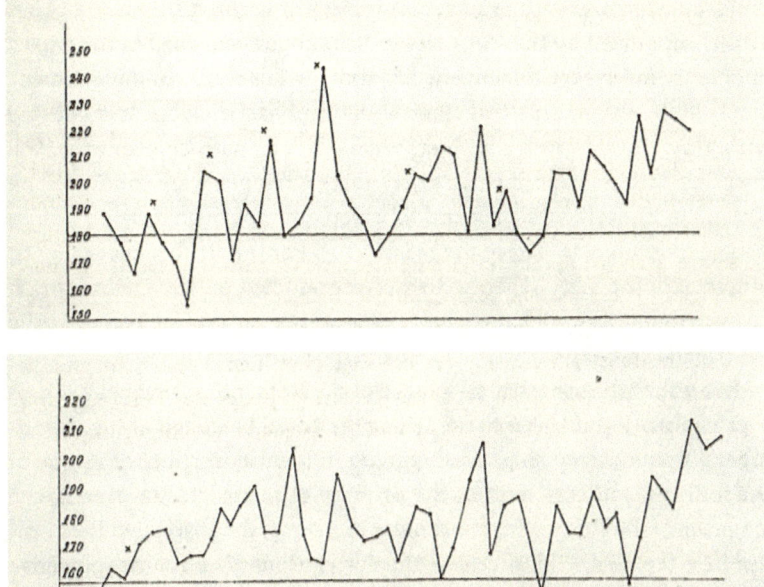

FIGURE 9. Alcohol's effect on simple reaction low dose (*top*) and high dose (*bottom*) (Kraepelin 1883c, 579, 581). Courtesy of the National Library of Israel.

arises from the "abolition of the ability to discriminate" (Kraepelin 1883c, 600–601).

The Experimental Frame and a Nosology of Intoxication

Employing psychological language to describe the effects of intoxicants in his study, Kraepelin clearly delineates the objectives of his research: this is an experimental rendering of the experience of embodiment, of the subject's existence in the milieu of the lifeworld. The scientificity of such a pursuit, a notion that underscored Wundt's entire research paradigm, relied on appeals to robust experimentalism. Kraepelin's earliest drug trials were in many cases exceptionally thorough, striving to outstrip the levels of experimental control seen in early efforts to measure physiological reaction. Where necessary, Kraepelin made every effort to control against confounding factors, from the possible influences of foul odors to the role of a full stomach. This was the experimental crucible that psychic phenomena would pass through in becoming constituted as scientifically valid.

Yet, in essence, Kraepelin's experiments were consistent with the radical empiricism that Wundt inherited from Fechner, and thereby the naturphilosophical, even Brunonian, empiricism of Fechner's time and place. This is supported by the identities of the subjects enrolled in the experiments themselves. Collectively, fifty-three trials were conducted using amyl nitrate, chloroform, and ether, with twenty-six trials being conducted with amyl nitrate, twelve with ether, and fifteen with chloroform. Well over half of those experiments, thirty-three attempts, were conducted on Kraepelin himself, while six and fourteen attempts were carried out on two willing subjects (Kraepelin 1883b, 430). For the alcohol experiments, nineteen of the forty-seven trials were conducted on Kraepelin himself (Kraepelin 1883b, 575). Kraepelin's experiments might more accurately be described as an experiential science of embodiment.

But unlike the experiential science of the Romantics, Kraepelin makes every effort to put the experiential in the distal language of the experimental frame. Kraepelin is forthcoming in identifying himself as one of the four test subjects, even as the primary test subject; however, in the language of the study, what most often remains is the abstracted *Reagirenden* (respondents). This is clear in the descriptions given for the subjective effects of the different intoxicants that are attributed to the experiences of all the test subjects, and yet upon review almost require that they be written by someone who had experienced these varied states of intoxication firsthand. The experiential components of his research are freely admitted, only to be obscured by the rhetorical expectations that underlie the scientific publication as a vehicle of epistemic permissibility. Where the findings of researchers working with vital substances had been eschewed by early neomechanistics on a methodological level, Kraepelin was working within an experimental methodology that could approach the phenomena of intoxication as an expression of embodiment, while still aspiring to scientificity.

The role of self-experimentation and experiential science has interesting implications with respect to Kraepelin's solution to the underlying problem of the uncertain dosage of an inhaled intoxicant, as well as the analogy Kraepelin draws between psychopathology and intoxication. As mentioned, an uncontrollable element in the initial experiments had been the ambiguity surrounding the dosage of an inhalant, and Kraepelin's solution had been to develop a method of categorizing *leichte* and *tiefere Narcose* (Kraepelin 1883b, 432). The categorization process necessarily relied on identification of a differentiating criteria of symptoms, which were then applied both to Kraepelin's own state of intoxication as well as those of the other participants. What Kraepelin had developed was a functional psychiatric nosology of intoxication, for the purposes of his experiment.

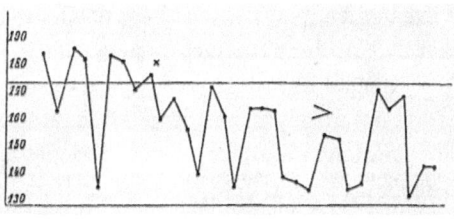

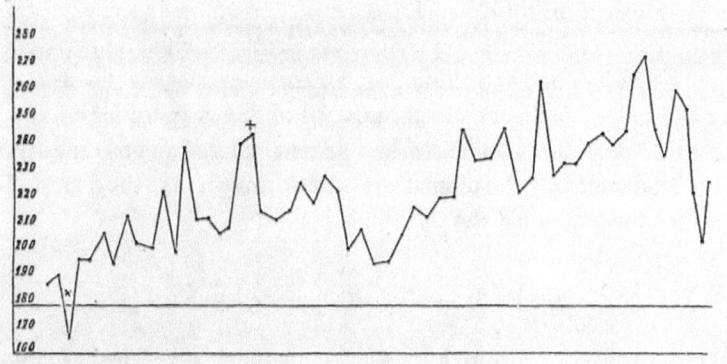

FIGURE 10. Alcohol's effect on discrimination reaction time low dose (*top*) and high dose (*bottom*) (Kraepelin 1883c, 585, 588). Courtesy of the National Library of Israel.

Though simple, Kraepelin's nosology of intoxication is a remarkable feature of his early study, in light of the overt significance Kraepelin ascribes to controlled experimentation with intoxicants for the understanding of the structure of mental states. Within this classification system, a given substance of intoxication serves as its own category, within which *leicht* and *tiefere* intoxication function as associated and yet distinct manifestations of mental disorder.

Experimentally, this variation is expressed through Kraepelin's organization of reaction times into representative tables and line graphs according to their classification as either *tiefere* or *leichte* intoxication. In the ether experiments, Kraepelin observed that under *leichte* intoxication, choice reaction times initially lengthened before shortening following the cessation of inhalation, while *tiefere* intoxication saw an immediate shortening (Kraepelin 1883b, 449–50). This structure further extends into the alcohol-centered section, despite the absence of the same dosing issues in the alcohol trials. Here, too, the results of trials with a diverse range in alcohol dosages are ultimately separated into high- and low-dose trials. Thus, *tiefere* and *leichte* function as separate diagnoses with their respective symptomatology, even when it is possible to approach degrees of intoxication as a continuum. Nosology emerges as an epistemic impulse from the very nascence of Kraepelin's study.

Intoxication here is not merely an analogy for mental states, but rather serves as a description of mental states as such. Kraepelin believes that substances of intoxication affect discrete psychological structures that may also be affected by a given mental illness (Kraepelin 1883b, 461–62). In this sense, intoxication, for Kraepelin, is a form of temporally and psychically localized mental disorder. Kraepelin's experiential reaction time trials then are to be understood as a self-imposed mental disorder, classified according to a reflective diagnostic process. All of this has profound implications for Kraepelin's parallelism, his understanding of mental illness, and, more fundamentally, the ultimate scientific conception of the relationship between the embodied and the bodily.

The Physical Made Real in the Psychical

In the identification of intoxication as an experimentally recognizable state of forced psychopathology, Kraepelin proffers his radical reconstitution of Wundt's psychological parallelism. Insofar as Wundtian psychology was borne out of alterity—both interpersonally and conceptually—with the neomechanical reduction of the mind to the body, Wundt's model was ill-equipped to integrate physiology as he had hoped. Wundt's parallelism was, as discussed, a secularization of the Fechnerian duality of the ephemeral body-world and the eternal God-mind. As if this were not enough, Wundt himself invoked comparison with Leibniz's dualism. Yet experimental psychology was flourishing. What might have been a minor shoot growing out of the cracks in neurophysiology had begun to bear fruit. It is clear to see, however, that this was a tenuous position. Neurophysiologists after the fashion of Meynert and Wernicke may have been visibly blundering their hasty equivalences between mental states and brain states, but brain research never stopped. In this sense, the dualism that had been the basis of experimental psychology's formation also threatened the validity of the entire research program. Wundt could exorcise Fechner's ghosts, but scientifically the psychomental remained in a state of liminality.

The answer, whether Wundt could ever see it or not, came in the form of substances of intoxication, from amyl nitrate to alcohol. By equating intoxicated states with mental states and passing this principle through the experimental crucible, Kraepelin renders the mind and body as a singular scientific object. It is in the context of Kraepelin's psychological reaction time trials that substances of intoxication intercede in the fundamental

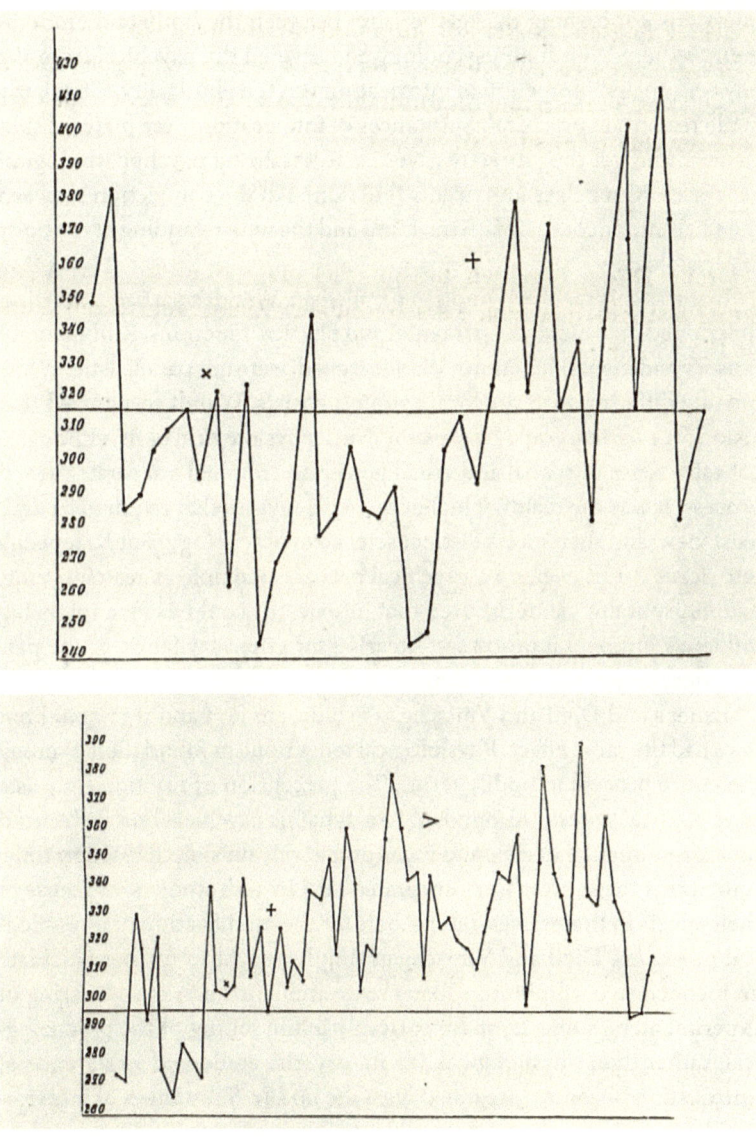

FIGURE 11. Alcohol's effect on choice reaction low dose (*top*) and high dose (*bottom*) (Kraepelin 1883c, 591, 592). Courtesy of the National Library of Israel.

questions concerning the relationship between the body and embodiment. Where Wundtian psychology had severed the psychical from the physical, the phenomenon of intoxication was the physicality of the body made real in the psychical. Substances of intoxication were material substances, and yet they directly gave rise to totalizing psychopathological processes. Kraepelin's intoxicants functioned as the connection between the scientific encounter with the mind and the understanding of the body in physiology.

Nor was this merely Wundtian parallelism. Wundt's partial parallelism functioned on a hierarchy of mental and physical functions. Rudimentary sensory and mental functions, Wundt argued, were not parallelistic. While broadly differing with the localization theorists, Wundt recognized that vision for example could be explained on a physiological basis, although a cohesive sense of spatial and visual perception implied a separate mental process. It was the realm of higher mental activities that required a parallelist view, and therefore a distinct science of psychology. But Kraepelin's intoxicant study, even on a psychical process as simple as reaction time, had indisputably demonstrated that intoxicants could exact a totalizing influence on mental processes—tearing into the very fabric of the perceptional lifeworld of embodiment.

Exner's and Dietl and Vintschgau's studies in 1873 and 1877 could not have had the same effect. Physiological reductionism found them framing the entire process in bodily terms. The perception of reaction time as a physiological process, as opposed to a dynamic psychical one, influenced their experimental decision to focus exclusively on simple reaction time. Intoxicants themselves were integrated into Exner's study not because of their effects in themselves, but as tools for the modification of physiological processes. Dietl and Vintschgau simply sought to further elucidate an inconclusive subpoint in Exner's research. Through the gathering of experimenters, subjects, and intoxicants in the setting of the psychological, rather than physiological, laboratory, the embodied experience of intoxication becomes integrated with the bodily. Substances of intoxication became testifying witnesses in a trial not of intoxicants themselves but of the ultimate nature of the relationship between the mind and the body. It is only in studying the effects of intoxicants on the mind, on the psychical process, that intoxication could be conceptualized as thinking with the body. Thus, it was on the testimony of intoxicating substances that mind and flesh, embodiment and the body, could be sublated within a unified physiological doctrine, without subjecting mental states to the reductionism of neural states. This was the beginning of a conception of the body that was not mechanical, but dynamically biological.

Simultaneously, the principle of psychological parallelism remains functionally intact as a working description of the perceptional dynamism of embodiment over and against the bodily reductionism of the brain psychiatrists. However, it functions solely as a framing principle—a conceptual model retained as a bulwark against the naive materialism of brain psychiatry. For this reason, Kraepelin has, at times, been labeled as ambivalent or blurry with respect to the overt stance on the mind-body divide (Hoff 2015, 34). This functional, as opposed to conceptual, parallelism still remains influential to this day, denoting a point of view rather than a "real" description of the mind-body relationship.

Drink and drug of all shapes and kinds had been changing the face of the medicine, of work, of leisure, and of the social order more generally in profound ways over the course of the nineteenth century. Now they interceded in the minds and bodies of experimenters and experiments alike. Kraepelin, alongside his fellow subjects, brought drugs into a radically new kind of "place" in the world: the psychological laboratory—a space implicitly characterized by a distinct horizon of epistemic possibility. Even as the early psychology lab employed methods similar to the physiologically oriented laboratories of Exner, Dietl, or Vintschgau, it existed within a very different constellation of theories and concepts. The apparatuses, cupboards, and work areas of Neumarkt, Leipzig, were themselves inscribed with the conditions for a distinct set of possibilities. Here, researchers and intoxicants gathered in a novel setting where intoxication could be seen as thinking with the body. Intoxication cast mind as flesh, not only by bringing substance to bear on mind but by bringing the experience of the mind to the forefront of bodily experience. This was a novel conception of the body and mind entirely at home in an age where all manner of cocaine tonics, chloral knock-out drops, and morphine pain pills not only existed but were ready-to-hand. But, for Kraepelin, this was only the beginning. In the ensuing period, Kraepelin would greatly extend the scope of his newfound pharmapsychological research program, while at the same time aspiring to develop a psychologically-founded science of psychiatry that could compete with the shortcomings of brain psychiatry.

10

Kraepelin's Nosology and an Intoxicated "Physiologie der Seele"

The First Compendium (1883)

Aspirations of completing his habilitation by further researching intoxication as a psychopathological phenomenon were cut short when Kraepelin lost his position working under Flechsig in the university's psychiatric hospital. Flechsig's reasons for firing Kraepelin appear justified: Kraepelin increasingly neglected the work he was being paid to do in the hospital, instead focusing on the work he was doing in Wundt's lab (Steinberg and Himmerich 2013, 249–50). Another possible factor was Flechsig's perception that his hopes of hiring an experienced clinical psychiatrist had been exploited by Kraepelin, in order to secure funding for Kraepelin's own research (Steinberg and Himmerich 2013, 249–50). Kraepelin's little "ploy" now threatened everything Kraepelin had worked toward. The firing left Kraepelin without a source of income, and so he made an urgent request to habilitate using work which he had already completed (Steinberg and Himmerich 2013, 249–50).

In lieu of a habilitation thesis, Kraepelin submitted three publications: "Ueber den Einfluss acuter Krankheiten auf die Entstehung von Geisteskrankheiten" (1881), "Ueber die Dauer einfacher psychischer Vorgänge" (1882), and—the paper just discussed—"Ueber die Einwirkung einiger medicamentöser Stoffe auf die Dauer einfacher psychischer Vorgänge" (1883) (Steinberg and Himmerich 2013, 250). His submission was very nearly rejected. One of his reviewers, the physiologist Carl Ludwig, who also happened to be Flechsig's foster father, decried the papers on reaction time as essentially derivative, lacking in creative ideas (Steinberg and Himmerich 2013, 250). To Kraepelin's good fortune, Ludwig handed the decision over to Erb, who ultimately granted Kraepelin his habilitation, bringing with it the rank of privatdozent and the ability to charge fees from students (Steinberg and Himmerich 2013, 250). By October 1882, just eight months after first getting the job with Flechsig, Kraepelin was officially part of the teaching staff in Leipzig's medical faculty (Steinberg and Himmerich 2013, 255).

Kraepelin had narrowly escaped a major set-back. Yet the premature completion of his habilitation requirements had been to his advantage. Kraepelin now had considerably greater flexibility in where and what he wanted to research. On the other hand, he was all but destitute and in serious need of an income. To help address Kraepelin's dire financial circumstances, Wundt encouraged Kraepelin to write a psychiatry textbook (Hoff 2015, 32; Heckers and Kendler 2020, 382). Published in 1883, this little textbook, Kraepelin's *Compendium der Psychiatrie*, was the infant form of the psychiatric nosology that would ultimately secure Kraepelin's place in the annals of medical history.

The first edition was simultaneously an effort to introduce Kraepelin's own ideas derived from Wundt's scientific psychology and a review of what Kraepelin perceived as being the best, most current literature in psychiatry (Hoff 2015, 32–33; Heckers and Kendler 2020, 381). Although Kraepelin acknowledges the influence of a number of existing psychiatry textbooks, the 1880 edition of Heinrich Schüle's textbook as well as the 1883 edition of Richard von Krafft-Ebing were the most directly formative (Heckers and Kendler 2020, 381). The nosologies proposed by both authors shared fundamental similarities, although they were slightly differently framed. Schüle classified pathologies as psychic, organic, or psychic-organic. Psychoses, meanwhile, were either psychoneuroses, mind diseases with no change in brain matter, or cerebropsychoses, which involved changes in the brain and were discernible by changes in motor function (Heckers and Kendler 2020, 381). This more or less arbitrary identification of the affectation of motor function as not only categorically distinct but also reflective of a structural change in the brain appears consistent with the relative success that neurophysiology had in localizing motor function. Krafft-Ebing's textbook instead proposed three different nosological approaches (anatomical, etiological, and clinical), which gave rise to three diagnostic categories: pathological illness, disease without pathological findings, and disorders of neurological development (Heckers and Kendler 2020, 381).

Kraepelin interpolated the ideas of Schüle and Krafft-Ebing, alongside those of Wilhelm Griesinger and Hermann Emminghaus, into the first edition of his compendium. However, Kraepelin had not intended for his psychiatric textbook to merely serve as a gloss on the existing literature, but rather to bring the existing literature into agreement with Kraepelin's own aspirations for psychiatry. In particular, Kraepelin hoped, on the basis of insights gleaned through experimental psychology, "to facilitate the understanding of mental disorders as much as possible and to indicate everywhere the roots of them in normal experience" (Kraepelin 1883a, viii). Kraepelin's compendium was, at least aspirationally, the herald of a new psychiatry, one in which the ultimate nature of psychopathological processes could be identified by direct experimentation on the mind.

At this point in his life, Kraepelin was far from the clinical nosologist he is known as today. He was more concerned about the effectiveness of diagnostic techniques, which implicitly, rather than explicitly, brought Kraepelin into confrontation with extant nosological systems (Engstrom 2015, 155). On this point, Kraepelin doubted the value of the anatomical and etiological nosodiagnostic categories identified by Schüle/Krafft-Ebing (Kraepelin 1883a, 189; Heckers and Kendler 2020, 382). Despite being incapable of discerning the anatomically healthy nerves from the pathological, neurophysiologists claimed it was brain science "which should be called upon to enlighten us about the true nature of the individual disturbances" (Kraepelin 1883a, 188). Such attacks leveled against the entire host of neurophysiologically-minded brain psychiatrists bordered on ridicule, but they also made Kraepelin's position overwhelmingly clear. Physiological pathology could not currently, and perhaps never could, form the basis of a valid diagnostic criterion. This was both because of the immature state of the study of neural anatomy as well as Kraepelin's concern "that were it possible to find constant structural changes [anatomically], they would only gain a real value through the relationship to the simultaneously observed functional disturbances" (Kraepelin 1883a, 188). It was hardly different for etiological criteria, where the vast majority of individual cases varied so greatly in the symptomatology and duration of their effects that it was nearly impossible to make firm deliberations with any certainty (Kraepelin 1883a, 188–89).

In contrast with the etiological and pathological criteria, "the clearest associations, in any case, are those which persist between the organically caused functional disorder and the clinical manifestations" (Kraepelin 1883a, 189). Here, too, distinctions between the presentation of symptoms across groups was exceedingly difficult, but the emphasis on visible symptoms overcame both etiological metaphysics and poorly supported physiological reductionism. The conclusion: "What it offers us are not diseases, but merely symptom complexes" (Kraepelin 1883a, 189). While Kraepelin provides independent reasoning for this choice of emphasis, it is difficult to ignore the potential theoretical and experimental influences behind this shift. For one, it brought Kraepelin's early psychiatry into agreement with Wundtian psychology. Wundt had approached everything outside of consciousness in physiological terms since the first edition of the *Grundzüge* (Engstrom 2015, 154; Araujo 2012, 41). It was thus consistent with Wundt's methodological philosophy for Kraepelin to favor the notion of symptom complexes—symptoms being the discernible rupturings of disorder into the conscious experience of embodiment.

Yet this is not a surprising turn for Kraepelin. He had applied this very same approach in his recently conducted research on the effects of intoxicants on the reaction time. There, Kraepelin had overcome the experimental problem of controlling for the variable dosage of an inhaled intoxicant by developing criteria for the classification of degrees of intoxication. The result had been a functional nosology of intoxication predicated on complexes of discernible symptoms. Although Kraepelin had the benefit of knowing which intoxicant had been consumed beforehand, the symptomatic and experimental expressions of deep and light intoxication were sufficiently distinct to warrant independent classification. This created a circumstance wherein Kraepelin, both reflexively (as examined) and 'clinically' (as examiner), was brought to classify states of mental intoxication solely on the basis of observable symptom complexes. To suggest that Kraepelin's intoxication trials had a direct influence on his diagnostic approach would seem dubious had Kraepelin himself not taken up intoxicants as experimental actors capable of inducing temporary psychopathological states. Kraepelin's decision to favor the clinical identification of symptom complexes as a diagnostic approach was shaped not only by his theoretical faith in scientific psychology but by recent experimental study on the effects of intoxicants on the human mind. The reaction time trials were intoxication as madness.[1] Thus, in the *Compendium*, madness was as Kraepelin had seen it, experienced it, through the haze of intoxication. Just as the intoxication trials had been a confirmation of the underlying somatic character of mental disorder, Kraepelin's early research on intoxicants had presaged his diagnostic approach. Nor was this the full extent to which intoxicants and the powerful mind-states they induced would affect his work—both in respect to his research on the effects of intoxicants on the body and mind, as well as in the psychiatric textbooks that ultimately shaped his legacy.

The Dorpat Psychological Society

For all that he had done between 1880 and 1883, Kraepelin was still eager to obtain a full professorship as soon as possible. A major motivator was his desire to finally be married. When Kraepelin got engaged in the summer of 1883, his mentor kindly advised him that it would be quite some time before he would be able to offer him anything approximating a full professorship, owing to the immature state of their shared field (Kraepelin 1983, 29). On the voyage home from a meeting in Freiburg, Kraepelin paid a visit to Gudden, who eagerly proposed a solution to Kraepelin's dilemma (Kraepelin 1983, 29). Kraepelin would return to Munich to fill a vacancy at the asylum there, while also assuming a lectureship at the university. By the fall of 1883, Kraepelin was once again in Munich (Kraepelin 1983, 29–30).

There Kraepelin was able to once again study the anatomy of nervous tissue, although the available staining and hardening technique greatly limited Gudden's aspirations of directly studying the relationship between groups of cells and nerve fibers (Kraepelin 1983, 30–31). Perhaps to distract from this time spent reviewing brain tissue, Kraepelin also bought a personal Hipp chronoscope and obtained all the necessary materials "to extend my experiments with medicines and stimulants even further, and then also to make time measurements on the mentally ill in order to gain a clearer idea of the mental changes" (Kraepelin 1983, 32).

However, the move to Munich had only been a temporary solution. In 1884, Kraepelin moved to Leubus to take a position at the local asylum. Upon accepting the position, he and his fiancée were married, and it seems that for a time, they lived a simple, pastoral existence in Leubus (Kraepelin 1983, 33). Research there was limited to studying differences in the composition of urine between groups of patients and occasional tests on mental chronometry. In May 1885, Kraepelin took up a job in the psychiatry department in Dresden (Kraepelin 1983, 36). Here, too, Kraepelin was limited in his scientific research, though he recalls being quite happy. That Christmas was spent with the Wundts, with whom he still retained a close relationship, and the following day he paid a visit to the eighty-five-year-old Fechner (Kraepelin 1983, 38).[2] In April 1886 Kraepelin finally found a way to make inroads into an academic career. Emminghaus wrote to him that he was leaving his professorship in Dorpat, and he had suggested Kraepelin as a possible replacement (Kraepelin 1983, 39).[3] Kraepelin was selected for the position and thus began his first professorship in Dorpat (Engstrom and Kendler 2015, 1191). On the way there, Kraepelin had hoped to introduce his wife to Gudden, only to discover that Gudden had drowned, possibly murdered, alongside King Ludwig II of Bavaria at Lake Starnberg (Freckelton 2012).

Though Emminghaus was brimming with glee to be returning to Germany from what Arthur Böhtlingk likened to "the desert" in an apparent allusion to the biblical wilderness, Dorpat would be formative for Kraepelin (Steinberg and Angermeyer 2001, 299). The teaching component of his position faced difficulty at almost every turn. Most of his students were

Estonian, Russian, or Livonian speakers and Kraepelin could only address them with an interpreter (Steinberg and Angermeyer 2001, 301–2). A policy of Russianization also created difficulties for German faculty members, restricting their academic freedom. Kraepelin would receive an order from the Livonian governor to sign documents in Cyrillic later in 1890 (Steinberg and Angermeyer 2001, 302).

Despite the difficulties, the status of a full professorship nevertheless also gave Kraepelin the latitude to finally continue his scientific work. Now securely in a professorship, Kraepelin was free to continue studying the effects of different intoxicants on mental processes. In 1887, Kraepelin founded the Dorpat Psychological Society, whose membership would conduct research in a regional laboratory for experimental psychology that Kraepelin set up in a small room offered by the university's president Alexander Schmidt (Steinberg and Angermeyer 2001, 305). Here, and elsewhere at Dorpat, Kraepelin oversaw experimental research on a wide range of psychological topics, including the perception of time and the depth of sleep. But it was also an opportunity for Kraepelin to continue research on the pet project he had started in Leipzig: the influence of intoxicants on mental processes.

Intoxicated Testimonies III: Dorpat and Heidelberg

Much of Kraepelin's research on this subject would not be published until 1892, when he collected his research from his time at Dorpat and, later, Heidelberg into the first book on the topic. A notable exception is Heinrich Dehio's dissertation in 1887, "Untersuchungen über den Einfluss des Coffeins und Thees auf der Dauer einfacher psychischer Vorgänge," which was written on research conducted in Kraepelin's Dorpat lab. Given Dehio's frequent reference to his mentor's own experimental objectives, it is worth briefly reviewing Dehio's dissertation before discussing the content of Kraepelin's book.

The objective of Dehio's dissertation was to expand upon Kraepelin's 1883 publication on the effects of amyl nitrate, ether, alcohol, and chloroform by conducting similar studies on the effects of a stimulant (Dehio 1887, 7). Dehio chose caffeine, on account of its availability, and made a comparison between the effects of pure caffeine and those of tea central to his experimental design (Dehio 1887, 8). This was additionally an opportunity to further overtake Exner and Dietl-Vintschgau, who had both studied tea and coffee, respectively, but done so with methods that Dehio deemed unreliable (Dehio 1887, 8–9). Kraepelin had published some cursory results on the effect of caffeine and other stimulants on reaction time in "Ueber psychische Zeitmessung" (1881), though these were far from extensive. Unlike these earlier trials, Dehio also administered his subcutaneous injections of pure caffeine to his subjects at doses of 0.5 grams, which elicited a light, albeit discernible, state of intoxication (Dehio 1887, 14). In other cases, subjects drank strong Russian tea (Dehio 1887, 14). The majority of the trials were conducted on Dehio himself, while the remaining attempts were made on Dehio's colleague, and fellow student of Kraepelin's, A. Sohrt (Dehio 1887, 15).

Though Dehio's thoughts are doubtlessly his own, his dissertation provides a snapshot of the methodological and conceptual developments that Kraepelin's own work had undergone while he was at Dorpat. Dehio's reaction time trials contained the simple and choice reactions, as Kraepelin's 1883 paper had, but he also tested "word reactions," which saw respondents reading monosyllabic words off cards, as well as "higher reactions" (Dehio 1887, 23–25, 36). These higher reactions included repeating the next number in a given sequence and doing basic arithmetic (Dehio 1887, 36–37). The methodological changes led to interesting results: Dehio found that caffeine and tea tended to shorten simple reactions, though it was inconclusive, and they had no discernible effect on choice reactions, but they definitely shortened word and "higher" reactions (Dehio 1887, 20, 31, 35, 37). In some cases, there was a slight lengthening in reaction times following the initial shortening, but generally the initial shortening was followed by a return to average reaction time and caffeine, as well as tea, had the effect of reducing variation in reactions overall.

From these observations, Dehio was able to draw two primary conclusions. The first was that, at the level of a strict comparison of caffeine dosages between tea and pure caffeine, tea was able to achieve comparable effects on reaction time to pure caffeine, but at a much lower caffeine dosage (Dehio 1887, 46). This raised questions about whether there was some additional alkaloidal component of the tea that was either directly affecting reaction time or amplifying the effects of the caffeine. The second was conceptual. Kraepelin had proposed that the intoxicating effects of alcohol arose from a slowing of the process of apperception, while the initial shortening in reaction time at low doses was a by-product of the stimulation of will time (Dehio 1887, 49–50). Dehio suggested that, in contrast, tea had no recognizable effect on will time (demonstrated by the negative effects on choice reactions) and that, although low-dose alcohol and tea both initially shortened reaction times, tea did so primarily through a moderate shortening of apperception time, a far cry from the disturbance caused by alcohol (Dehio 1887, 49–51). Any elongation in reaction times after tea or caffeine was very moderate, and Dehio attributed it either to fatigue or to a slight elongation in the process of apprehension (Dehio 1887, 50–51). Dehio, and presumably Kraepelin as well, understood these findings as yet another tentative confirmation of Kraepelin's initial hypothesis that different substances of intoxication modified different psychic processes, and in different ways (Dehio 1887, 48–49). Caffeine demonstrably elicited psychological effects of an entirely different nature than those previously seen in alcohol or other sedatives. Such confirmation was, however, only cautiously acknowledged: there were still so many different forms of intoxication to research. Nonetheless, Dehio finished his text on a hopeful note, confident "that here we have found a tool that promises to open up a completely new, previously unrealized path of scientific research" (Dehio 1887, 55).

Heidelberg and Arzneimittels

Dehio's dissertation reflected a development in the theoretical and methodological approach to the study of various intoxicants on mental process in Kraepelin's laboratory during his time in Dorpat. But Kraepelin would not be in Dorpat much longer. On November 9, 1890, Kraepelin received his appointment to a chair at Heidelberg (Steinberg and Angermeyer 2001, 316). Little is known about how Heidelberg became aware of Kraepelin as a candidate. Surviving letters between Kraepelin and Wundt implied prior contact and spoke hopingly about the opportunity to bring scientific psychology into an entirely new region (Steinberg and Angermeyer 2001, 316–17). Kraepelin had always intended Dorpat to be a stepping stone, though the ever-strengthening influence of Russianization may well have helped push Kraepelin out the door (Steinberg and Angermeyer 2001, 317). The Russian ministry officials had repeatedly complicated Kraepelin's scientific and clinical work, with Kraepelin going as far as personally overseeing the furnishing, renovation, and installation of linoleum in the clinic, even enlisting the help of his assistants in installing electrical wiring, in order to avoid meddling from Petersburg (Kraepelin 1983, 42–45). Notably, Dehio—the student who had so judiciously assumed the Kraepelinian study of caffeine as his doctoral project—came with him.

Once at Heidelberg, Kraepelin continued his research on experimental pharmapsychology, compiling his work into the book *Ueber die Beeinflussung einfacher psychischer Vorgänge durch einige Arzneimittel* in 1892, the first monograph-length publication on the subject. Here, Kraepelin discussed the effects of alcohol, tea, morphine, chloral, amyl nitrate, chloroform, ether, paraldehyde, caffeine, nicotine, and cocaine on the wide array of mental processes, including reading, writing, arithmetic, recall, and many more (Kraepelin 1892a, vii–viii). Concerned with how the mercurial nature of mental processes might interfere with his results, Kraepelin had intended for these to be "substances of such an energetic effect that [concerns about interference caused by normal variation] had to take a back seat to all other causes" (Kraepelin 1892a, 3). The subjects had to get high enough that there were no questions about the nature of any variations from baseline. This was not merely research on the effects of various medicines; it was the self-conscious study of the intoxicated "Seelenleben" (soul-life) of embodied consciousness (Kraepelin 1892a, 3).

Alcohol, Chloral, Morphine, Paraldehyde, Ether, and Amyl Nitrate

Kraepelin's earlier work on the effects of alcohol have been discussed at some length, and Kraepelin reviews his earlier work in *Ueber die Beeinflussung*. New additions from his unpublished research from the period between 1881 and 1883 are his association studies on alcohol, which used the experimental method for measuring simple reaction time but in response to sound association (Kraepelin 1892a, 51). When it came to testing the effects of alcohol using the post-1883 methods, Kraepelin had a far greater number of test subjects to pick from. Apart from himself, Kraepelin identified Dehio, "Herren Dannenberg (Da.), Hahn (Ha.), Heerwagen (He.), Michelson (M.), and Oehrn (O.)" as respondents (Kraepelin 1892a, 68). This may have been to his detriment, however, since it exaggerated individual differences that were less visible in a smaller cohort and would have been diffused in a larger study. Subjects were given either 20 or 30 grams of pure alcohol, depending on when the trial was conducted. The experiments in question took place between December 1888 and December 1889 and amounted to a total of twenty-seven runs: seven on reading, ten on arithmetic, and ten on memorization of figures (Kraepelin 1892a, 68–69).

In the trials that saw subjects work on adding strings of numbers, alcohol elicited a generalized decrease in work performance, which was immediately seen after consumption (Kraepelin 1892a, 72–73). The secondary effect on the adding was an increased variability in the rate with which the work could be completed (Kraepelin 1892a, 74). Kraepelin found that the task that required subjects to memorize sets of twelve digits varied greatly between individual subjects. In the first quarter of an hour, there was an overall decrease in their subject's abilities to complete the task, though some subjects initially performed the task better than they had prior to consuming alcohol (Kraepelin 1892a, 76). Nevertheless, the consistent pattern found in nearly all cases was an initial drop in performance followed by a consistent improvement toward baseline starting thirty minutes after ingestion (Kraepelin 1892a, 77). Kraepelin emphasized that any initial increase in work performance was certainly a fleeting phenomenon, though he personally found that it was consistent with subjective experience (Kraepelin 1892a, 77). He related this to his findings in the low-dose simple reaction trials, where he also found an initial improvement in performance. On this point, Kraepelin offers the same possible explanation: those subjects who initially performed better possibly had some increased resistance to alcohol's effects, whether by natural constitution or familiarity with the substance (Kraepelin 1892a, 77–79). Another factor that Kraepelin was constantly aware of was the difference in individual learning methods, which likely accounted for some degree of variation between individuals but not enough to alter the trend (Kraepelin 1892a, 86).

Reading times, too, were generally slowed under the influence of alcohol, with at least one case of an initial acceleration (Kraepelin 1892a, 87–88). On these grounds, Kraepelin suggested that under certain conditions alcohol caused an acceleration of speech. Kraepelin also noted, however, that this only occurred once in trials using a 30-gram dose, which appeared consistent with his findings elsewhere about the stimulating effects of lower doses of alcohol (Kraepelin 1892a, 88). Reading did seem more sensitive to this initial acceleration effect, but also appeared to function at a deficit for a longer period of time as well (Kraepelin 1892a, 88). He did further tests on muscular flex response using a dynamometer, as well as alcohol's effect on the prediction of time intervals, where alcohol's effects were consistent with the other tests (Kraepelin 1892a, 91, 99, 105).

For Kraepelin's trials using paraldehyde, he ran sets of word and association reactions, using a hand-operated contact key (Kraepelin 1892a, 148–149). Of the fourteen observations, seven were on Kraepelin himself, four on Alfred Lehmann, and three on Ernst Rehm (Kraepelin 1892a, 149). Contrary to Kraepelin's expectations that paraldehyde would merely lengthen reaction times, paraldehyde produced an array of seemingly paradoxical results (Kraepelin 1892a, 149–50). Instead, Kraepelin observed fluctuations between increases and decreases in response times (Kraepelin 1892a, 150). This behavior appeared to scale with dosage, showing the highest magnitude in fluctuation at 5-gram doses and lower variation at 2 grams (Kraepelin 1892a, 150). Kraepelin proposed that these were not merely irregularities: there was a recognizable acceleration of some psychic processes. The onset of this acceleration effect was recognizable starting approximately five minutes after ingestion, and it peaked at twenty-three to twenty-seven minutes (Kraepelin 1892a, 150). Combined with the negative results, Kraepelin realized that the shortened intervals were the result of a tendency toward premature reaction under the influence of paraldehyde (Kraepelin 1892a, 149–52). An explanation proposed by Kraepelin is the appearance that paraldehyde elicited an initial elongation in reactions, to which the subject then unconsciously attempts to compensate for (Kraepelin 1892a, 152). Where word reactions were used in place of discrimination reactions, paraldehyde appeared to have similar effects as other reactions but on an enlarged scale (Kraepelin 1892a, 153).

These findings suggested to Kraepelin that paraldehyde had a strong stimulatory effect on the will impulse, apparent not only from the subjective feeling of a quicker reaction, but also from the frequency with which premature reactions were registered in the experiments (Kraepelin 1892a, 155). Simultaneously, there was a marked lengthening in times when premature reactions did not occur. This combination of prolonged reactions with a tendency toward premature responses suggested to Kraepelin that paraldehyde had a twofold effect: at once unfettering the will process and suppressing psychical apprehension (*Auffasung*) (Kraepelin 1892a, 157–59).

Kraepelin had previously done a limited series on the effects of chloral hydrate, just two sets of experiments. On these two occasions, he gave himself a dose of 2 grams of chloral and tested simple and choice reaction times (Kraepelin 1892a, 161). Simple time slowed very rapidly after ingestion, but the resultant lengthening was relatively small, seeing a return to baseline after approximately thirty minutes (Kraepelin 1892a, 161). Subjectively, Kraepelin experienced that chloral took a long time to induce feelings of fatigue, which was more apparent in the results from the choice reactions, where it took approximately six to fourteen minutes to show an effect (Kraepelin 1892a, 162). The response was a visible slowing of the election process, an elongation of far greater magnitude than was seen in simple reactions. Notably, there was no shortening or premature reactions, as had been seen in paraldehyde (Kraepelin 1892a, 162). Kraepelin also conducted two additional sets of trials with chloral closer to 1892, now using 1 gram of chloral during sets of choice and word reactions (Kraepelin 1892a, 164). Both experiments saw significant lengthening in reactions, with results that were fundamentally similar to the earlier chloral trials (Kraepelin 1892a, 165).

Morphine, along with tea, exerted what Kraepelin felt was one of the most distinct affectations of the psychological process, and he considered it to be of the highest practical and theoretical interest (Kraepelin 1892a, 166). However, Kraepelin was also extremely wary of inducing morphinism in any of his subjects, so he limited trials to himself (Kraepelin 1892a, 167). In the 1881–1883 period, Kraepelin did two sets of experiments with morphine, both of which were choice reactions conducted under the influence of a 0.01-gram subcutaneous injection of morphine muriaticum (Kraepelin 1892a, 167). What Kraepelin measured was a gradually increasing moderate elongation in reactions, which reached its highest value after eighty minutes (Kraepelin, 1892a, 167). In the second set of experiments, there was a shortening below baseline in the first half hour that was too consistent to be a mere fluctuation (Kraepelin 1892a, 167–68). Kraepelin conducted two additional sets of experiments on morphine while in Heidelberg, this time taking 0.01 grams in an oral solution prior to conducting trials on choice and word reactions (Kraepelin 1892a, 168). Immediately after ingestion, Kraepelin registered a moderate prolongation in choice reactions that peaked after thirty-five to forty minutes (Kraepelin 1892a, 168). The word reactions produced a completely different result. Here, Kraepelin observed a sudden shortening in reactions that reached peak magnitude below baseline levels at thirty to thirty-five minutes and then slowly returned to normal levels (Kraepelin 1892a, 168). Both responses were so distinct, and relied on so many individual measurements, that Kraepelin had absolute confidence in the validity of these observations (Kraepelin 1892a, 168–69). Kraepelin proposed that this twofold effect of morphine was likely because it excites the perception of external impressions, while choice acts are made more difficult. Kraepelin also repeated his earlier work on amyl nitrate and ether and found that the observations in his 1883 paper remained reliable (Kraepelin 1892a, 170–71).

From the midst of this flurry of measurements, two more or less distinct classes of drug effects emerged with respect to their discernible effects on the conscious process. Both sets intoxicated, albeit by affecting consciousness at different levels of the conscious process. Alcohol, amyl nitrate, and other sedatives notably lengthened reactions, with the exceptions of erratic, anticipatory jolts (Kraepelin 1892a). Morphine and, to a far lesser degree, tea slowed choice reactions, reflecting their effect on the will, while markedly shortening simple reaction time, discrimination reactions, as well as the time it took to do arithmetic (Dehio 1887; Kraepelin 1892a). In keeping with the definitional strictures of Wundtian psychology, it was clear to Kraepelin that the former group had an inhibitory effect on the sensory/intellectual processes by undermining the normative function of apperception and the latter had an excitatory effect on sensory/intellectual processes by stimulating external impressions (Kraepelin 1892).[4] These represented two relatively distinct modes by which particular intoxicants elicited pathological effects on the conscious subject.

Intoxicated Psychosis

While created only through great effort and with considerable assistance, it is difficult to deny that Kraepelin's 1892 *Ueber die Beeinflussung* remained consistent with the objectives laid out in his initial letter to Wundt, back when he was proposing a habilitation project.[5] Further still, Kraepelin appears hopeful about the value of his research, as he says:

> A brief review of the whole path we have traveled up to this point shows, as I believe, that through the methods applied here we are in a position to express in definite numerical values those changes in our soul-life which we are otherwise only able to describe in very general outlines through the deceptive aid of introspection, and to now trace them back with certainty to very simple elementary disturbances. (Kraepelin 1892a, 227)

Kraepelin's psychological research on the effects of intoxicants on the mind had not only developed a method of experimentally encountering "simple elementary disturbances" but had also established a more or less fixed relationship between constellations of symptoms and the disorder of certain mental processes. Each and every substance has a discrete effect on mental life, expressed through the combination of various affectations of basic mental processes (Kraepelin 1892a, 228). This was sufficiently predictable for Kraepelin to speculate about the action of taking various drugs at once. For example, Kraepelin suggested these results demonstrated why alcohol helps with morphine abstinence, while also exhibiting how the underlying psychological character of alcohol and morphine euphoria are different (Kraepelin 1892a, 226). With regard to mental processes, this research brought Kraepelin to the realization that motor disturbances were almost always accompanied by sensory and intellectual disturbances, while sensory and intellectual disturbances accompany one another so closely that they appear inseparable (Kraepelin 1892a, 228).

Given the scope of the maturation of Kraepelin's research into the effects of intoxicants on the mind relative to the period surrounding his habilitation and the first *Compendium*, it is worth considering whether these new developments correspond with the changes Kraepelin made to further editions of his textbook. The second edition of the *Compendium*, now called *Psychiatrie: Ein kurzes Lehrbuch für Studirende und Aerzte*, was published in 1887 and can be distinguished from the first edition by the expectation that symptom complexes will converge on discernible patterns (Heckers and Kendler 2020, 382–83; Hoff 2015, 36). The 1887 edition also immediately followed Dehio and Kraepelin's research on caffeine (Dehio

1887; Steinberg and Müller 2005, 140–41). There, Dehio suggested that in caffeine they had found some degree of confirmation for Kraepelin's initial hypothesis that different intoxicants affected different psychic processes in different ways, especially as Kraepelin's 1883 paper focused on substances with roughly similar effects (Dehio 1887, 48–49). This hopeful albeit ambiguous position is consistent with the changes that Kraepelin made in 1887 with respect to the suggestion that symptom complexes would likely converge on a pattern, much as the caffeine trials had tentatively associated specific psychic disturbances with particular substances. Structural tendencies and impulses realized in the experimental encounter with intoxicated states were reflected in Kraepelin's shifting nosological application of symptom complexes.

A far more remarkable shift came with the 1893 edition, which, once again, shortly followed Kraepelin's 1892 *Ueber die Beeinflussung*. The major theoretical development of *Ueber die Beeinflussung*, as stated by Kraepelin, was the realization that distinct forms of mental disturbance, such as those triggered by intoxicants, could be associated with the affectation of specific psychological processes. This was a shift from the position seen in 1883 where the finite distinctions between substances of intoxication were so poorly defined that low and high doses functionally operated like independent symptom complexes. By 1892, the associations made with certain forms of intoxication had been largely clarified and the behavior of particular substances at higher or lower doses was given a consistent definition. Intoxicants could even be grouped according to their direct affectations of discrete psychological subprocesses.

It is this subtle, yet precise, theoretical shift that one finds reflected in the nosological concept underlying the 1893 edition of *Psychiatrie*. Here, Kraepelin advocated for the classification of mental disturbances on the basis of long-term, careful observation of the entire course of a given mental disturbance, which could then be grouped on the basis of these broader similarities (Kraepelin 1893, 242–43; Heckers and Kendler 2020, 383). Crucially, there is the impression that relatively stable disease categories could emerge through this method of classification, something more rigid than mere symptom complexes. He also introduced the symptomatological terminology of "dementia praecox," which would come to constitute a diagnostic category in 1896 and, in time, become schizophrenia (Decker 2007, 339). This is nothing if not a perfect mirroring of the theoretical development seen in Kraepelin's research on intoxicants, where earlier ambiguity surrounding symptomatology had been overcome by the realization that patterns of mental disturbance could be understood as the observable affectation of specific psychic processes. Methodologically, many of these ambiguities, for example those pertaining to alcohol, were ultimately explained through the use of a wider array of subjects and greater breadth in experimentation, just as Kraepelin had proposed in the 1893 edition of *Psychiatrie*. By the 1896 edition of *Psychiatrie*, the classification scheme proposed in 1893 would concretize into the basis of what would become his psychiatric nosology (Heckers and Kendler 2020, 383; Kraepelin 1896). It becomes conceivable that many of the foundations of Kraepelin's psychiatric nosology were not derived solely from the few clinically focused moments in Dorpat and his first years at Heidelberg, and were instead greatly influenced by his hard-won experimental research on the effects of intoxicants on the mind.

Of Diagnostic Cards and Intoxicants

Much has been made of Kraepelin's *Zählkarte* (diagnostic cards) method of collecting, organizing, and—subsequently—grouping essential patient information and its role in the formation of his psychiatric nosology around natural disease entities. Yet, by Kraepelin's own account, the dizzying scope of clinical labor involved in not merely gathering but sorting the necessary information, realizing the sheer volume of new patients, well exceeded the energies of Kraepelin and his busy colleagues (Kraepelin 1983, 142; Moskowitz and Heim 2011; Weber and Engstrom 1997). Instead, Kraepelin elected to initially carry out the process "only for individual, particularly important groups" (Kraepelin 1983, 143). Even then, many diagnostic cards contained little if any information of note. As Weber and Engstrom suggest, "it is difficult to imagine how the textual content of the diagnostic cards could have been the exclusive basis of Kraepelin's nosological concepts" (Weber and Engstrom 1997, 383). The question is not whether Kraepelin really made use of, even relied upon, diagnostic cards. He certainly did, particularly from 1896 onward. It is a matter of the identity of the conceptual a prioris around which clinical observations could be signified, "particularly important groups" could be recognized, and large collections of clinical observations could gain organizational significance.

Given that Kraepelin's famous organization of psychosis into dementia praecox and manic depression began in 1893 (for dementia praecox) and 1896 (for the earliest collective grouping of periodic insanities), there are enduring questions surrounding whether experimental psychology might have provided just such an a priori basis. The topic has garnered some discussion in the recent past. Helmut Hildebrandt focuses on the framework found in the Wundtian modeling of the conscious process as critical to engendering the possibility of Kraepelinian categories, even proposing that Kraepelinian nosology all but singularly relied on the foundations wrought by Wundt's experimental psychology (Hildebrandt 1993). In a similar vein, Voelker Roelcke attributes Kraepelin's readiness to disregard the causative significance of fluid historical influences on disease classifications to his formative engagement with experimental psychology (Roelcke 1999).

But this is likely only part of the picture. After all, when we speak of Kraepelin's experimental psychology, we are speaking of an experimental science of intoxicated embodiment, one deeply enmeshed with intoxication as a way of knowing. The (provisionally designated) upper-tier theories at hand in Kraepelinian psychology are clear enough. Kraepelin makes no explicit claim that the purpose of his experiments with intoxicants was to identify natural disease structures. Psychometric measurements of intoxicated states served to define the border between normal and pathological psychical processes. Attending exclusively to these surface claims, one concludes that the explicit character of endogenous psychoses had no place in Kraepelin's experiments at all, and the stakes were set at the nevertheless ambitious goal of delineating the limits of pathological processes. It is on such grounds that the historiography has tended to alienate Kraepelin's psychology from his nosology.

However, it is difficult to come by this notion without feeling it to be a tad disingenuous.

Kraepelin's psychometric research on the effects of intoxicants was located at the intersection of enduring questions surrounding the basic underlying nature of psychical and physiological phenomena, the viability and necessity of psychology as a distinct experimental science, and the quantifiability of normal versus pathological states. It further engaged with the long-standing questions surrounding the relationship between intoxication and madness, in addition to the ambiguous psychological character of intoxication itself. These questions, trailing behind them a snaking baggage trains of methodologies, apparatuses, concepts, associations, and participating entities, filled the experimental space of his psychological laboratory. Möbius's 1882 ascriptions of intoxication to temporary insanity were silently present, as were his 1886–1892 dualities of exogenous and endogenous psychoses. The multitudes of physicalist experimental physiology were there too. Some standing in abeyance while others were called to testify, a host of seen and unseen actors were in every sense as constitutive as the furnishings, equipment, and human agents who outwardly facilitated the experimental encounter. Enjoined with the apparatus, not merely in body but in embodied mind, was Kraepelin himself.

In the rapturous state of intoxication, the sum of Kraepelin's learning and aspirations coalesced in his ownmost embodied experience. For all his efforts to shroud the immanence of his own subjectivity behind the discursive frame of the experiment, Kraepelin was dually positioned as subject and observer in most, sometimes all, of the observations pertaining to the psychological effects of particular intoxicants. The morphine trials, recognized above all others for the distinct importance of their findings, were constituted by a series of events that took place solely in the perceptional lifeworld of Kraepelin himself (Kraepelin 1892a). Intoxication was, for Kraepelin, a way of experientially encountering the discrete forms of mental disturbance that would later become the subject of psychiatric nosology.

Like a soldier at his post, awaiting, with bated breath, that break in the perfect stillness that betrays the nearness of danger, one can imagine Kraepelin readying himself for the incoming stimulus. And it does come. The experiment is prepared again, except, this time, it is punctuated by the sting of a subcutaneous injection, a smack of raspberry syrup in pure alcohol, the fruity pungence of amyl nitrate. Here, the emotive state of intoxication brought lower-tier concepts to the fore in a mercurial flux of liminal associations, clinging to the perceptional immediacy of embodiment. All the more so because, for Kraepelin, the embodied experience of intoxication itself, as a temporally circumscribed manifestation of psychopathology, was the object of experimental inquiry. Confined by the definitional strictures of Wundtian psychology, the nature of psychological phenomena and the organizational validity of preexisting nosologies, even the nature of psychoses themselves—all supposedly secondary, lower-tier concepts—became associable with the range of psychical disturbances engendered by potent intoxicants. Intoxication was a way of knowing and making, constituting the conditions of possibility for the candidacy of certain claims about the world and the place of the body in it.

Even in those cases where Kraepelin did not explicit inhabit a dual position within the experimental apparatus, Kraepelin was nevertheless *liminally* positioned within the apparatus. Only fragments of the private, emotional experiences of the other participants, carrying within them a quiet cacophony of conceptual associations, could be interpolated with the experimental process, since, after all, it was Kraepelin himself who commanded the experimental process. Kraepelin was, in this sense, analogously positioned within the apparatus, insofar as new observations and measurements were inescapably synthesized in accordance with the perceptional contingency of Kraepelin's own experience. This is not merely to banally point out that it was ultimately Kraepelin who published the research, and thus necessarily structured the significance of experimental measurements in keeping with his subjective partiality. It was the experience of intoxication itself that was implicated in making real the measurements, observations, and experiences of other respondents, a notion that Kraepelin himself timidly acknowledges in remarking that external measurements align with his *subjective experience* (Kraepelin 1892a).

I have already intimated that intoxicated ways of knowing figured centrally within Kraepelin's experimental psychological studies, that this dynamic may have influenced the hardening of "exterior" symptom complexes into "interior" diagnostic categories. It is worth considering whether this same dynamic, of intoxication as a way of knowing and making, helped lay the foundations of Kraepelin's nosology directly. It was Kraepelin, after all, who—on the basis of sustained experimental encounters with intoxicants—identified two distinct groups of temporal mental disturbances, the sensory-intellectual inhibition of alcohol and sedatives contrasting with the sensory-intellectual excitation of caffeine and morphine.

A similar dualism is found in the nosological demarcation of endogenous psychoses into dementia praecox and manic-depressive illness, which would later comprise the nosological legacy celebrated by neo-Kraepelinians to this day. Dementia praecox would be framed as a degenerative illness of the apperception, much as inhibitory intoxicants such as alcohol affected the apperceptive process. Manic-depressive insanity would likewise come to be characterized by the acute forcefulness with which it affected the integration of external impressions, a quality Kraepelin had already attributed to morphine and caffeine.

It seems possible that the organizational a prioris absent from Kraepelin's *Zählkarte* might instead be found in Kraepelin's experimentation with intoxicating substances, in the symptomatology of intoxication itself. Though Kraepelin never explicitly states this connection, the parallel symptomatologies of endogenous and exogenous psychoses gesture at intoxication's structural significance for Kraepelinian nosology, at the existence of an organizing a priori borne out in the experiential lifeworld of intoxication itself. Intoxicants, the vivaciousness with which they assail day-to-day pretensions of volition, of will, jolt us from our dogmatic slumber. They illuminate the extents to which that hidden continent, the domain of high-brained conceptual thought, is imminently, inextricably participatory in the experiential constitution of knowledges. They make the world.

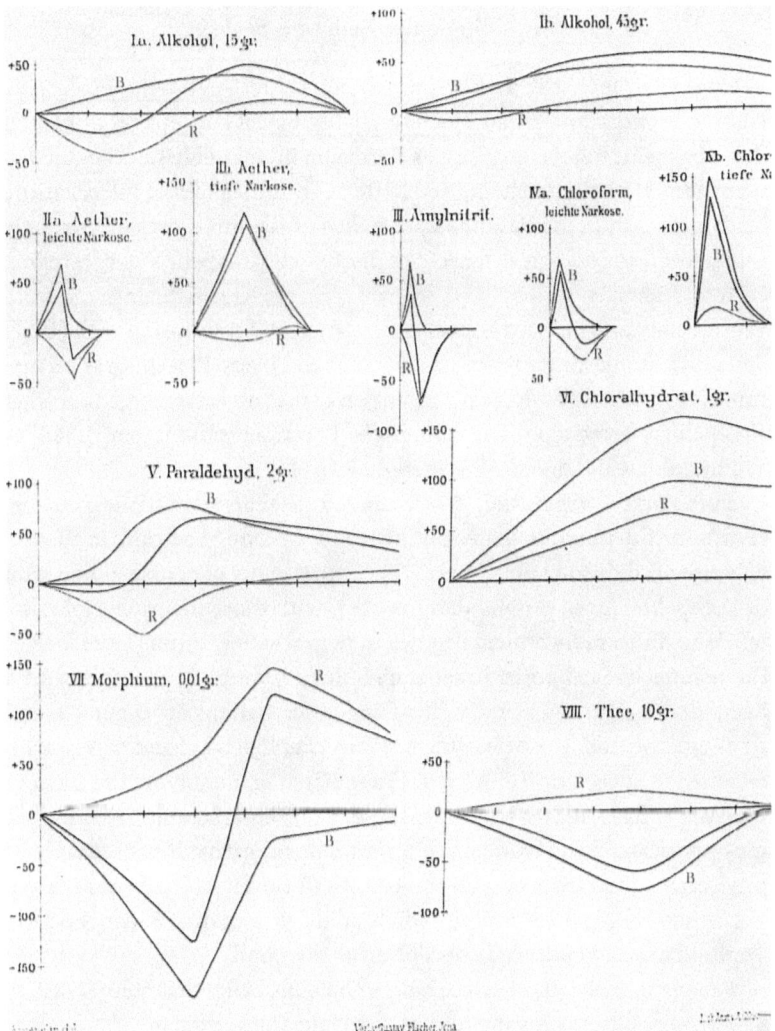

FIGURE 12. Collection of reaction time graphs from Kraepelin's 1892 work. Line (R) corresponds to motor function and line (B) corresponds to sensory/intellectual functions (Kraepelin 1892a, unnumbered appendix). Note the placement of morphine and tea at the bottom and how the sensory-intellectual measurements diverge from the graphs above them, suggesting that morphine and tea are sensory-intellectual stimulants. Courtesy of the National Library of Israel.

Physiologie der Leib und Seele

What is clear is that Kraepelin's psychological research with intoxicants laid the foundations for an increasingly biological conception of mental disturbance. It follows that just as Kraepelin increasingly saw mental disturbances as analogously inducible through substances of intoxication, his conception of mental illness, and thus of the mind, was increasingly understood in biological terms. As discussed, Kraepelin's 1883 research on intoxicants demonstrated that, even early on, Kraepelin diverged from Wundtian parallelism to propose a more dynamic conception of the body and mind. But it was here, in the late 1880s and 1890s, that this notion matured into a strictly biological conception of the subject, emergent of what Kraepelin referred to in "Psychologische Forschungsmethoden" (1888) as a "Physiologie der Seele" (Kraepelin 1888, 13).

In the first *Compendium*, Kraepelin left no secret concerning his disregard for the possible benefit of neurophysiology/-anatomy in the development of diagnostic criteria. This position was of course a reflection of Kraepelin's more general disagreement with those neurophysiologists who sought to reduce mental states to neural states, through the lens of the neomechanical body. Kraepelin's influence on his students has even been identified as the origin of the famous accusations of "brain mythology" against the likes of Meynert, Wernicke, Flechsig, and so on, with Kraepelin's successor in Munich, Franz Nissl, first applying the term to Flechsig (Hlade 2021, 4).[6] In this sense, Kraepelin remained steadfast in his opposition to the brain psychiatrists, a position that his students happily carried forward well into the twentieth century (Hlade 2021, 4–5). Yet, as has already been established, this did not mean that Kraepelin truly opposed a unified physical conception of the mind.

Kraepelin, himself, seems to have upheld his belief that mental disturbances were ultimately somatic illnesses (Engstrom 2015, 156). The error of the neurophysiologists, brain psychiatrists, and neomechanists had been to derive clinical theory from anatomical research, where Kraepelin saw no possibility of dissecting the complexities of the human mind by way of anatomical research alone (Engstrom 2015, 155–56). It had been an a priori recognition of the body as the locus of the real, of thinking about the mind via the body, that had led others toward neuropathology in the first place. The issue for Kraepelin had been that there was no place in these perspectives for the dynamic complexity of the psyche. A totalizing biologistic conception of the mind and body was tenable if it was inclusive of the perceptional diversity of mental life.

To this point, Kraepelin had taken up the methodologies of scientific psychology to bring the testimonies of substances of intoxication to bear on the physical character of the mind, and body. Kraepelin had used intoxication to think about the mind with the body, rather than describe the mind through the body. This had allowed him to understand and uphold the mind in all of the dynamic vigor of embodied perceptional existence while simultaneously grounding it in the physical. It was the physical made real in the psychical. It was only in conjunction with an approach to the mind grounded in experimental psychology and the clinical encounter that neuropathology had its place, and there is evidence of this in Kraepelin's work.

Though the 1883 *Compendium* was firm in its position on the questionable value of neuropathology for psychiatry, this attitude softened in subsequent editions. The 1887 edition of *Psychiatrie* saw Kraepelin advocating for the importance of including pathological research in the development of psychiatric classifications, particularly the postmortem findings of patients with similar symptom complexes or clinical pictures (Kraepelin 1887, 211; Heckers and Kendler 2020, 383). Into the late 1890s, Kraepelin continued his research on the effects of intoxicants on the mind, with several of his students at Heidelberg publishing on their work. Arnold Löwald wrote his dissertation on bromides, Hans Hänel on trional, August Hoch on tea, A. Oseretskowsky on the ergography of alcohol and caffeine, and Martin Meyer on alcohol's effect on writing tasks, in addition to further, albeit limited, studies on tobacco, morphine, and cocaine (Hänel 1898; Kraepelin and Hoch 1896; Kraepelin and Oseretskowsky 1901; Kraepelin 1983; Löwald 1896; Meyer 1901). Kraepelin's own journal would publish many of these findings. With the first edition being published in 1896 as *Kraepelins Psychologische Arbeiten*, the pages boldly attest to Kraepelin's renewed, or perhaps expanding, enthusiasm for the relationship between pharmapsychology and psychopathology. It was then in the late 1890s that Kraepelin reports of witnessing "Nissl's beautiful animal experiments with subacute maximum poisoning, in which it was possible to follow with surprising clarity the quite different influence of the individual poisons on the nerve cells" (Kraepelin 1983, 121; Nissl 1898, 613). Nissl's experiments demonstrated the specific effects of morphine, sulfonal, alcohol, morphine, trional, potassium bromide, chloral hydrate, and nicotine on nerve cells, further extrapolating that changes in the cell elicited by drugs offered insights into the nature of psychopathology (Nissl 1898, 613-614). Further illustrating this point, the 1896 edition of *Psychiatrie* introduced a category of metabolically induced mental disorder understood as "autointoxication" (Heckers and Kendler 2020, 383; Noll 2004). Far from being stymied by disinterest after the publication of his 1892 book, research on the effects of intoxicants continued in the form of a significant research program, with Kraepelin adopting a leading role. States of intoxication not only provided an experimental foundation for the development of Kraepelin's psychiatric nosology but increasingly also served as the bridge between the psychical and the physical.

Kraepelin's physiology of the soul was not in contrast to, or even separate from, the physiology of the body. Owing to the functional irreducibility of mental states to neural states, they would necessarily serve as two distinct, albeit interdependent, sciences. They ultimately became the shared foundations of a holistic conception of the body and mind, one that was inclusive of the psychical experience of embodiment and of the body as object. Where the neomechanists encountered the body within a framework predicated on the a priori exclusion of the psychical, the mind—the psyche—remained an anomaly. Where strict parallelism undermined the potential for closeness between the psychical and the physical, the body itself—the vehicle of lived activity—remained an anomaly. By upholding distinct scientific encounters with the body and the mind as the basis for any reckoning of mental life and deriving it from the experimental testimony of intoxicating substances, Kraepelin demonstrated what is now meant by biologism. This was the emergence of the biological subject, a being for whom the sum of physical and mental existence was understood as biochemical in origin, and yet the study of said being was irreducible to physiology.

The biological subject turns out to have been an intoxicated subject, emergent of the emotional experience of intoxication as a way of knowing, itself taken as an object of experimental encounter. Though Kraepelin's work would inspire a number of his students to continue research in the effects of intoxicants around the world, the impact of Kraepelin's intoxicated studies on the mind was not limited to the scientific community. Arguably, any notion of a cohesive biologism, one that participates in the dominant conceptual a prioris underlying an array of subjectivities, can expectedly be found in other discursive communities, as far as they are concerned with the body and mind. To this point, it is important to also look to the emerging body philosophies of the period. For Nietzsche, Freud, and Weber, thinkers whose work would leave indelible marks on our sense of the modern, it would be not only the work of Wundt and Kraepelin that would shape their conceptions of the body, but intoxicants themselves. In the mingling of the scientific encounter with intoxicants with their own intoxicated lives, each would, in their own way, argue for a foundational conception of the biological subject.

✴ 11 ✴
Drunken Songs of Tomorrow

NIETZSCHE, FREUD, WEBER, AND INTOXICATION

Toward a Transpyschological Biology of the Subject

There is an illusory gap between the world of philosophy and that of the sciences. Such perceptions somehow even persist when it comes to the formations of subjectivities, biological or otherwise. There is no denying that physiological and biomedical research directly intercedes in one's conception of what it means to be. We might resist, striking out against the biomedical categories of experts failing to describe the uniqueness of our experience. And yet these points of resistance become enfolded in biologism itself, changing it without negating it. Medicine, institutional policies, civil health ordinances, and economic regulatory practices, to name but a few, are constantly negotiated, and renegotiated, at the sites where they cross into our lives—they all help make us up as subjects. This diversity of inputs, given unique expression through the radical particularity of place and time, entails a certain expectation that conceptual transmission will almost always be partial, fractured, and distorted. Yet there is a peculiar degree of conceptual cohesion surrounding the emergence of the biological subject toward the end of the nineteenth century, one that persists to this day. As has hopefully been made clear, this cohesiveness is irreducible to any presumed facticity of being in a body. It could, however, be expected that such cohesiveness would entail commonality between seemingly conflicting points of view, that the horizon of possible questions that could be asked apropros the body and mind, both within and without the sciences, had radically shifted.

The answer to this query can be found by turning to the work to those late nineteenth-century social theorists and philosophers who abandoned spirit in favor of a purely biological understanding of the human being. One need look no further than some of the period's most influential personalities: Friedrich Nietzsche, Sigmund Freud, and Max Weber. Peering not only into their work but their lives, a common thread emerges: drunkenness, narcosis, and inebriation pervaded their bodies and informed some of their most influential ideas. For each, intoxication figured centrally in their lives at the critical junctures of their interface with the emerging experimental interest in substances of intoxication. For Nietzsche, this would culminate in biopsychiatric philosophy of intoxication. For Freud, his cocaine-charmed life would mark his only real turn to experimental psychology, followed directly by the emergence of psychoanalysis. And, for Weber, his personal struggles with intoxication would translate into a fetishization of sobriety, one that would figure centrally within his conceptions of capitalist modernity, scientific epistemology, and the modern body. Each would, in turn, develop their own languages, their own vocabularies, with which to voice a dynamically biologistic perspective that was both consummately organic and yet inclusive of a distinct, scientific concept of embodiment.

Nietzsche: Psychiatrist of the Will

The scholarship surrounding the potential sources of Nietzsche's ideas and his possible influences is greatly indebted to Nietzsche's efforts to obfuscate their origins. Even in his central published works, Nietzsche writes for a learned reader. He constantly levels highly focused critiques against the works of individuals he rarely, if ever, names, instead leaving it to the reader to be well-enough versed in what Nietzsche himself had read. It is perhaps a great irony then that today Nietzsche is mostly wasted on the youth. Even after his vast personal library is cataloged, the contents of his letters and journals are considered, and records of his lifetime borrows from the library are queried, what we're left with is still a picture of negative space. That is, the blank space left by the impression of what Nietzsche would seem to have read, and yet lacks direct evidence. This dilemma is the product of both the simple reality that people do not generally leave a record of everything they read and Nietzsche's own self-conscious efforts at self-mythologizing. Nietzsche actively obfuscates his origins and, where an association of his is well known, he misrepresents the nature of others' influence. A very obvious example of this is his friendship with Paul Rée, with whom Nietzsche shared a profound personal and intellectual closeness until their falling out, for which Nietzsche would overcompensate in attesting to the complete lack of influence that Rée ever had on Nietzsche's thought (Holub 2015, 157–60).

One domain of great interest concerning Nietzsche's philosophy, the subject of considerable effort in coming understand its influence, is physiology. Early on, Nietzsche began to advocate for some kind of material doctrine of the body. Some have suggested this was emergent of Nietzsche's lifelong poor health, plagued by a condition that would leave him periodically blinded and assailed by near constant indigestion, Nietzsche's awareness of his body was repeatedly conditioned through illness (Dahlkvist 2014, 138). Nietzsche was thus primed to have an interest in both medicine and physiology, not only in order to understand these subjects but, in a sense, to make use of them himself (Dahlkvist 2014, 138).

Another factor was his friendship with Paul Rée. Nietzsche and Rée met in May 1873 and became close friends by 1876, at the latest. It was Rée's passion for the natural sciences, and especially the life sciences, that inspired Nietzsche to broaden his reading in the late 1870s and early 1880s. Prior to their friendship, Nietzsche did not have a more than passing interface with serious scientific publications. An important starting point was Nietzsche's first encounter with Friedrich Lange's *Geschichte der Materialismus* in 1866, which was not long after his introduction to the works of Schopenhauer (Brobjer 2017, 26). After reading Lange, Nietzsche produced a science reading list in 1868 that included Helmholtz's *Über die Erhaltung der Kraft*, Wundt's *Über die Menschen- und Thierseele*, Lotze's *Streitschriften*, and Oken's *Zeugung*, among others (Brobjer 2017, 28). It is not clear which books he ultimately got around to reading. In 1872–1873, Nietzsche became acquainted with Zöllner's *Über die Natur der Kometen: Beiträge zur Geschichte und Theorie der Erkenntnis*, a text that would have a tremendous impact on Nietzsche's thinking (Brobjer 2017, 32). As early as 1874, Nietzsche made a passing reference to du Bois-Reymond in *Schopenhauer als Erzieher*, and at least two more references are found in the *Nachlaß* from that spring/summer (Nietzsche NF-1874, 35[12]). It was around this time that Nietzsche may have come to own du Bois-Reymond's 1874 *Über Geschichte der Wissenschaft* (Campioni et al. 2003, 199). In 1875, Nietzsche purchased and read Julius Bernstein's *Die fünf Sinne des Menschens* (Campioni et al. 2003, 141). *Unsere Körperform und das physiologische Problem ihrer Entstehung* by Wilhelm His was purchased in 1876 (Campioni et al. 2003, 299).

Needless to say, Nietzsche developed a sustained interest in the life sciences, before and into the year 1876. The year 1876 is noteworthy, as it marks the divestment of Nietzsche's thought from the influences of Schopenhauer and Wagner, which Brobjer has argued was at least partially due to the influence of some of the books listed above (Brobjer 2017, 34). The year 1876 was also the beginning of Paul Rée and Nietzsche's close collaboration. Rée had studied philosophy at Leipzig and even had the opportunity to study science in Berlin (Holub 2015, 157). Although he would not become a physician until later in life, it appears Rée was always deeply influenced by scientific research and especially physiology, medicine, and Darwinism. That Rée's scientific proclivities would leave an impression on Nietzsche is all but a foregone conclusion. The two spent a winter together in Sorrento, and all the surviving documentary evidence paints the clear image of two men who partook in an extremely free exchange of ideas (Holub 2015, 157). For a while, it appears that Nietzsche was even something of a follower of Rée, in both his scientific and historiopsychological interests (Holub 2015, 157).

Under the influence of Rée, Nietzsche himself relays that in the period between 1876 and 1883 he read almost nothing but books of scientific interest (Brobjer 2017, 35). Among many others, Nietzsche acquired James Johnston's *The Chemistry of Common Life*, Michael Foster's *A Text Book of Physiology*, Bain's *Mind and Body: The Theories of Their Relation*, and Adolf Fick's *Ursache und Wirkung*, as well as various scientific periodicals (Brobjer 2017, 37, 39). Thus, when Nietzsche wrote "What is a word? The representation of a nerve stimulus in sounds," or equated bodily activity with intellectual virulence, this was not informed by popular materialism but rather by a relatively learned awareness of the current scientific literature (Nietzsche 1873 [1975], WL-1, §1; Nietzsche 1881 [1975], M-18, §18).[1] Nietzsche was arguably even a positivistic materialist during the period between 1873 and 1883, jotting in a 1876 note that "positivism [is] totally necessary" (Nietzsche 1876 [1975], NF-1876, 20[19]). Even as Nietzsche himself grew critical of positivistic approaches and developed a broadly skeptical attitude, Nietzsche's rejection of metaphysical notions and his focus on the somatic character of mental phenomena remained demonstrably clear.

An additional, and perhaps more crucial, influence further exacted via Nietzsche's friendship with Rée was that of psychology. Perhaps more than anything else, Nietzsche understood his philosophical work to be psychological in nature, at times lamenting that he was still not popularly referred to as a psychologist (Golomb 2015, 1). First identifying himself as a psychologist as early as 1878, Nietzsche would sporadically renew the label for the remainder of his life (Nietzsche 1878, BVN-1878 [1975], 762). The subtitle of his final publication was nothing less than "Aktenstücke eines Psychologen" ("From the Case Files of a Psychologist") (Nietzsche 1889 [1975], NW-Titelblatt). The question is: In what sense did Nietzsche understand himself to be a psychologist, and how did he figure his approach in to the psychologies that were blooming in the late 1870s and 1880s?

Here, too, Paul Rée was a formative influence. Nietzsche had been exposed to Rée's moral psychology as early as 1875, when Nietzsche read Rée's 1875 *Psychologische Beobachtungen* (Holub 2015, 168). Two years later in 1877, Rée debuted *Der Ursprung der moralischen Empfindungen*, a psychological genealogy of the moralistic sense.[2] Though these were precisely the subjects that would make Nietzsche famous, it appears these interests originated in Nietzsche and Rée's interactions. Nietzsche's psychophilosophical interests in a broader sense do, however, predate his close friendship with Rée. As early as "Über Wahrheit und Lüge im außermoralischen Sinn" (1873), Nietzsche expanded on Kant's critical work in arguing that any perceived categories of conscious themselves are mere leaps of metaphor. "True" knowledge of self and the world could not be independently verifiable, and yet we encounter the world knowingly. Metaphysical frameworks, speculative systems, transcendent principles—these were valuable not on account of their epistemic durability, but because of the mythic purposes they served for individuals and for society (Nietzsche 1873 [1975], WL-1, §1; Nietzsche 1874 [1975], HL-Vorwort-1). Nietzsche's understanding of the psychological implications of this argument was clearly articulated in "Vom Nutzen und Nachtheil der Historie für das Leben," where the shared cultural imaginary, given expression through a culture's historical sense, could function either as a force for stagnation or for excitation, anticipating the conflict between life-celebrating and life-denying values as a central facet of Nietzsche's own moral psychology (Nietzsche 1874 [1975], HL-1).

It already becomes clear what Nietzsche intends when, shortly after becoming close friends with Paul Rée, Nietzsche begins to write about the need for a "new psychology" (Golomb 2015, 1; Liebscher 2014, 362). It would be psychology, Nietzsche suggests, that would provide the instruments with which to diagnose and understand the great crises of the age. However, Nietzsche scarcely makes explicit how he would ground such a psychology methodologically. A strong hint can be found in Nietzsche's discussion of experimentation in *Morgenröte: Gedanken über die moralischen Vorurteile*. Here, the principle of experimentalism is identified with the generation of new possibilities, futures, and ways of being (Bamford 2016, 13). Nietzsche muses about experimentation not only in a laboratory but on the social, moral, and perspectival structure of ourselves and our societies, proclaiming "we are experiments" (Nietzsche 1881 [1975], M-453). Contrasting sharply with philosophies of compassion and security, Nietzsche identifies experimentation with risk, potentiality, and new beginnings (Bamford 2016, 13–14). Though any experimentation can only ever be preliminary for Nietzsche, it is clear that experimenting *psychologically*, in concert with experimentation in the natural sciences, is a significant way in which Nietzsche's new psychology would function.

Of course, what Nietzsche here refers to as experimentation falls outside of the understanding that was typical within the purview of the physical sciences, and certainly Nietzsche's psychophysically minded contemporaries. In this sense, Nietzsche's words here betray the influence of scientifically minded psychological-philosophers, notably Zöllner, Lange, and Hartmann (Lehrer 2015, 182). Yet, given the respect, even admiration, Nietzsche expresses for the natural sciences and his self-identification with psychology, it is worth nonetheless considering the possible influences of scientific psychology in the tradition of Wundt and Kraepelin.

The Wundtian Nietzsche

As with many of Nietzsche's influences, Nietzsche never explicitly names the likes of Wundt, nor his followers, in his published works. Yet there is fairly substantial evidence that Nietzsche had more than a passing interest in Wundtian thinking. To begin with, the works of Eduard Hartmann are found in Nietzsche's personal library, and there Nietzsche would have been exposed to Wundt's ideas in considerable detail (Campioni et al. 2003, 276; Lehrer 2015, 182). Nietzsche also listed one of Wundt's works, at the very least aspiring to read Wundt, in his earlier attempt at a scientific reading list. Then there are the contents of Nietzsche's letters and surviving papers. In the summer of 1877, Nietzsche wrote to Rée with some urgency about his efforts to arouse the editor of *Mind*'s interest in their work, very casually remarking that an essay by Wundt was being published there (Nietzsche 1877 [1975], BVN-1877, 643). On a note written in 1879, Nietzsche simply writes "Wundt 'Aberglaube in der Wissenschaft,'" without further context (Nietzsche 1879 [1975], NF-1879, 47[15]). Paul Rée also wrote to Nietzsche about Wundt in 1881. Taken together, it is difficult to deny that Nietzsche had a sustained awareness of Wundt's publications as far back as the late 1860s, and that this sustained awareness translated into discussions with Rée until at least the early 1880s. The pièce de résistance though is a letter sent by Nietzsche to his publicist, Constatin Georg Naumann, on November 8, 1887, requesting that free copies of his upcoming book, *Zur Genealogie der Moral*, be sent to "Herrn Professor Dr. Wundt, Leipzig[;][...] Herrn Professor Dr. Leuckart, Leipzig[;] [...] Dem Geheimrath Prof. Dr. Helmholtz[;][...][and] Herrn Professor Dr. Du Bois-Reymond, Berlin," among several others (Nietzsche 1887 [1975], BVN-1887, 946). This extends the timeline of Nietzsche's interest in Wundt until the final year of his conscious life, while affirming his sustained interest in physiology.

The evidence overwhelmingly suggests that Nietzsche had an awareness of Wundt's work and even aspired to capture the interests of the experimental psychologist. This in itself intimates a great deal about Nietzsche's philosophical legacy. But there is further evidence of not only Wundt's influence on Nietzsche, but perhaps Nietzsche's influence on Wundt. This is particularly clear in Nietzsche's notion of will, the beating heart of Nietzsche's mature psychological-philosophy. It is first in *Jenseits von Gut und Böse* that Nietzsche introduces his fully developed notion of the *Wille zur Macht*. Here, Nietzsche outlines his conception of will in contrast to a dualistic or spiritual idea, instead identifying not a single will but a ceaseless combat of unseen wills bubbling beneath the surface of all psychological processes (Cowan 2005, 48). In this sense, the *Wille zur Macht* referred not to a particular will but to the process of conflicting wills behind all organic life. Crucial to this idea is both the noncausal nature of Nietzsche's will and the opacity of the willing process relative to human awareness. As Nietzsche poetically pondered in *Also Sprach Zarathustra*, "if I wanted to shake this tree with my hands, I would not be able to do it; but the wind, which we do not see, twists and bends it wherever it wants" (Nietzsche 1883 [1975], Za-I-Baum). We are, everywhere, twisted and bent by unseen hands.

Nietzsche's idea of the will as diffused throughout every facet of embodied experience, rather than an explicit faculty or volitional force, draws direct parallels with the idea of will found in Wundt's *Grundzüge*. In the 1874 *Grundzüge*, Wundt had already described that "all organic and psychic functions as more or less complex expressions of a basic form of 'will' at the origin of all physiological and psychic activity" (Cowan 2005, 50). All psychical and physical actions, both intended and unintended, are "modalities" of the willing process (Cowan 2005, 51). Whereas instinctual reactions were acts of certain sets of wills, decisions—such as those tested in choice reactions—involved a more sophisticated "struggle of conflicting motives," a combat of wills (Wundt 1909, 225; Cowan 2005, 51).[3] Although Nietzsche never identifies Wundt as an influence, the evidence establishing Nietzsche's awareness of, and interest in, Wundt, as well as the clear parallels between Nietzsche's psychological model and Wundt's, leaves little room to deny the possibility of influence. This is also true of Nietzsche's interest in experiment as a central component of his concept of a new psychology, one that was foundational to what Nietzsche even referred to in *Jenseits* as a "Physio-Psychologie" (physiopsychology), a striking allusion to Wundt's "physiologisch[e] Psychologie" (physiological psychology) (Nietzsche 1886 [1975], JGB-23). Far more compellingly, shifts in Nietzsche's understanding of intoxication may align with the concepts "baked into" the Wundtian research program of experimental psychology.

In Opio Veritas

Intoxication had been a focal point of Nietzsche's symbolic lexicon as early as *Die Geburt der Tragödie aus dem Geiste der Musik*, where the constrictive, logocentric impulse of the Apollonian was contrasted with the boundless, dynamic creativity of the Dionysian. The scope of the available literature on Nietzsche's Dionysus as both signifier and signified is effectively boundless. It is a dominant fixture of Nietzsche's corpus. But there is a clear shift in how Nietzsche takes up intoxication, and intoxicants, in his symbolic language as he moves from the so-called middle period into his mature philosophy.

By the writing of the *Fröhliche Wissenschaft* (1882), intoxication plays a central, albeit still symbolic, role in Nietzsche's thought (Ciaccio 2018, 119). As Nietzsche wrote of the failings of modern theater, "the strongest thoughts and passions are made available to those who are incapable of thought and passion—but of intoxication [...] theater and music are the hashish smoking and betel chewing of Europeans" (Nietzsche 1882 [1975], FW-86). This had been the Dionysian principle at play behind Nietzsche's temporary infatuation with Wagner—music, art, the aesthetic experience had long been the premier European narcotic. Narcosis and stupefaction are expressive of the deleterious potentials of excesses in a given form of intoxication, while intoxication more generally is simultaneously the much-needed antidote to constrictive metaphysics. Nietzsche reflected on this very principle in *Ecce Homo*, remarking that "if one wants to do away with an unbearable pressure, hashish is necessary; thus, I had a need for Wagner" (Nietzsche 1889 [1975], EH-Klug-6). Within his emerging, physiopharmacological lexicon, the narcotic excess becomes representative of the complacent immovability of what was once the Apollonian, the antidote to which is joyful, invigorating intoxication.

That Nietzsche would use the symbolic language of intoxicants to denote the aesthetic is not at all surprising. By this time, Nietzsche was already long adrift in a sea of very real intoxicants, a situation that would only intensify over the course of his life. Opium, chloral, potassium bromide, hashish, at times even alcohol—all were consumed frequently and at incredibly high doses (Anderson 2011, 110; Ciaccio 2018). In one letter to Paul Rée and Lou Salome dated December 20, 1882, Nietzsche wrote, imaginably in an opium-induced stupor, "*in opio veritas*: long live wine and truth" (Nietzsche 1882 [1975], BVN-1882, 360). Famously, he would acquire many of these intoxicants by signing fraudulent prescriptions written by "Dr. Nietzsche" (Anderson 2011, 110). His own mother relayed that Nietzsche would buy chloral in bulk and relied on it for sleeping, and Nietzsche's sister would attribute his descent into madness to Nietzsche's use of a still-unknown Javanese narcotic (Ciaccio 2018, 118). Thus, intoxication was not merely metaphorical for Nietzsche. It was a literal expression of Nietzsche's perspectival encounter with his own everydayness. Far from an occasional notion, it was a constant companion in Nietzsche's daily life. *Rausch* was, in this sense, both literal and emblematic. There is another dimension to Nietzsche's thought, one touched upon in many of the references cited up to this point and of the utmost importance to Nietzsche's understanding both of intoxicants and of his identity as an experimenting psychologist of the age, that of *degeneration*.

Dégénérescence, Entartung, Degeneration

It is clear that when Nietzsche writes "vice is not a cause; vice is a consequence," or characterizes the narcotizing cultural decadence of the belle epoque as the "hashish smoking and betel nut chewing of the European," that he is drawing on the late nineteenth-century metanarrative of degeneration theory (Nietzsche 1882 [1975], FW-86; Nietzsche 1888 [1975], NF-1888, 14[113]). This is as true of his grand cultural analysis of a European culture on the precipice of ruin as it is of his complex relationship with intoxication. From the outset, the Morelian systemization of existing hereditizing impulses in French psychiatry realized in the 1857 *Traité des Dégénérescences* identified intoxicants as a significant, if not the most significant, cause of degeneration (Bynum 1984; Morel 1857). Mid nineteenth-century French psychiatry was deeply intertwined with the vibrant tradition of French orientalism. Interest in hashish, coupled with the orientalization of opium, was influenced by the Commission des Sciences et des Arts and their scientific encounters during Napoleon's campaign, providing the basis for Moreau de Tour's Middle Eastern voyage, subsequent interest in hashish, and the related activities of the Club des Hashischins (Black 2022; Kamieński 2016). Coupled with then-living memory of the French, German, and English moral panics of the late eighteenth and early nineteenth centuries, which saw intellectuals and legislators wringing their hands over the popular consumption of alcoholic spirits, intoxicants such as opium, hashish, and alcohol served as tangible vehicles of moral degeneracy, economic disruption, and cultural otherness.

Nietzsche drew heavily on not only Morel, but also the swelling continental discipline of degenerationist thought, which included Charles Féré, Caesare Lombroso, and—possibly, though the matter remains hotly debated—Arthur de Gobineau (Owen 2021; Martin 2004; Moore 2000).[4] It should be acknowledged the Nietzsche was far from a typical degenerationist. Nietzsche appropriates the vernacular of degeneration and looses it upon the ostensibly Christian moralism implicit to the values of degeneration theory, finding in criminality, intoxication, and deviance hints of a vigor for life at odds with bourgeois decadence. We can catch glimmers created by this tension in the Nietzschean prose of the bank-robbing anarchist philosopher Renzo Novatore. Degeneration, for Nietzsche, was inseparable from the liberal bourgeois society that sought to pathologize it. David Owen and Ken Gemes go as far as proposing that Nietzsche seeks to interpolate agents of social infection into a higher culture rather than to divest society of degenerative elements, a heuristic that echoes the Nietzschean call to affirm *all* of history (Gemes 2021; Owen 2021). It is on these grounds that the influential degenerationist Max Nordau would identify Nietzsche not as a kindred spirit, but as ego-maniacal degenerate par excellence (Nordau 1892).

Pharmapathologies of the Will

The space in which questions of intoxication, physiology, the will process, and degeneration intersect is in Nietzsche's late pharmapathology of the will, a movement playfully—à la Nietzsche—identified with the advent of Nietzsche the diagnosing *psychiatrist*. Accompanying his longstanding discussion of intoxication, a major conceptual shift occurs in Nietzsche's mature works. Nietzsche had previously provided an aspirational identification of his philosophical project with an "experimental" physiopsychology of culture. Intoxication, and intoxicants, were lexically diffused throughout his early work, and yet this psychological project rarely directly engaged with his symbolic language of intoxication. However, from the time of *Jenseits* onward, Nietzsche moves to invoke the psychiatric language of psychopathology (Cowan 2005, 53). It is only here that Nietzsche begins to speak of "weakness of the will" (*Willensschwäche*) and "loss of will" (*Willensverlust*) (Cowan 2005, 53). Moral impulses become the product of a pathology of the will process, which informed all manner of mental and physical activity, as he recorded in a fragment from his *Nachlaß*:

We say, for example [...] one becomes a proper person because one is a proper person: that is, because one is born as a capitalist of good instincts and prosperous conditions . . . If one is born poor, from parents who have done nothing but squandered in everything and saved nothing, then one is "irredeemable," one wants to say ripe for the penitentiary and the lunatic asylum. . . . Today we no longer know how to think of moral degenerescence as separate from physiological degenerescence: it is a mere symptom complex of the latter. Bad: the word here expresses certain incapacities, which are physiologically connected with the type of degeneracy: e.g., the weakness of the will, the insecurity and personal being-in-the-majority of the "person," the impotence to suspend the reaction to any stimulus and to "control" oneself, the bondage to any kind of suggestion of an alien will. Vice is not a cause; vice is a consequence . . . . Vice is a rather arbitrary conceptual delimitation to summarize certain consequences of physiological degeneration. (Nietzsche 1888 [1975], NF-1888, 14[113])

The moral behavior that Nietzsche had opposed for over a decade now is no longer merely emergent of the psychohistorical process; they are "symptom complexes," just as Kraepelin would say—the discernible effects of a disordering in the psychological process. This would be a defining change in Nietzsche's approach, one which marked the transition from Nietzsche's ambiguous, early middle-period metaphysics of the will to a fully biological conception of the subject. In *Götzen-Dämmerung*, Nietzsche goes as far as identifying how "every mistake, in every sense, follows from *instinctual degeneration*, from disintegration of the will: one almost uses it to define the bad" (Nietzsche 1888 [1975], GD-Irrthuemer-2).[5] Each example of perceived decline, of decadence, in both the individual and in society becomes a "Symptom" (Nietzsche 1888 [1975], GD-Vernunft-6, GD-Moral-2, GD-Moral-5, GD-Deutsche-6). Nietzsche the psychologist becomes Nietzsche the psychiatrist, diagnosing the psychical diseases of the age and reframing them as disorders of the mental process.

Some scholars, notably Cowan, Lampl, and Haaz, have also sought to credit this pathological turn in Nietzsche's later work to the influence of Ribot and French medical science, seeing them as central to Nietzsche's late development of a body-philosophy (Cowan 2005, 56; Lampl 1987; Haaz 2002). They make a strong case, and Nietzsche's knowledge of the French tradition is clear. To their further credit, Nietzsche's pathological turn drips with the influence of the—largely French—tradition of degeneration theory. But a closer look at the interactions between Nietzsche's Wundtian conception of the will, his long-established symbology of intoxication, and his pathological turn suggests a further dimension to his late body-psychology.

It is, above all, in Nietzsche's later remarks on the nature of intoxication that his newfound pathology of the will comes to light. In a note from 1888 titled "Zur Physiologie der Kunst," Nietzsche appears to have listed in what way, physiologically speaking, the effects of art are likenable to those of intoxication, writing, "1. intoxication as a prerequisite: causes of intoxication 2. typical symptoms of intoxication 3. the feeling of strength and fullness in intoxication: its idealizing effect" (Nietzsche 1888 [1975], NF-1888,17[9]). Here, Nietzsche appears to draw on the familiar symbolic language of intoxication, only now he grounds it even more explicitly in the embodied psychophysical experience of intoxication. This dynamic becomes all the more apparent in *Götzen-Dämmerung* (also 1888), where Nietzsche the psychiatrist makes his first major debut and where Nietzsche introduces the language of "intoxication of the great will" (Nietzsche 1888 [1975], GD-Streifzuege-11). Now intoxication is explicitly identified as a temporal modification of the will. However, intoxication is not merely a modification of the will; it is a pathology of the will process, as Nietzsche says in another fragment from 1888:

There are two states in which art itself appears in man like a force of nature, affecting them whether they want it or not: on the one hand as a compulsion to vision, on the other hand as a compulsion to orgiasm. Both states are also present in normal life, only weaker, in dreams and in intoxication. (Nietzsche 1888 [1975], NF-1888, 14[36])

Intoxication, and intoxicants, cease to merely serve as a representation of or metaphor for the effects of art. Intoxication now refers to a temporal pharmapathology of psychological processes, giving voice to orgiastic expression.

Nietzsche's identification of intoxication as a pharmapathology of the psychological process is not merely an accident of the characteristic looseness of Nietzsche's symbolic language. It figures heavily in his diagnoses of the age. Speaking on the failures of German culture, Nietzsche remarks that they should come as no surprise, as "nowhere are the two great European narcotics, alcohol and Christiandom, being abused more licentiously" (Nietzsche 1888 [1975], GD-Deutsche-2). He goes as far as pondering "how is it actually possible that young men, who dedicate their existence to the most spiritual of goals, do not feel the first instinct of spirituality, the instinct of self-preservation of the spirit, in themselves— and drink beer" (Nietzsche 1888 [1975], GD-Deutsche-2). Beer drinking here is both symptom and cause, directly disordering the psychological process. This is a radical departure from the metaphorical equivalence between drugs and stupefecation in Nietzsche's earlier work, scarcely more than an ambiguous echo of the familiar Marxist comparison of religion and opium. This is even a departure from a simple attribution of societal narcosis to intoxicants. Intoxication is identified with explicit physiopsychological effects on Nietzsche's overtly Wundtian conception of the psychological process. Such notions reek of nothing less than Kraepelin's work in the 1880s interpreted through the lens of Nietzsche's psychohistorical moral theory.

In this way, intoxicants are located at the point of transition from Nietzsche's metaphysical, or at least linguistic, conception of the will to an overtly biological subject. By Nietzsche's final years, the mores of society that Nietzsche had spent most of his public life condemning had become pathologies. Pathologies of the physiopsychological process, of the will, of the will to power. Better put, they were pharmapathologies of the will: psychological disturbances both caused by and analogous to the effects of intoxicants on the body and mind. It appears that Nietzsche's own considerable, career-long use of opium, potassium bromide, chloral, and hashish initially provided the symbolic language through which he could express what he perceived as the crises of his day, only for the symbolically signified to collapse into the signifier. Nietzsche was not a scientific researcher. What Nietzsche's physiological and psychological research provided was a grammar, a language, and finally a framework with which to translate the experience of intoxication into a dynamically biological conception of subjectivity. The newly discovered biological subject, for all its pathologies, was an intoxicated subject.

Could Nietzsche have been influenced by Kraepelin? Matters of influence are forever fluid, insolvent, undecided. Even where influence is freely admitted, the historiography is free to ply away at questions surrounding the scope and nature of conscious and unconscious transmission. As for Nietzsche and Kraepelin, it has hopefully already become clear that the two fin de siècle diagnosticians shared more than one might at first expect. Much like Nietzsche, Kraepelin broadly engaged with the degeneration theories of his day. The better part of Kraepelin's explicitly degenerationist thinking would materialize from 1903 onward, a shift coinciding with his travels in Dutch colonial Java and which saw Kraepelin interested in preserving the health of the *Volkskörper* (Engstrom 2007). Kraepelin certainly engaged with earlier theories of degeneration, such as in the work of Krafft-Ebing, whose work on the pathology of criminality influenced the writing of a textbook of formative significance for Kraepelin's 1883

Compendium. Degenerationist thinking would increasingly figure into Kraepelin's macroscopic conceptions of mental pathology, a deepening trend in Kraepelin's thought hinted at as early as his 1893 description of dementia praecox as a "psychical degeneration process" (*psychische Entartungsprocesse*) (Hoff 2008; Kraepelin 1893, xiii). At the same time, Kraepelin was in keeping with Nietzsche in his general ambivalence concerning the specific claims marshaled by degeneration theorists such as Lombroso (Hoff 2008). This is, above all, exemplified in their respective understandings of intoxicants as, at once, a medium of psychical degeneration as well as a literal and figurative avenue of profound mental stimulus.

Of the possibility of Kraepelin's direct influence, Nietzsche, as is typical, leaves little explicit evidence to go on. But from the language of symptom complexes to the association between intoxicants and pathologies of a Wundtian psychological process, the late Nietzsche certainly appears to be rehearsing Kraepelin. Having established the likelihood of a Wundtian influence, it becomes easier to question whether Kraepelin, in particular his work on intoxication, figured into Nietzsche's conception of the subject as well. Kraepelin's work, including "Ueber die Einwirkung einiger medicamentöser Stoffe auf die Dauer einfacher psychicher Vorgänge," was published in Wundt's journal, *Philosophische Studien*, alongside many of Wundt's own works, where Nietzsche had the will and the means to access them. Kraepelin was also a serious contributor to Wundt's foundational research program of reaction time. Those interested in Wundt's early experimental work would have almost inevitably crossed pathes with the work of Kraepelin and Wundt's other students, as well as their experimental encounters with intoxicants as pathological states. Even outside of Wundt-adjacent publications, Kraepelin interacted with texts and individuals with which Nietzsche had a strong association. Kraepelin was a regular contributor to *Literarische Centralblatt für Deutschland*, what was then the most widely read literary review in Germany and certainly well known to Nietzsche (Steinberg 2000). In 1885 and then in 1886, Kraepelin wrote reviews for the *Literarisches Centralblatt* of Paul Rée's *Die Entstehung des Gewissens* and then *Die Illusion der Willensfreiheit: Ihre Ursachen und ihre Folgen* (Kraepelin 1885, 1697–98; Kraepelin 1886a, 41). Not only might Nietzsche have read Kraepelin's reviews of Rée's work, but Kraepelin's review of *Die Entstehung des Gewissens* connected its central argument to some topics in criminal psychology—a further consideration of Nietzsche's (Kraepelin 1885).

Ultimately, possible influences can only be assessed on the basis of changes in Nietzsche's thought, especially on the symbolic and polemic function of intoxication and intoxicants. If we are to assume there was no interaction between the two figures, it can nevertheless be said that Kraepelin and Nietzsche danced to similar music: the steady percussion of degeneration, the crooning brass of physiology, and the swelling strings of the psychical—all bending to the bittersweet melody of intoxication. For Nietzsche as for Kraepelin, it is the marriage of intoxicants and psychology that ultimately gives rise to the conception of the biological subject. It is, perhaps, a fitting coincidence that Erwin Rohde Jr., the son of Nietzsche's dear friend Erwin Rohde, would go on to be a pupil of Kraepelin's, even becoming an experimental pharmacologist (Kraepelin 1983, 136, 242). At that, let us now continue on to consider the case of Sigmund Freud, a figure for whom even his limited foray into the psychometrics of intoxication appears to have been central to the emergence of the mind.

Freud: The Cocainization of Mind

The legacy of Freud is sometimes contrasted with that of Emil Kraepelin. Both born in 1856, their divergent impacts on the shape of mental health care over the course of the twentieth century have even led to calls of a Freud-Kraepelin dualism (Trede 2007, 237). This characterization is undoubtedly at least partially grounded in a mischaracterization of both figures: Freud as metaphysician of the psyche and Kraepelin as hard-nosed brain scientist. Yet both figures understood all mental disturbances as ultimately biological in character, while also upholding the practical necessity of a nonphysiological science of the mind. There is no need to look further than Freud's early neurophysiological research or his lifelong endorsement of evolutionary ideas, first after the fashion of Clausian Darwinism and later after that of Haeckelian neo-Lamarckianism (Perkins-McVey 2022). The real point of departure between these two men was the nature of *the cure*, though that too was not always the case. For at the turning point in Freud's shift toward his psychoanalytic method, one finds not only substances of intoxication but a brief foray into experimental psychology. Such was the substance of Freud's 1884/1885 *Über Coca* and related publications. In cocaine, Freud had sought and, for a moment, found a radical cure for a host of then barely treatable psychiatric maladies. How he came to these conclusions was not only through self-experimentation with cocaine intoxication, but through psychophysical research into cocaine's influence on reaction time and muscular output. In this sense, Freud's psychological cocaine studies would mark the first serious break with neomechanist neurophysiology and pave the way for Freud's later psychoanalytic work. Combined with a possible Kraepelinian influence, Freud's shift toward a composite, biological conception of the psyche relied, ultimately, on the cocainization of the mind.

During the period surrounding the publication of *Über Coca* and the subsequent papers on cocaine, Freud was an assistant in Meynert's psychiatric clinic, starting in 1883 (Dalzell 2011; Perkins-McVey 2024).[6] Though they would later have a quite famous falling out, Freud evidently at least initially held Meynert in high regard. There is evidence that Freud had, for a time, found inspiration, even cause for emulation, in the aspirations of his supervisor and teacher, drawing on Meynert's work in his early description of free nerve cells and the later development of his topographical model (Perkins-McVey 2024). Although the Austrian neurologist Gabriel Anton (1858–1933) did not regard Meynert's teaching very highly, Freud found Meynert to be a venerable teacher and "the greatest brain-anatomist of his time" (Freud 1899; Dalzell 2011, 68). The depth of Meynert's initial impact on Freud is clearly reflected by changes in the contents of Freud's publications, as he moved from Brücke's academic physiology lab to Meynert's psychiatric clinic.

Under Brücke, Freud published "Über den Bau der Nervenfaser und Nervenzellen beim Flußkrebs" (1882), a study of the anatomical structure of the nerve cells and fibers found in freshwater crayfish. "Über den Bau" had been a commendable piece of research on the neurophysiology of invertebrates, although it was arguably reserved as regards its possible theoretical implications. After taking the assistanceship under Meynert, the content of Freud's publications changed dramatically. "Die Struktur der Elemente des Nervensytem" (1884) was, by comparison, a highly theoretical text that attempted to translate Freud's earlier histological studies on the structure of nervous tissue in crayfish into a model for understanding the cellular structure of the nervous system in general. In this way, "Die Struktur" clearly reflected the influence of Meynert's theoretical emphasis on nerve fiber superstructure, and perhaps even shared in some of Meynert's boldness. "Ein Fall von Hirnblutung mit indirekten basalen Herdsymptomen bei Skorbut" (1884), meanwhile, was an impressive neurological case study after the Meynertian fashion, which tracked the rapid decline of a young man who arrived at the clinic with a case of scurvy (Freud 1884b). Both publications demonstrated the formative effect of not only Meynert's mentorship, but also the transition to a clinical setting.

Freud attained the post of lecturer in 1885, which would provide an opportunity to secure funding for his sought-after trip to Paris in order to intern with Charcot (Grzybowski and Żołnierz 2021). This trip has often been marked as a breaking point in Freud's personal, and conceptual, relationship with Meynert—with Charcot introducing Freud to hypnosis, male hysteria, and psychic trauma (Grzybowski and Żołnierz 2021; de Marneffe 1991, 71; Miller et al. 1969, 608). Freud left for Paris a neurophysiologist and returned to Vienna a psychoanalyst—or so the story goes. But what if the sea change in Freud's life was something else altogether? In Meynert's psychiatric clinic, Freud became well acquainted with the employment of chloral, morphine, and other intoxicants to suppress, or mask, the worst symptoms of their patients, for whom they could offer little cure. Potent intoxicants suffused the literal and figurative space of Freud's clinical life. But this quiet collection of actants were not as silent as at first appears. Before Freud turned to the "talking cure," cocaine would represent Freud's first attempt to find a lasting cure for mental disturbance (Ciaccio 2018, 118). Could it be that Freud's interstitial 1884–1885 self-experimental studies on the effects of cocaine intoxication could account for Freud's sudden transformation, could prefigure Freud's shift from physiology to psychoanalysis?

Let us begin with the nature of Freud's cocaine research and the extent to which it represented an effort to finally cure mental disturbance. Freud contextualized his own cocaine study within the broader history of European coca research, reaching back to the earliest days of Spanish intervention in the Americas (Freud 1885c, 1–5). Early European accounts of coca use among the indigenous populations of South America served as a quasi-mythological backdrop for many of the claims Freud would go on to make, relaying dubious stories of elders untouched by illness and whole populations who had used coca their whole lives without repercussion (Freud 1885c, 4–6).

Studying the effects of cocaine "in repeated experiments on myself and others," the chief concern of *Über Coca* was the benefits of cocaine's euphoric properties, which Freud relayed with intoxicating exuberance (Freud 1885c, 11):

> The psychic effect of Cocainum mur. in doses of 0.05–0.10 gr. consists of an exhilaration and enduring euphoria, which differs in no way from the normal euphoria of the healthy person. There is a complete absence of the feeling of being altered that accompanies the exhilaration caused by alcohol, and there is also an absence of the characteristic urge to immediate activity that accompanies the effects of alcohol. One feels an increase in self-control, feels more vital and able to work; but when one works, one also misses the noble excitation and increase of mental powers caused by alcohol, tea or coffee. (Freud 1885a, 12)

These euphoric highs were not without their drawbacks. Freud reported how his early, oral consumption of 0.05 grams of cocaine in solution elicited some gastrointestinal effects, numbing of the mouth and lips, a quickening of the pulse, and a drying of the mucous membranes that persisted for hours after consumption, most of which abated with repeated consumption (Freud 1885c, 11–12). But it was worth it, for the exhilarating effects of cocaine allowed Freud to complete long-lasting mental and physical labor without any sign of fatigue (Freud 1885c, 13). Better yet, Freud found that cocaine's euphoria was not followed by depression or fatigue, nor did high doses lead to a clouding of one's conscious awareness (Freud 1885c, 14, 15).

Such miraculous effects led Freud to propose cocaine as a powerful cure for select forms of mental disturbance. One such benefit of cocaine, as Freud understood it, was that it had the potential to fill a substantial hole in the pharmacopoeia of psychiatry, as he suggested: "To many physicians, cocaine seemed to be called upon to fill a gap in the pharmacopoeia of psychiatry, which, as is well known, has enough means to reduce the increased excitation of the nerve centers, but knows no means of increasing the reduced activity" (Freud 1885c, 16). In this way, Freud had found a risk-free "Zaubermittel," a magic formula, of immediate import to psychiatrists around the world (Reicheneder 1988, 172). The 1885 "Ueber die Allgemeinwirkung des Cocains" was even put to print by Freud with the main intent of spreading awareness of cocaine's potential for psychiatrists (Freud 1885d, 49). Through repeated small doses over a long time span, Freud suggested that the course of these illnesses could be reversed—melancholia, neurasthenia, sexual asthenia, hysteria, even hypochondria had reportedly been cured through coca therapy (Freud 1885c, 16–17). Freud had found in cocaine not only a treatment for psychical disturbances "which we interpret as states of weakness and depression of the nervous system without organic lesions," but an actual cure (Freud 1885d, 51).

Best of all, Freud was confident that at moderate doses even chronic use had no negative effects on one's health, though he at least acknowledged that extreme use led to symptoms similar to "Alkoholismus und Morphinismus" (Freud 1885c, 4, 6). Yet combating precisely these two modern afflictions was one of the potential applications Freud envisioned for cocaine therapy. Freud reported that it had been shown to reduce withdrawal symptoms in those reliant on morphine or alcohol, and reduce the need for them (Freud 1885c, 21, 22). Freud wrote of having read and heard of, at times, miraculous reversals in not only the reliance on morphine but actual rejuvenation of the general health of the patient when morphine was substituted for cocaine (Freud 1885c, 22). Once patients switched to cocaine they could then far more easily wean off it (Freud 1885c, 20–22). Freud spoke to this with the theory that cocaine had "a direct antagonistic effect against morphine" (Freud 1885c, 22).

Where chloral, morphine, or paraldehyde could only hide a patient's symptoms, bury them in a doss of droopy eyelids and quiet hearts, cocaine cured—bringing light to dreary minds, and maybe their guts too. The potential Freud saw in cocaine therapy would ultimately be deemed misguided. But it may very well have been the first major step toward the emergence of psychoanalysis, and to understand that development it is important to work through the psychophysical turn at the heart of Freud's cocaine study and discern what it owed to Kraepelin.

The Psychometrics of Cocaine

While *Über Coca* already ranked as perhaps the most substantial publications on cocaine's therapeutic benefit in its time, one of the most remarkable elements of the study is Freud's often overlooked turn to psychophysical measurement. Freud's measurements of cocaine on muscular strength and simple reaction time were published in his 1885 "Beitrag zur Kenntnis der Cocawirkung," with some brief remarks on dynometrics and reaction time making their way into the addenda of *Über Coca* (Freud 1885a, 7; Freud 1885c, 25). This would be Freud's only real foray into the world of psychophysics or—in the Wundtian sense—psychological experimentation, and it centered around cocaine.

For the reaction time experiments, Freud borrowed Exner's "Neuramoebimeter," which was in many ways comparable to the apparatus Exner had used in his past experiments on simple reaction time in 1873 (Freud 1885a, 7; Freud 1885c, 25). Though Freud clearly took some methodological guidance from Exner, the conceptual foundations of their experiments were radically different. As discussed, Exner had experimented with intoxicants as controlled modifiers of fatigue, while Freud designed his experiment with the objective of understanding the psychological and physical effects of his miracle cure, cocaine. Freud's simple reaction time trials produced clear results: the introduction of cocaine consistently produced simple reaction times that were shorter than the baseline, and more stable (Freud 1885a, 7; Freud 1885c, 25). The tests with the dynamometer were consistent with the reaction time experiments. Here, Freud found that there was a marked increase in motor power starting fifteen to twenty minutes after ingestion, which only gradually decreased over the following four to five hours (Freud 1885a, 5–6; Freud 1885c, 25).

Most remarkably, Freud observed that the psychophysically measurable effects of cocaine were directly preceded by the onset of *euphoria* (Freud 1885a, 7). Both the shortening of reaction time and increased muscle output run "parallel to the coca euphoria and also seems to originate more from the central readiness to work, from the elevation of the general condition, than from a direct influence on motor apparatuses" (Freud 1885c, 25).[7] Freud the neurophysiologist appeared to be suggesting that the curative potential of cocaine relied on the uplifting euphoria of cocaine intoxication, rather than the direct effects of cocaine on the nervous system (Reicheneder 1988, 173; Freud 1885a, 7). Specifically, it is attributed to cocaine's apparent stimulation of the subject's willingness to work, which could be understood as the will more generally (Freud 1885c, 25).

This represented a startling departure from Freud's previous publications, otherwise an ever-abiding series of increasingly careful anatomical studies. It is tempting to interpret Freud's observations in neuroanatomical terms. Freud's argument that cocaine had "a direct antagonistic effect against morphine" might, in this sense, be framed pharmacologically, with Freud understanding the relationship between cocaine and morphine as analogous to how Balthasar Luchsinger described the antagonism between atropine and pilocarpine in 1877 (Freud 1885c, 22; Luchsinger 1877). And yet Freud's remarks on cocaine's therapeutic potential and psychophysical effects are far more readily couched in psychological terms. Much as cocaine's euphoric effects entail its dynametric effects, the antagonism between cocaine and morphine or alcohol derive from Freud's conception of cocaine as a general stimulant of the will. Freud, it would seem, had buried the lede, intentionally or unintentionally concealing the psychological character of his cocaine study.

Given the content of Freud's psychophysical research on cocaine intoxication, it is worth considering the extent to which Kraepelinian ideas figured into Freud's experimental concept and theoretical conclusions. Much like Nietzsche, Freud would actively and passively, through both omission and misrepresentation, obfuscate some of his formative influences. To this point, Freud and Kraepelin almost certainly had some form of—now lost—written correspondence and were aware of each other's work (Dalzell 2011, 116). Exner, Freud's colleague, would have known a great deal about the reaction time research being conducted by Wundt and Kraepelin, and potentially could have directed Freud to their work (assuming Freud was not already aware).

Most of the justification for speculation concerning a possible influence by Kraepelin lies with the conceptual shift embodied in Freud's singular turn to psychophysics, and how that is reflected in his publications. There is a strong similarity between Kraepelin's 1883 description of alcohol as affecting consciousness by disordering apperception, while also dysregulating the will, and Freud's identification of cocaine as a stimulant of the will, without affecting consciousness (Freud 1885c, 14, 15). Since the euphoria, in Kraepelinian terms, is merely the subjective experience of the drug's effect on psychological subprocesses, it could be argued that Freud's proposal of an antagonistic effect on morphine is best understood in a Kraepelinian formulation, where pharmacological antagonism occurs at the level of opposing influences on psychological subprocesses. The degree of Freud's indebtedness to Kraepelin's 1883 research on intoxicants cannot be definitively established with the evidence available at this time. There remain, however, serious questions regarding the extent to which Freud's turn to psychophysics, his study of an intoxicant, and a shift into the realm of psychology coincidentally overlap with Kraepelin.

Cocainizing the Psychoanalytic Turn in Freud

Freud's experimental research on cocaine appears to have had a clear effect on the course of his later psychoanalytic work. It was here, in his work with cocaine intoxication, that Freud firmly breaks with his neurophysiological education and instead attributes physiological expression to psychological causes. One possible way of orienting this development in Freud's thinking, before even the shift in his therapeutic method under Charcot, is to identify this development as a case study in Freud's emerging interest in the psychosomatic character of illnesses yet unexplainable through brain lesions (Springer 2002, 21–22). Scheidt understood Freud's cocaine studies as illuminating Freud to an unseen subconscious, which then developed while he was in Paris with Charcot (vom Scheidt 1973).

While it is difficult to align *Über Coca* with a wholesale discovery of the unconscious, there is something to the suggestion that Freud's cocaine studies mark the site of one of history's forsaken crossroads, the well-trodden path to Freudian psychoanalysis and the one less taken. It has already been suggested that the genesis of this short-lived project is likely attributable to the influence of Wundt and Kraepelin, whose methodologies, as well as the conceptual structure of their findings, strongly accord with those of Freud. An apparent disciplinary critique of this assessment could point to a simple Exnerian influence. It was from none other than Exner himself that Freud acquired the apparatus employed in his experiments. Yet Freud's progressive mentorships with Claus, Brücke, Meynert, and Exner meant that Freud was—in the words of Clark Glymour—"reared to think that psychology should be a neurophysiology of the mental" (Glymour 1991, 51). Sure enough, this statement accurately reflects Freud's thinking up to the time of his cocaine studies. In asserting the ontological priority of the perceptual experience of euphoria over the outwardly physiological process of motor response, Freud's cocaine studies are an open affront to the underlying premises of the Exnerian research program of physiological reaction time.

Freud's intoxicated dabbling in experimental psychology further marked a clear point of departure in Freud's conceptual understanding of the body and mind. This is made apparent by Freud's subsequent transition from the physiological basis of Meynert and Brücke to a partial parallelism, where "everything has an organic basis but not everything is organic" (Panhuysen 1998, 20–21). As much as the Freudian project of psychoanalysis is often erroneously characterized as an effort to chart the troubled interiority of the unconscious, this reputation owes a great deal to the reception of Freud, particularly in France, which was decidedly less attracted to biological and neurological postulates. This is plainly exemplified, above all, by Freud's enduring commitment to some form of evolutionary theory, largely of a neo-Lamarckian bent, as a formative influence on sexual development, in addition to forming the basis for a wide range of psychological disturbances (Perkins-McVey 2022).[8] Psychoanalysis was to psychiatry what histology was to anatomy, essentially a hyperspecialization in the psychical frame of reference, a dialing of the mental microscope, bringing into focus the unseen contours of the psyche (Trede 2007, 239).

In this sense, Freud's intoxication trials formed the basis of his conception of the biological subject. Cocaine shaped the rhetorical bridge between strict neurophysiology and the partial parallelism that defines his personal and professional perspective. Where he had once seen nothing but the body, Freud's passage into psychoanalysis via an experimental psychological encounter with cocaine opened unto new vistas in how the somatic duplicity of psychological embodiment and the physiological body could be upheld. The subject he saw sitting across from him was biological through and through, a seething sea of not only unconscious drives but a chimeric unity of blood, brains, and embodied life. As Freud reportedly liked to say, "behind every psychoanalyst stands the man with the syringe" (Campbell 2007).

Max Weber: Work, Sobriety, and the Struggle of the Will

In the cases of Nietzsche and Freud, the argument for the significance of intoxication in their work is almost freely given. As an employee of the psychiatric clinic and a neurophysiologist, Freud had been a direct interlocutor in many of the same discussions as Kraepelin and conducted direct psychophysical research on cocaine. Nietzsche, meanwhile, had been nothing less than the self-proclaimed emissary of a wine god. But Max Weber, a principal figure in the history of sociology, was a patron of sobriety. Or so it would seem. Weber's language of sobriety lay at the crossroads of his conceptions of objectivity, science, capitalism, power, and the will—in short, the essential questions of European modernity. However, the basis of Weber's interest in sobriety was not solely as a historical-sociological analysis of sobriety as a social phenomenon. It was emergent of the tensile interplay between intoxication and sobriety in his personal life, given categorical form by the work of Kraepelin, occasioned by Weber's own foray into the psychophysics of labor. Intoxication and sobriety increasingly appear to be central to Weber's conception of subjectivity in capitalist modernity. To make sense of this, it is important to first sketch out the context surrounding Weber and Kraepelin's dispute concerning the extension of experimental psychiatry into the factory, before exploring the finer elements of Weber's conceptions of sobriety and intoxication.

The Psychophysical Factory

In the decade after 1900, Kraepelin made several attempts to extend the psychiatric clinic into broader society, most notably in the classical degenerationist domains of criminality, alcoholism, and labor (Brain 2001, 658). This was emblematic of Kraepelin's difference with Wundt. Where Wundt was content to focus on the development of psychology as a legitimate science, Kraepelinian psychiatry had, since its inception, been grounded in an aspiration to apply experimental psychology—to bring it into the lives of others. Back in 1880, Kraepelin's *Die Abschaffung des Strafmasses— Ein Vorschlag zur Reform der heutigen Strafrechtspflege* had advocated for a reckoning on society's approach to criminality, instead aspiring to identify and treat the criminal as a kind of psychopathology (Kraepelin 1880; Engstrom 1991, 112). Evidently drawing on the degenerationist preoccupation with the pathological nature of criminality, *Die Abschaffung des Strafmasses* framed the moral quandary of crime as a psychological, even psychiatric, problem, although Kraepelin's interest appears to have initially waned in favor of his psychological research. By 1906, Kraepelin—bolstered by his post-1903 commitment to the broader social effects of degeneration—had renewed and remobilized his earlier interest in the topic of criminality, manifesting in the 1906 publication of "Das Verbrechen als soziale Krankheit" (Kraepelin 1906a). That very same year, Kraepelin published "Der Alkoholismus in München," going on to publish "Die Schildknappen des

Weinkapitals an der Arbeit" and "Die Psychologie des Alkohols" in 1911. In connection with the degenerative effects of alcohol, Kraepelin sought to develop a scientific explanation for the political upheavals rooted in working-class life, exemplified by his psychiatric interest in the aversion to work—neurasthenia (Engstrom 1991, 129). The experimental foundations of nearly all these ideas are found in their nascence in Kraepelin's research on intoxicants, then still being done in Kraepelin's lab (Kraepelin 1899; Kraepelin and Hoch 1896).

What united these topics was a shared etiology emergent of what Kraepelin understood as a disorder, or dysregulation, of the will (Brain 2001, 658; Engstrom 1991, 111–13). Framing the social ills posed by the criminal, by reluctance in labor, by drunkenness as emergent of a disruption or weakening of the will is, on its face, typical degeneration theory. Just such a closeness between degeneration and a loss of will can be seen throughout the *grandes oeuvres* of mainstream degeneration. Nietzsche's identification of a *Willenschwäche* readily comes to mind. But what distinguished Kraepelin's approach was the rootedness of this association in his experimental encounters with intoxicants. It was in the lab that an ill-defined palsy of the will was experimentally realized and patterned onto the minds of the masses, constituted in reflection of the measurable psychological effects of alcohol and other drugs on the will process. By way of an equivalence between pathological mental states and intoxication, Kraepelin had taken the testimony of a wide array of substances of intoxication to find experimental validity in the concept of a disturbance of (Wundtian) will. It was these experiments that had laid bare the "mechanisms of mental disturbance" (Kraepelin 1883a, 187). Thus, just as a young Kraepelin had hoped to reform psychiatry on the empirical foundations of experimental psychology, the scientific validity of Kraepelin's psychiatry justified, perhaps even demanded, the extension of the clinical enterprise into the broader social order.

The relationship between Kraepelin's extensive research on the effects of alcohol and his crusade against alcoholism, which he waged with fanatical zeal, is immediately comprehensible (Engstrom 1991, 116–17). Alcohol and other intoxicants inhibit the psychological will process (Kraepelin 1892a). Even Kraepelin's 1892 *Ueber die Beeinflussung* briefly analyzed the long-term effects of repeated alcohol use, and there is a certain consistency to the suggestion that repeated disordering of the mental processes disorders the mind. Kraepelin's intoxication trials had also raised a host of auxiliary questions and observations, however. Particularly important for Kraepelin were the phenomena of fatigue from repetitive tasks and improvement through practice—the effects of both of which were constantly considered in order to discern the "real" effects of intoxication (Kraepelin 1892a, 239, 246, 249). Simultaneously, efforts to mitigate the methodological problem of fatigue and practice in psychometry alluded to the role of fatigue and practice in the domain of labor, the worker's psychological existence in the factory. This would transform into a series of publications on mental labor titled *Ueber geistige Arbeit*, with the foundations of Kraepelin's experimental encounter with the phenomena of practice and fatigue stemming directly from his study of the effects of intoxicants on psychological processes. In 1901, Kraepelin and his student Oseretzkowsky expanded on these vague research questions, publishing the results of a series of experimental ergographic studies where they studied the effects of workplace intoxicants like coffee, alcohol, and tea (Kraepelin and Oseretzkowsky 1901). The conceptual foundations of their project likewise drew on the rudimentary dynamometric research included in Kraepelin's 1892 *Ueber die Beeinflussung* (Kraepelin 1892a, 95).

Synthesizing these various elements, Kraepelin transformed his ergographic studies on repetitive physical activities into a "work curve," a means of measuring a participant's, or worker's, personal fatigue equation for psychophysical labor (Brain 2001, 660, 665; Kraepelin 1903, 6). This, Kraepelin argued, allowed for the determination of an individual's "Arbeitkraft" (work power), relative to their personal rates of fatigue following repeated completion of a mundane task and mitigating factors, such as practice (Kraepelin 1903, 6–7). It was the mental processes underlying the phenomena of fatigue, repetition, and practice that gave way to the neuroses of the working class—just as constant pressures wore down machines in a factory, so too did the monotonous vigor of labor wear down the minds of men (Brain 2001, 659). Further still, alcohol, tea, and tobacco—psychoemotional lubricants that flooded the factory floors like so much machine oil—had the potential to either stymie or hasten the dissipation of work power. The parallel degenerationist paths of reluctance to work, alcoholism, and—in the extreme case—criminality thus converged at a crossroad, that of the factory as a cultural nexus of working-class intoxication.

All of this was absolutely crucial for Max Weber. When Weber went to conduct empirical field research on the lives of factory laborers—the very epitome of Weber's conception of rational modernity—he relied both on Kraepelin and, as will soon be discussed, on his own intoxicated encounters. A point of almost immediate contention for Weber was the legitimacy of Kraepelin's "work curve" (Brain 2001, 665, 666). Much as Kraepelin himself had come to recognize, the work, or performance, curve represented the outward trend of an unrecognized array of subprocesses

(Weber 1995, 168–70; Brain 2001, 664). This was in part an entailment of the intended utility of the work curve. In formulating his work curve, Kraepelin hoped to abstract the localized mental labor seen in the laboratory into the remote specificity of the factory. But, as Weber suggested, the fluid contingency of industrial labor was hardly equivalent to the isolated mental work studied in the laboratory, an incongruence concealed by the seeming instrumentality of the work curve (Weber 1995). In this sense, the work curve, at best, did little to meaningfully represent the psychophysical phenomena of industrial labor. At worst, its development as a metric actively obfuscated Kraepelin's intended object of study. It was a square peg trying to fit in a round hole. Kraepelin's solution had been to break the work curve down into its constitutive subprocesses, though Weber suggested this only accentuated the issues apparent with the initial work curve (Brain 2001).

A further point of criticism raised by Weber centered on Kraepelin's duality of recovery and fatigue, where Weber questioned both the psychological unity of fatigue and its differentiation from tiredness (Brain 2001, 666). The distinction between the perception of tiredness and the measurable phenomena of fatigue had been a significant auxiliary focus of Kraepelin's earlier intoxication trials, one rooted in the originary constitution of psychological reaction time. Through the lens of Kraepelin's intoxicated research, Exner's inability to make sense of the effects of morphine, wine, and tea on physiological reaction time rested on Exner's failure to discern between tiredness and fatigue. Where Exner had hoped to elicit fatigue using drugs that instilled a sensation of tiredness, Kraepelin's work with various intoxicants had demonstrated that the subjective perception of tiredness (*Müdigkeit*) could exist unaccompanied by any signs of psychological fatigue (*Ermüdung*) (Exner 1873, 627; Kraepelin 1883c; Kraepelin 1892a, 198). They were, in fact, separate phenomena, an idea that, for Kraepelin, proffered experimental support for the primacy of psychological, as opposed to physiological, reaction time (Kraepelin 1892a, 198). This distinction, Weber suggested, "opened a window onto some of the conceptual problems of industrial psychophysics" (Brain 2001, 666–67). Even if experimentally valid, the readily apparent interrelations of fatigue and tiredness exemplified the way in which the apparatus of experimental psychology led to theoretical distinctions of dubious purchase outside of the four walls of the laboratory, into something like a worker's fatigue rate.

Outwardly, Weber's analysis broke with Kraepelin. The endeavor to extend the laboratory into the factory was confounded by the particularity of factory life, so much more than the mere locale of abstracted labor processes. The factory was a constellation of labor practices, of material circumstances, of dynamic motivators, and of housing conditions (Brain 2001, 668). None of the factory's radical particularity could be captured in a single laboratory. Nevertheless, though Weber appeared to dissent with Kraepelin on methodological grounds, Weber still based his conclusions on Kraepelinian categories (Lazarsfeld and Oberschall 1965, 189). Weber was actually hopeful that experiments similar to Kraepelin's could be conducted in the factory itself but was deterred by the exorbitant cost (Lazarsfeld and Oberschall 1965, 189; Weber 1995). The work curve might have been lost in translation, but the foundational premise of abstracting labor processes by way of psychophysical methods went unassailed. Nor did Weber's criticism of the fatigue/tiredness distinction constitute an outright critique. Weber merely problematized the epistemological validity of Kraepelin's direct translation of the experimental artifact of fatigue into as dynamic a setting as the factory floor. In this way, the fundamental psychological categories associated with Kraepelin's experimental interactions with intoxicants of all kinds were essentially untouched by Weber's analysis.

Between Drunkenness and Sobriety

But this was not the full extent of Weber's interface with intoxication. Weber's fetishization of mental sobriety entailed a twofold fetishization of intoxication, one that struck at the very heart of Weber's epistemology and conception of modernity. Though Weber was hesitant to become mired in the debates raging between historicists and positivists, Weber nonetheless understood that the work he hoped to do in the social sciences necessitated a clear epistemology. By Weber's appraisal, the scientific study of cultural life differed significantly with that of the physical world, particularly concerning the intersubjective validity of their conclusions, as he wrote in *Die "Objektivität" sozialwissenschaftlicher und sozialpolitischer Erkenntnis*:

> There is no such thing as an absolutely "objective" scientific analysis of cultural life or—what perhaps means something more specific, but for our purposes certainly nothing essentially different—of "social phenomena" independent of special and "one-sided" points of view, according to which they are—explicitly or implicitly, consciously or unconsciously—selected, analyzed and representatively structured as objects of research. (Weber [1904] 2018, 174)

The social sciences, or the scientific analysis of cultural existence, can never be objective (in the way natural science can be), explicitly because there is no transcultural, or sufficiently totalizing, perspective through which to properly take in and organize the minutiae of human life. Perhaps more importantly, while the natural sciences sought to make knowable the laws of the physical world, even identifying the laws of the social existence would tell you nothing about the radical particularity of social reality (Weber 2018, 174–75).

However, the impossibility of objectivity does little to undermine the historical expectation, or need, for objectivity. On the contrary, the historical naturalization of the requirement for clarity, objectivity, and scrutability is precisely what Weber meant when he said that "it is the fate of our times, to be characterized by rationalization and intellectualization, by, above all, the disenchantment of the world" (Weber 1917, 22). The conceptual a prioris behind the systematization and rationalization of the cosmos was, for Weber, what it meant to be modern, manifested in, for example, the emergence of the industrial factory and all that it entailed. Objectivity in the social sciences was thus an impossible, but nevertheless desired, ideal—a professional value to strive for, rather than a conceptual category.

How then does Weber characterize the ideal quasi-objectivity of the social scientist? He employs the language of sobriety. For Weber, sobriety represented an essential quality of an industrial capitalist's comportment toward social and material relations, as he wrote in *Die protestantische Ethik und der Geist des Kapitalismus*: "And in the same way, of course, it is one of the fundamental characteristics of capitalist private enterprise that all is rationalized on the basis of strict arithmetical calculation, planned out and *soberly* directed toward the desired economic result" (Weber 2016 [1905], 206–7).[9] With the emergence of industrial capitalism, of rationalization, "the old, sedate, and comfy way of life gave way to cold, hard sobriety," where it is "sober self-control and moderation, which immensely increases efficiency" (Weber 2016, 189, 182). The character of sobriety is identified with the rationalization of modern capitalist society, the very same perspectival comportment behind the emergence of the social sciences. Weber had even identified the duties of the university academic with upholding a "pitiless sobriety of judgment" as early as *Die Evangelisch-sozialen Kurse in Berlin im Herbst dieses Jahres* (1893) (Hennis 1991, 36–37). It thus becomes clear that, for the social scientist, the pursuit of the unattainable ideal of objectivity finds expression in the epoch-defining aspiration toward sobriety.

Sobriety itself, however, is a negative state. Without intoxication, sobriety ceases to be sobriety: it is merely the baseline condition.[10] The language of sobriety in this sense simultaneously implies intoxication, and intoxication was something Weber knew a great deal about. Drinking had been "the primal pleasure for the young Weber" (Radkau 2009, 29). Ever a willing drinking buddy, "alcohol-tinged male company, even without deeper friendships, seems to have been what excited [Weber] *par excellence*" (Radkau 2009, 37). Weber's drinking even came to affect Weber's mental well-being, saying "that he suffered from obsessional thoughts and, especially after nights of drinking, sometimes imagined for the whole of the next day that he was Jumbo the elephant and lived in a zoo" (Radkau 2009, 46). Opium, too, was a frequent companion of Weber's into the twentieth century, specifically to help combat his insomnia (Radkau 2009, 155). But the youthful pride that Weber once took in holding his drink and drugs would turn to regret as his life progressed. In 1919, Weber's "anger at the years of self-intoxication burst out of him, and his fight against insomnia suddenly turned into a fight against the 'torture' of a life 'which has to be beaten down with a thousand drugs and poisons, from opium to cocaine, in the form in which it moves'" (Radkau 2009, 154).

It appears arguable, if not readily apparent, that Weber's operative metaphor of sobriety was emergent of his own preoccupation with intoxication. Sobriety becomes both a symptom and a biological description of the prolific rationalization that Weber saw as defining the world around him, one that he had fought with his intoxicated body and productive mind. It was undeniably an appropriate metaphor. The self-same rationalization behind the disenchantment of the world was also the root of the tendency to think of the subject biologically, and biorationalization found embodiment everywhere in a society that Weber recognized was run on intoxicants.

The Weberian modality of sobriety can further be contextualized through the earlier discussion of Weber's interface with Kraepelinian ideas, particularly concerning the will and intoxication. Weber clearly expressed how sober-mindedness was an essential quality of capitalist enterprise that made possible the finite planning and execution necessary to achieve economic goals (Weber 2016, 206–7). This evaluation is a prime example of Weber's own conception of power, as the capacity to "carry out his own will despite resistance" (Warren 1992, 19). In this sense, power is as dependent on the capability to exert one's will as it is on the capacity to effectively direct one's will. It was Kraepelin who, as discussed, experimentally developed the conception of intoxication in general and drunkenness in particular as a disordering of the psychological processes involving the will (Kraepelin 1883c, 600–601; Kraepelin 1892a, 157–59). To be sober then both figuratively and literally meant to guard your psychological constitution, to care for the will. It was for such a reason that Kraepelin himself swore off alcohol (Kraepelin 1983, 79). Sobriety, as Weber's crucial metaphor for perceptional capitalist modernity, was ultimately then about the struggle of biological society against the Kraepelinian, degenerationist disordering of the will rooted in intoxication. Capitalism's successes depended on stone-cold sobriety. Peering through a Weberian lens, the emergence of teetotalism alongside industrial capital takes on a radically new dimension.

Intoxication as Conduct and Way of Knowing

At its core, Weber's perspective on the sober modern was emergent of his own intoxicated way of knowing: intoxicants, and the struggle with them, shaped the grammar of Weber's diagnosis of the age. Intoxication not only lent itself to the vocabulary with which Weber described the world around him, but the scientific discourse surrounding intoxicants made the distinction between intoxication and sobriety available as both a figurative and literal description of the relationship between industrial capitalism and the will. And yet it is this tension—this alterity between the epochal sobriety of capitalism and the diffusive revelry of intoxication—that simultaneously elevates intoxication into a form of conduct, a way of *stepping outside the iron cage*.

Lawrence Scaff makes the convincing case that Weber, despite his transparent obsession with Nietzsche, was far from simply a relativistic commentator, another prophet of fate in a historically given world (Scaff 1989). Weber is instead identified with the project of coming to understand the modern subjectivist interplay between competing life orders, or spheres of conduct. The fin de siècle culture of intoxication, with which Weber frequently participated, might be understood as just such a sphere of conduct, at once caught within the givenness of capitalist modernity and yet in competition with the aspirational sobriety of industrial rationalization. Intoxication represented a temporal life order agonally oriented toward the hegemonic conduct of the given age. In opposing the "sober self-control and moderation" underling industrial capitalism, one steps outside the iron cage of modernity, realizing alternate spheres of conduct and forms of life (Weber 2016, 182). At the same time, such avenues of "escape" are enmeshed with the continued possibility of mass industrial production, recourse for weary bodies and minds. The antitheticalism of intoxication is, in this way, implicated as participatory in engendering the possibility of industrial sobriety. Intoxication and sobriety thus become contrasting conducts or forms of life that simultaneously constitute the emerging cultural-historical identity of capitalism. At the foundation of this approach, one finds the experience of intoxication itself, just as central as the tension between intoxication and sobriety was in Weber's private life. Quite appropriately, Weber's ownmost lived experience, that of intoxication, was crucial to the constitution of his subjectivist appraisal of the modern condition.

For Weber, as it had been for Nietzsche and Freud, it was not only merely his engagement with psychological research on the nature of intoxication that shaped his understanding of the modern self, but his own intoxicated life. The interpositional conflict between aspirational sobriety and intoxication, a contrast of subjectivist life orders in which Weber was personally implicated, further converged on his evaluation of the historical givenness of modernity. It was the experience of intoxication that brought these concepts to the fore, forcing him to feel. Each figure, in turn, impressed upon their work an intoxicated way of knowing—a tacit, embodied encounter of the self with the body and the world, which is subsequently concealed by the rhetorical dance of knowledge making. It is worth pondering if, after all, the entirety of the biological subject rests on the unseen influences of substances of intoxication.

A Horizontal Fall

Opium leads the organism to death in a euphoric mood. The torture is the process of returning to life against one's desires. An entire spring excites the veins to madness, bringing with it ice and fiery lava.

I recommend the patient who has gone without for eight days bury his head in his arms, stick his ear to those arms, and wait: catastrophe, riot, factories exploding, flood; the ear can detect the entire apocalypse in the star-lit night of the body. (Cocteau 1930, 23)

In 1930, the French writer Jean Cocteau recorded his musings about life, love, and intoxication in a diary accounting his second attempt to break free of the requirement of smoking opium, a fashionable habit among artist, poets, and philosophers in twentieth-century interwar France. More poetry than diary, Cocteau's recollection is far from prosaic, fraught with aphoristic declarations, among them "life is a horizontal fall" (Cocteau 1930, 37). Cocteau's *Opium: Journal d'une désintoxication* has been called "the diary of a lost love" (Barnes 1970, 58). But, more than that, the quote above, just one of many throughout the memoir, speaks to the text as a rumination on the relationship between the embodied experience of intoxication and the biological subject. Cocteau's *Opium* reflects the mind of an individual who, despite being far from the biomedical sciences himself, conceives of and subjectively experiences the body as biological, a confluence of nerve centers, tissue, and pumping blood. *Opium* is the narcotic memoir of a biological subject. Intoxication becomes a conflagration, illuminating the darkling sky of the body.

Our narrative has centered on the varied histories of science and medicine in their intersections with intoxicants. Maybe this has been at the expense of the ordinary experience of intoxication. And yet the ordinary experience of intoxication, if there is such a thing, is forever in contention with the scientific and medical concepts that surround them. Intoxication is never an experience in itself. Intoxication is value laden, pregnant with meaning. Much as contemporary drug culture is enjoined—even consciously—with concepts of addiction, bioavailability, and neurotransmission, the recipient of Brunonian stimuli was never completely removed from the scientific enterprise of Brunonianism itself. The knowing imparted by the living experience of intoxication might give way to something surprising, even unanticipatable, but it is always emergent of the experiential convergence of conceptual associations made available in the particularity of place and time. Intoxication is a way of knowing and making the world, one long overlooked in light of its purchase in the otherwise sober-minded annals of scientific investigation. In this case, intoxication has been a way of knowing and making a new group of subjects, a subject stretched across parallel sciences of the body and mind, a biological subject.

This has been the story of just such a subject—at least in part. More specifically, this has been the story of the possibility of such a subject, the historical emergence of the biological subject through various expressions of intoxicated ways of knowing. At each step, intoxicated ways of knowing made particular conceptual associations real in the perceptional experience of the intoxicated. Even sure-footed efforts to resist the influence of intoxication as a way of knowing served to engender the possibility of their eventual inclusion in the developing conceptions of the body and mind. By its end, biological modernity can be seen reflected in the inebriated lives still being led to this day.

Fueled by colonization and driven by experimentalism, the explosive development of the early modern pharmacopoeia dovetailed with the Newtonian pursuit of a scientific medicine. Though there were several eighteenth-century experiments in medical system building, it was John Brown's quantifiable principle of excitability, a vital conception of the body borne out of the experience of intoxication, that took hold. Central to his therapeutic model were vital substances, remedies, in particular intoxicating remedies, that affected the organism by stimulating the vital principle.

In the Germanies, Brunonianism, for a moment, enjoyed tremendous success, owing in large part to Andreas Röschlaub's intrepid efforts to bring the concept of Brunonian excitability into agreement with the language of Romantic philosophy. This placed intoxicated excitability, and vital substances, at the forefront of *Naturphilosophie*, with the effects of opium serving as Schelling's primary example of cause and effect. The vital substance concept even figured into the appearance of alkaloids as a new kind of thing to be in the world, via Sertürner's own intoxicated encounters. Through Oken, the vital substance concept made its way into Johannes Müller's *Handbuch*, where, even as Müller distanced himself from Romanticism and Brown, he nevertheless pursued anatomical justification for the vital substance concept.

So inextricable were vitalism and vital intoxicants in mid nineteenth-century Germany that when Müller's students, Brücke, du Bois-Reymond, and Helmholtz, attempted to divest physiology of vitalism they expelled any residual conceptions of the vital substances. At its most extreme, it appeared that this position on the part of the "neomechanists" expressed itself in the form of a complete disassociation with researching the effects of intoxicants on the body. Physiochemical reductionism meant privileging the experimental encounter with the physical world as the singular source of meaningful scientific knowledge. For the next generation of neomechanists, this translated into an interest in physiologically establishing the theoretical equation of mental states with brain states, the most prominent examples being the cortical theories of psychiatrists Meynert and Wernicke. They, too, maintained critical distance from any scientific encounter with intoxicants, outside of clinical practice.

Yet the world of the late nineteenth century churned with intoxicants. Morphine, alcohol, and opium had been made readily available for mass consumption, but there were other intoxicants then too. There were novel alkaloids, like cocaine, which promised effects then unseen on the European continent, as well as the rapidly expanding domain of synthetic drugs, among them chloral and amyl nitrate. Intoxicants only became more present; advertised in newspapers, available in sundry stores—substances of intoxication were a critical facet of society. Thus, in hindsight, it comes as little surprise that, out of the budding science of experimental psychology, substances of intoxication would intervene in the unfolding deliberations concerning the nature of the body and mind to concretize a new form of subjectivity. By taking the experience of intoxication as the object of experimental study, Kraepelin imparted scientific validity to the embodied, perceptional encounter between intoxicated subjects, the body, and the world. It made possible the scientific identification of intoxication as the physical made real in the psychical—upholding a subject that was fully biological, and yet only studiable through a parallel science of conscious embodiment. Further still, intoxication would help shape Kraepelin's nosologies of mental illness, by providing an experimental model for understanding psychopathological states. Intoxicated ways of knowing directly intervened in the experimental process to make recognizable not only a biological subject but a dynamic conception of the biological mind, then beyond the reach of neurophysiology.

But this does not capture the significance of the developing interest in the psychology of intoxication, nor its import for the biological subject. Foundational to the cohesiveness of the biological subject is the realization of a biologistic conception of the mind and body as a conceptual a priori—both expressed through, and furthered by, the concept's place in the late nineteenth-century philosophies and social theories of the body. To this point, both the influence of psychological research on intoxication and their own most drunken, stimulated, and narcotized experiences were found to be central to Nietzsche's, Freud's, and Weber's own concepts of the body and its place in modernity.

In this sense, the history of the biological subject is a history of physiology, of psychiatry, of chemistry, of neuroscience, of pharmacy, of philosophy—but, above all else, it is a history of the epistemological structures, both conceptual and perceptual, that make up the body and mind. What I have striven to establish is that intoxicated ways of knowing should be included in such a history. It may very well be the case that intoxicated ways of knowing serve a further role, perhaps another equally as impactful one in the history of the biomedical sciences. The biological subject is in a horizontal fall. It is more apparent today than it was at the beginning of the twentieth century. Biological subjects careen through modern society, always changing, yet faithfully biological and fideliously intoxicated. More questions come to the fore than have just been answered, specifically: What does this mean for the biochemical identity of the modern brain? How does this factor into a society that increasingly employs the language of neurotransmitters and "chemical imbalance" to describe basic psychological wants, needs, and disturbances? Into the enduring legacy of the "opioid epidemic"? An answer may very well be found in further explorations of intoxication as a way of knowing. But that's for another story.

Acknowledgments

I must first thank my wonderful wife. Your belief spans continents and stands the innumerable tests of time. None of it would be possible without you.

I would like to give heartfelt thanks to Gordon McOuat, without whose continuing support neither this project, nor I myself, could possibly stand where it does today. You have been indispensable as a teacher and a friend. I am grateful to Robert Brain for being a superlative reader of my manuscript, whose reflections on the text at every level helped me realize what it is today. His support and mentorship continue to drive and inspire my research. I am deeply appreciative of Richard Brown, from whom—over long hours spent amid endless stacks of papers—I learned to translate fleeting ideas into enduring projects, and whose impassioned drive continues to inspire me. I would also like to thank William Barker for seeing the potential in this project from its nascence to its ultimate completion, as well as Hans-Günther Schwarz, who emboldened this manuscript's philosophical ambitions.

I owe a big thanks to Johannes Chan for offering your time to review an earlier version of this manuscript. Your generous comments greatly informed my sense of which sections needed further development.

Although not directly funding this project, significant revisions of the manuscript were completed while supported by the Social Sciences and Humanities Research Council of Canada (SSHRC). I would further like to thank Tamar Kravetz at the Bar-Ilan University Library; the Staatsbibliothek zu Berlin—Preußischer Kulturbesitz; the Bayerische Staatsbibliothek; Vicky Pohlen, Charlie Achuff, and Laura McNulty at the United States National Library of Medicine; and the National Library of Israel for all their assistance.

Finally, I must express my heartfelt appreciation to Karen Darling, executive editor, for shepherding this manuscript through the review process with patience, grace, and no small amount of faith.

Additional material can be found online at https://press.uchicago.edu/dam/ucp/books/pdf/IWK_Translations.docx.

Notes

Convergences

1. All translations are the author's own.

Chapter 1

1. It should be noted that, insofar as extreme excitement also led to indirect debility, the cure for both extremes of the scale was powerful stimulants. See Perkins-McVey (2023) for a detailed discussion of how the interplay between stimulus and indirect debility influenced the early disease theory of alcoholism.

2. This has been copied from the 1795 translation, as prepared by Thomas Beddoes, although much of the preface, including the story of Brown's conversion by opium, are taken word for word from Brown's own 1788 translation from the original Latin (Brown 1790).

Chapter 2

1. Citations reference poem number. The translation is literal as opposed to poetic.

2. The relationship between opium and Coleridge's life and work is a vast topic of scholarly inquiry. For a cursory overview of the literature, see L. Wagner's "Coleridge's Use of Laudanum and Opium" (1938), A. Hayter's *Opium and the Romantic Imagination* (1968), M. H. Abrams's classic, albeit dated and controversial, *The Milk of Paradise* (1971), M. G. Cooke's "De Quincey, Coleridge, and the Formal Uses of Intoxication" (1974), and R. Holmes's *Coleridge: Early Visions, 1772–1804* (2011), to name just a few.

3. As with all Brunonians, De Quincey is not blind to the dangers of opium, an awareness on full display in *Confessions*. For all the Brunonian passion for the curative good represented by opium, it was nevertheless dangerous, precisely because it was such a potent, uniquely positioned remedy.

Chapter 3

1. Christoph Heinrich Pfaff, a student of Kielmeyer's before being made a professor of medicine, chemistry, and physics in Kiel, published a second German translation of Brown's *Elements of Medicine* alongside a commentary in 1798. As early as 1798,

Pfaff remarked on how "[Brown] is compared with Newton" (Pfaff 1798, vii). For, as Newton "established the laws of dead matter, [...] Brown brought the grand basic laws of life out of the obscurity into which they had been reduced through prejudice and ignorance" (Pfaff 1798, vii). Further drawing comparisons with Newton, Brown had "built the science of living matter on solid foundations with a kind of mathematical certainty, in a science which embraces every modification of life that exists on this globe, the entire animal and vegetable kingdom" (Pfaff 1798, vii). Pfaff was relatively critical, particularly of the stimulus doctrine, which Pfaff regarded as outwardly true, though he questioned its underlying nature.

2. "gleichsam wie ein Cocon."

3. Hoffman should be distinguished for her attention to the positive aspects of Kant's discussion.

4. Lenoir, in particular, identifies Kielmeyer and Blumenbach with the development of a novel biological science, centered around morphology. Lenoir's argument is, in part, in contention with Robert Richard's. My own thesis extends Lenoir's, while also drastically limiting its farthest reaching conclusions.

5. Ritter tested the effects of the electric pile on all manner of bodily function, from sneezing to defecation, and even ejaculation (Strickland 1998, 457). The latter experiment involved connecting the electrical pile to his own reproductive organs (Strickland 1998, 457). Ritter wrote to his publisher about his desire to marry his electrical pile (Strickland 1998, 455).

6. Ironically, pure opium may well have worked, as the effect of opiates as a cure for diarrhea is now well known.

7. Note the similarity with Brown's language.

8. Brunonian physicians would term this reciprocal activity "contrastimulus," later a central focus of Brunonianism's detractors.

9. Emphasis added.

Chapter 4

1. This is in reference to Thomas Kuhn's suggestion that it is often younger scientists, less integrated into the community of the current paradigm, whose research engenders paradigm shifts.

2. "schiest dagegen in prismnatischer Form unter einem Winkel von 30 bis 40 Grad an."

Chapter 5

1. Tom Broman argues this incongruence is at least partially explained by the disintegration of the Holy Roman Empire, which brought down with it a relatively united body of cultural practices, traditions, and approaches in education (Broman 2002).

2. See Nicolaas Rupke's *Alexander von Humboldt: A Metabiography* for an in-depth account of how von Humboldt's life and work has historically been reframed and reoriented in light of ever-changing presentations of German science. Part of this is the shifting importance attached to *Kosmos*, relative to his other work and his activity as a social promoter of science (Rupke 2008).

3. Steigerwald's remarks concern Alexander von Humboldt's and Johann Ritter's galvanic experiments, but they further apply here.

Chapter 6

1. Moritz Romberg was a Berlin neurologist who, as Müller had in physiology, contributed greatly to the study of neurology through the collection, translation, and synthesis of neurological research from across Europe, primarily active between 1820 and 1850 (around the same time as Müller).
2. Virchow understood physiology, and particularly cell theory, to be politically consequential. Politics, Virchow argued, was nothing less than the medicine of the *polis*, referring to both the environmental origins of many ailments and the analogous relationship between cells in a body and people in a society. Just as a small number of cancerous cells can undermine the entire body, strife in one sector of society compromises the well-being of society as a whole. Such conclusions led Virchow and, to a lesser degree, other physiologists to advocate for sweeping political and economic reforms (Weindling 1993).
3. This issue ultimately resolved in 1875 when the psychiatric clinic was split in two, with Meynert leading the second psychiatric clinic (Dalzell 2011; Guenther 2015).
4. *Associations* here denotes the relations between ideas, which is distinct from how I use the term with respect to intoxicated ways of knowing.
5. Notably, the text itself uses Brunonian concepts to explain the function of the various drugs on the nervous system (Kane 1881).

Chapter 7

1. "Bildungsbürgerlich" refers to the social and moral values of the "Bildungsbürgertum," the educated, bourgois upper middle class, who were typically employed as doctors, academics, or civil servants in nineteenth-century Germany. They were strongly associated with humanistic education and German Idealist philosophy.
2. Rudolf Lichtenfels and Rudolf Frohlich's paper "Ueber den Puls als ein Symptom, sowie als numerisches Maass der physiologischen Arzeneiwirkung" is cited here as being published in 1857 as opposed to 1851. It attempts to associate visible changes in pulse with specific conditions, including narcotic intoxication.
3. Henning Schmidgen argues that Wundt was greatly influenced by the extensive references to Liebig's work in Mill's *Logic*, suggesting that Wundt's model of the conscious process is likenable to a chain of chemical reactions (Schmidgen 2003).

Chapter 8

1. Everett Mendelsohn (1965, 203) argues that questions of vitalism are secondary to the influence these ideas had in their scientific theories/methodologies.
2. Wöhler and Liebig had previously corresponded in the resolution of a disagreement over the characteristics of a silver compound with fulminic acid and silver cyanate.
3. There is an interesting discussion to be had concerning the rhetorical significance of the decision to include these colonial histories of coca in a paper concerned with the identification of a chemical isolate.
4. Niemann's cause of death is allegedly lung damage, possibly a result of his experimentation with mustard gas, which also begun in 1860.
5. The German chemical dye industry and pharmacy science are not only

co-original, but share in their origins. Adolf von Baeyer, for example, began his career in pharmacology but would win the 1905 Nobel Prize in Chemistry for his work on synthetic indigo dye.

6. De Quincey is the most famous ninteenth-century literary promoter of narcotics, but he was far from the only one. Just as quickly as morphine became an iconic component of medical practice, morphine captured the imaginations of poets, artists, and broader society. Heinrich Heine would even name a poem after morphine before his death in 1856 and appears to have developed a serious reliance on morphine late in life. Several authors have suggested this may have been his actual cause of death (Auf der Horst and Labisch 1999).

7. Karch (2006, 38) provides an alternative date of 1868.

8. A somewhat dishonest case could also be made that the opium pipe deserves this title. It is nevertheless important to make an ontological distinction between morphine and opium.

9. This is itself an interesting vignette, reflecting the broader significance of the vital substance concept for medical theory and practice.

10. There is considerable debate about the extent to which the oft-cited link between the American Civil War and widespread morphine use is a myth, an exaggeration, or a justified historical narrative.

11. Heinrich Byk's Berlin-based chemical factory later greatly expanded and now produces additives and measuring instruments under the name BYK Additives & Instruments.

Chapter 9

1. This is an interesting observation because it complicates the perception of Wundt as a failed physiologist. To this end, it is worth considering that although Wundt's physiological work never received the praise Wundt hoped it would, Wundt was already working in the upper tier of academic physiology and seeking praise from its foremost representatives. Wundt's self-perceptions of failure as a physiologist may have been little more than a matter of his ambitious goals.

2. Note the inclusion of strychnine, which, though it is primarily used as a rat poison today, was counted as a narcotic into the twentieth century.

3. In English: "Experimental Investigation of the Simple Psychological Process, First Part" and "The Behavior of the Physiological Reaction Time under the Influence of Morphine, Coffee, and Wine."

4. See figures 9–11 for a panel of Kraepelin's graphs comparing reaction times with alcohol.

5. Kraepelin was deeply concerned about the effect of practice, or conditioning, on his results; however, in cases such as these it was clear to Kraepelin that the subject's reaction far exceeded baseline deviations emergent of prior conditioning.

Chapter 10

1. In reference to Kraepelin's concept of intoxication as artificial insanity, exogenous psychosis, or model psychosis.

2. Kraepelin and Wundt sustained a friendship for the remainder of their lives.

There is a particularly famous photograph of Kraepelin and Wundt at Wundt's seventieth birthday.

3. Dorpat was the German name for the modern Estonian, formerly Russian, city of Tartu. Culturally, the city had a long historical association with Baltic Germans, but in the eighteenth century it came under Russian control and subsequently went through a gradual process of Russianization. In the time of Emminghaus and Kraepelin, Dorpat was already part of Russia.

4. See figure 12 for context.

5. The reference to effort involved in the creation of the text is a paraphrase of Kraepelin's own words in the text's introduction. Here, Kraepelin relays the tremendous difficulty involved in not only developing controlled experiments, but also in making the many individual observations and measurements. Given that Kraepelin conducted most of the measurements on himself, this is inescapably also in reference to all of the instances in which Kraepelin drugged himself in different ways, particularly with soporifics, which was surely "exhausting."

6. As discussed earlier, this was also the origin of Karl Jaspers's use of the term.

Chapter 11

1. All citations for Nietzsche refer to the digitized version of the standard critical edition of Nietzsche's complete works. As such, all citations are formatted thus: (author year of publication [year of digitization], reference abbreviation in the *Kritische Gesamtausgabe Werke und Briefe* (KGWB), page number).

2. In typical fashion, Nietzsche claims that, upon reading *Der Ursprung der moralischen Empfindungen*, he disagreed with just about every premise.

3. "Kampf solcher widerstreitender Motive."

4. As Martin suggests, the extent of Nietzsche's direct contact with Gobineau's thought is the subject of sustained scholarly dispute, ranging from the assertion that Nietzsche was worlds apart from Gobineau to the contentious suggestion that Nietzsche owed everything to the man (Martin 2004). It is the case that Nietzsche mentions Gobineau, albeit only twice, in letters from 1865 and 1888, and that Gobineau was an influential friend to Wagner and his Bayreuth circle (Martin 2004). This would mean that Nietzsche writes of Wundt more often than Gobineau.

5. Emphasis mine.

6. When *Über Coca* is cited the text being referred to is the 1885 reprint with the addition of supplementary addenda (*Nachträge*).

7. "also der Coca-Euphorie parallel und scheint auch eher von der centralen Arbeitsbereitschaft, von der Hebung des Allgemeinbefindens, als von einem directen Einflusse auf motorische Apparate herzurühren."

8. A classic, and in many ways troubling, example of this is Freud's characterization of hereditary (neo-Lamarckian) Jewish neuroses in *Der Mann Moses und die monotheistische Religion*.

9. Emphasis added.

10. This is, of course, only theoretical, so centrally has intoxication figured in human history, if sometimes only in the shadows.

Bibliography

Abraham, J., and S. Mathew. 2019. "Merkel Cells: A Collective Review of Current Concepts." *International Journal of Applied and Basic Medical Research* 9 (1): 9–13.

Abrams, M. H. 1971. *The Milk of Paradise: The Effect of Opium Visions on the Works of DeQuincey, Crabbe, Francis Thompson, and Coleridge*. Octagon Books.

Adler, H. 1991. "Gustav Theodor Fechner: A German Gelehrter." In *Portraits of Pioneers in Psychology*, edited by G. Kimble et al. American Psychological Association / L. Erlbaum Associates.

Aldea, A. S., and A. Allen. 2016. "History, Critique, and Freedom: The Historical A Priori in Husserl and Foucault." *Cont Philos Rev* 49:1–11.

Alderwick, C. 2016. "Nature's Capacities: Schelling and Contemporary Power-Based Ontologies." *Angelaki* 21 (4): 59–76.

Allik, J., and E. Tammiksaar. 2016. "Who Was Emil Kraepelin and Why Do We Remember Him 160 Years Later?" *TRAMES* 20 (70/65), no. 4: 317–35.

Alturkistani, H. A., Tashkandi, F. M., and Z. M Mohammedsaleh. 2015. "Histological Stains: A Literature Review and Case Study." *Global Journal of Health Science* 8 (3): 72–79.

Amrein, M., and K. Nickelsen. 2008. "The Gentleman and the Rogue: The Collaboration between Charles Darwin and Carl Vogt." *Journal of the History of Biology* 41 (2): 237–66.

Anctil, M. 2015. *Dawn of the Neuron: The Early Struggles to Trace the Origin of Nervous Systems*. McGill-Queen's University Press.

Anderson, M. 2011. "Telling the Same Story of Nietzsche's Life." *Journal of Nietzsche Studies* 42 (1): 105–20.

Anonymous. 1722. *Pharmacopoeia Edinburgensis*. 2nd ed.

Anonymous. 1802. "Lob der allerneusten Philosophie." *Allgemeine Literatur-Zeitung* 225, 339.

Anonymous. 1866. [Review:] "On the Speedy Relief of Pain and Other Nervous Affections by Means of the Hypodermic Method." *The British and Foreign Medico-Chirurgical Review* 37 (74): 395–401.

Anonymous. 1878. "Ernst Heinrich Weber." *Nature* 17:286.

Anonymous. 1938. "Gustav Fritsch (1838–1927)." *Nature* 141:403–4.

Araujo, S. 2012. "Why Did Wundt Abandon His Early Theory of the Unconscious?

Towards a New Interpretation of Wundt's Psychological Project." *History Of Psychology* 15 (1): 33–49.
Araujo, S. 2014. "Bringing New Archival Sources to Wundt Scholarship: The Case of Wundt's Assistantship with Helmholtz." *History of Psychology* 17 (1): 50–59.
Araujo, S. 2016. *Wundt and the Philosophical Foundations of Psychology: A Reappraisal*. Springer International.
Auf der Horst, C., and A. Labisch. 1999. "Heinrich Heine, der Verdacht einer Bleivergiftung und Heines Opium-Abuses." *Heine-Jahrbuch* 38:105–32.
Bache, F. 1980. *A System of Chemistry for the Use of Students of Medicine*. Edited by T. Thomson. Arkose Press.
Bacon, F. 1844. *Novum Organum: Or, True Suggestions for the Interpretation of Nature*. London.
Bailey, W., and L. Truong. 2001. "Opium and Empire: Some Evidence from Colonial-Era Asian Stock and Commodity Markets." *Journal of Southeast Asian Studies* 32 (2): 173–93.
Bamford, R. 2016. "The Ethos of Inquiry: Nietzsche on Experience, Naturalism, and Experimentalism." *Journal of Nietzsche Studies* 47 (1): 9–29.
Ban, T. A. 2006. "Academic Psychiatry and the Pharmaceutical Industry." In *Progress in Neuro-Psychopharmacology and Biological Psychiatry* 30 (3): 429–41.
Barfoot, M. 1993. "Philosophy and Method in Cullen's Medical Teaching." In *William Cullen and the Eighteenth Century Medical World*, edited by A. Doig, J. Ferguson, I. Milne, and R. Passmore). Edinburgh University Press.
Barnes, C. 1970. "Stage: 'Opium,' Farewell to Addiction." *The New York Times*, October 6, 58.
Bartmann, W. 2003. *Zwischen Tradition und Fortschritt: Aus der Geschichte der Pharmabereiche von Bayer, Hoechst und Schering von 1935–1975*. Steiner.
Beck, H. 1992. "The Social Policies of Prussian Officials: The Bureaucracy in a New Light." *The Journal of Modern History* 64 (2): 263–98.
Beer, M. 1995. "Psychosis: A History of the Concept." *Comprehensive Psychiatry* 37 (4): 273–91.
Behrens, P. J. 1980. "An Edited Translation of the First Dissertation in Experimental Psychology by Max Friedrich at Leipzig University in Germany." *Psychol. Res* 42:19–38.
Benitez-Rojo, A. 1996. *The Repeating Island*. Duke University Press
Bergman, N. A. 1991. "Humphry Davy's Contribution to the Introduction of Anesthesia: A New Perspective." *Perspectives in Biology and Medicine* 34 (4): 534–41.
Bernstein, J. 1868. "Zur Theorie des Fechner'schen Gesetzes der Empfindung." *Archiv für Anatomie, Physiologie and wissenschaftliche Medicin*, 388–93.
Bernstein, J. 1870. "Ueber die physiologische Wirkung des Chloroforms." In *Untersuchungen zur Naturlehre des Menschen und der Thiere*, ed. J. Moleschott. Giessen.
Bernstein, J. 1871. *Untersuchungen über den Erregungsvorgang im Nerven- und Muskelsysteme*. Heidelberg.
Berridge, V. 1999. *Opium and the People: Opiate Use and Drug Control Policy in Nineteenth- and Early Twentieth-Century England*. Free Association, London.
Biagioli, F. 2016. "Helmholtz's Relationship to Kant." In *Space, Number, and Geometry from Helmholtz to Cassirer*, special issue of *Archimedes (New Studies in the History and Philosophy of Science and Technology)* 46, 1–21.

Bilal, M., B. Edwards, M. Loukas, R. J. Oskouian, and R. S. Tubbs. 2017. "Johann Gaspar Spurzheim: A Life Dedicated to Phrenology." *Cureus* 9 (5): e1295.
Bistricky, S. L. 2013. "Mill and Mental Phenomena: Critical Contributions to a Science of Cognition." *Behavioral Sciences* (Basel, Switzerland) 3 (2): 217–31.
Black, S. 2022. *Drugging France: Mind-Altering Medicine in the Long Nineteenth Century*. McGill-Queen's University Press.
Boring, E. G. 1950. *A history of Experimental Psychology*. 2nd ed. Appleton-Century-Crofts.
Bos, Gerrit. 1996. "'Balādhur' (Marking-Nut): A Popular Medieval Drug for Strengthening Memory." *Bulletin of the School of Oriental and African Studies, University of London* 59 (2): 229–36.
Bowler, P. J. 2005. "Revisiting the Eclipse of Darwinism." *Journal of the History of Biology* 38 (1): 19–32.
Brain, R. 2001. "The Ontology of the Questionnaire: Max Weber on Measurement and Mass Investigation." *Studies in History and Philosophy of Science* 32:647–84.
Breidbach, O. 2004. "Schelling and Experiential Science." *Sudhoffs Archiv* 88 (2): 153–74.
Brobjer, T. 2017. "Nietzsche's Reading and Knowledge of Natural Science: An Overview." In *Nietzsche and Science*, edited by G. Moore and T. Brodjer. Routledge
Brock, W. 2002. *Justus von Liebig: Chemical Gatekeeper*. Cambridge University Press.
Broman, T. H. 2002. *The Transformation of German Academic Medicine, 1750–1820*. Cambridge: Cambridge Univ. Press.
Brown, J. 1790. *The Elements of Medicine: Or, A Translation of the Elementa Medicinae Brunonis Vol. I*. Philadelphia.
Brown, J. 1791. *The Elements of Medicine: Or, A Translation of the Elementa Medicinae Brunonis Vol. I*. Philadelphia.
Brown, J. 1795. *The Elements of Medicine: Or, A Translation of the Elementa Medicinae Brunonis Vol. I*. London.
Brown, J. 1804. *The Elements of Medicine: Or, A Translation of the Elementa Medicinae Brunonis Vol. I*. Portsmouth.
Brown, T. 1987. "Medicine in the Shadow of the Principia." *Journal of the History of Ideas* 48 (4): 629–48.
Brühl-Cramer, C. 1819. *Ueber die Trunksucht und eine rationelle Heilmethode derselben*. Berlin.
Brunton, D. 2000. "A Question of Priority: Alexander Wood, Charles Hunter and the Hypodermic Method." *Proceedings of the Royal College of Physicians of Edinburgh* 30:349–51.
Buchholz, Ch. F. 1800. "Versuche die Zerlegung des Opiums in seine nähere Bestandtheile betreffend; nebst einigen dahin gehörigen Bemerkungen." *Journal der Pharmacie* 8 (1): 24–62.
Bürgy, M. 2008. "The Concept of Psychosis: Historical and Phenomenological Aspects." *Schizophrenia Bulletin* 34 (6): 1200–10.
Butler, T. 1970. "The Introduction of Chloral Hydrate into Medical Practice." *E. Bulletin of the History of Medicine* 44 (2): 168–72.
Bynum, W. F. 1984. "Alcoholism and Degeneration in 19th Century European Medicine and Psychiatry." *British Journal of Addiction* 79 (1): 59–70.

Bynum, W. F., and Roy Porter. 1988. *Brunonianism in Britain and Europe.* Welcome Trust.
Cahan, D. 1993. *Hermann von Helmholtz and the Foundations of Nineteenth-Century Science.* University of California Press.
Campbell, N. D. 2007. "The Man with the Syringe: Pain and Pleasure in the Experimental Situation." In *Discovering Addiction: The Science and Politics of Substance Abuse Research.* University of Michigan Press.
Campioni, G., P. D'Iorio, M. Fornari, F. Fronterotta, and A. Orsucci. 2003. *Nietzsche's Persönliche Bibliothek.* De Gruyter.
Canales, J. 2010. *A Tenth of a Second: A History.* University of Chicago Press.
Cartwright, F. 1972. "Humphry Davy's Researches on Nitrous Oxide." *Brit. J. Anaesth.* 44:291–96.
Cayleff, S. 2016. *Nature's Path: A History of Naturopathic Healing in America.* Johns Hopkins University Press.
Chvátal, A. 2017. "Discovering the Structure of Nerve Tissue: Part 3: From Jan Evangelista Purkyně to Ludwig Mauthner." *Journal of the History of the Neurosciences* 26 (1): 15–49.
Ciaccio, J. 2018. "Between Intoxication and Narcosis: Nietzsche's Pharmacology of Modernity." *Modernism/Modernity* 25 (1): 115–33.
Clark, W. 1999. "The Death of Metaphysics in Enlightened Prussia." In *The Sciences in Enlightened Europe,* edited by W. Clark, J. Golinski, and S. Schaffer. University of Chicago Press.
Cocteau, J. 1930. *Opium: Journal d'une désintoxication.* Stock, Delamain & Boutelleau.
Coleman, W. 1999. *Biology in the Nineteenth Century: Problems of Form, Function, and Transformation.* Cambridge University Press.
Cooke, M. 1974. "De Quincey, Coleridge, and the Formal Uses of Intoxication." *Yale French Studies* 50:26–40.
Cooper, A. 2017. "Kant and Experimental Philosophy." *British Journal for the History of Philosophy* 25 (2): 265–86.
Corley, T. 1987. "Interactions between the British and American Patent Medicine Industries 1708-1914." *Business and Economic History* 16:111–29.
Courtwright, D. T. 1978. "Opiate Addiction as a Consequence of the Civil War." *Civil War History* 24 (2): 101–11.
Courtwright, D. T. 2002. *Forces of Habit: Drugs and the Making of the Modern World.* Harvard University Press.
Cowan, M. 2005. "'Nichts ist so sehr zeitgemäss als Willensschwäche': Nietzsche and the Psychology of the Will." *Nietzsche-Studien: Internationales Jahrbuch Fur Die Nietzsche-Forschung* 34:48–74.
Cramer, T. 2015. "Building the 'World's Pharmacy': The Rise of the German Pharmaceutical Industry, 1871-1914." *Business History Review* 89 (1): 43–73.
Cranefield, P. F. 1966. "The Philosophical and Cultural Interests of the Biophysics Movement of 1847." *Journal of the History of Medicine and Allied Sciences* 21 (1): 1–7.
Crawford, M. J. 2019. "An Imperial Pharmacopoeia? The Pharmacopoeia Matritensis and Materia Medica in the Eighteenth-Century Spanish Atlantic World." In *Drugs on the Page: Pharmacopoeias and Healing Knowledge in the Early Modern*

Atlantic World, edited by M. Crawford and J. Gabriel. University of Pittsburgh Press.
Cullen, W. 1789. *Materia Medica, Vol. 1*. Edinburgh.
Dahlkvist, T. 2014. "Nietzsche and Medicine." In *Handbuch Nietzsche und die Wissenschaften*, edited by H. Heit and L. Heller. De Gruyter.
Dalzell, T. 2011. "Freud and Kraepelin." In *Freud's Schreber between Psychiatry and Psychoanalysis*. Karnac/Routledge.
Danziger, K. 1980. "Wundt's Theory of Behavior and Volition." In *Wilhelm Wundt and the Making of a Scientific Psychology*, edited by R. W. Rieber. Springer.
Darmon, I. 2011. "No 'New Spirit'? Max Weber's Account of the Dynamic of Contemporary Capitalism Through 'Pure Adaptation' and the Shaping of Adequate Subjects." *Max Weber Studies* 11 (2): 193–216.
Daston, L., and P. Galison. 1992. "The Image of Objectivity." *Representations* 40:81–128.
Daston, L., and P. Galison. 2007. *Objectivity*. Zone Books.
De Kock, L. 2014. "Hermann von Helmholtz's Empirico-Transcendentalism Reconsidered: Construction and Constitution in Helmholtz's Psychology of the Object." *Science in Context* 27 (4): 709–44.
de Marneffe, D. 1991. "Looking and Listening: The Construction of Clinical Knowledge in Charcot and Freud." *Signs* 17 (1): 71–111.
De Quincey, T. 1885 [1821]. *Confessions of an English Opium-eater*. New York.
Decker, H. 2007. "How Kraepelinian Was Kraepelin? How Kraepelinian Are the Neo-Kraepelinians?—from Emil Kraepelin to DSM-III." *History of Psychiatry* 18 (3): 337–60.
Dehio, H. 1887. "Untersuchungen über den Einfluss des Coffeins und Thees auf der Dauer einfacher psychischer Vorgänge." PhD diss., Universität Dorpat.
Deming, S. 2011. *The Economic Importance of Indian Opium and Trade with China on Britain's Economy, 1843–1890*. Whitman College Economics Working Papers, 25 (1).
Derks, H. 2012. "Tea for Opium Vice Versa." In *History of the Opium Problem: The Assault on the East, ca. 1600–1950*. Brill.
Derosne, J.-F. 1803. "Mémoire sur l'opium." *Ann.Chim* 45:257–85
Derosne, J.-F. 1804. "Ueber das Opium." *Journal der Pharmacie für Aerzte, Apotheker und Chemisten* (Leipzig) 12:223–53.
Descartes, René. 1993. *Meditations on First Philosophy: In Which the Existence of God and the Distinction of the Soul from the Body Are Demonstrated*. Translated by Donald A. Cress. Hackett Publishing Company.
Dettelbach, M. 2001. "Alexander von Humboldt between Enlightenment and Romanticism." *Northeastern Naturalist* 8:9–20.
Diamond, S. 2001. "Wundt before Leipzig." In *Wilhelm Wundt in History: The Making of a Scientific Psychology*, edited by R. Rieber and D. Robinson. Springer Science.
Dierig, S. 2006. *Wissenschaft in der Machinenstadt: Emil Du Bois-Reymond und seine Laboratorien in Berlin*. Valstein Verlag.
Dietl, M., and M. Vintschgau. 1877. "Das Verhalten der physiologischen Reactionszeit unter dem Einfluss von Morphium, Caffée und Wein." *Pflügers Archiv* 26:316–408.

Digby, A. 1999. *The Evolution of British General Practice, 1850–1948*. Oxford University Press.
Dilthey, W. 1901/1927. "Friedrich der Große und die deutsche Aufklärung." In *Studien zur Geschichte des Deutschen Geistes*. Springer.
Domanski, C. W. 2004. "A Biographical Note on Max Friedrich (1856–1887), Wundt's First PhD Student in Experimental Psychology." *Journal of the History of the Behavioral Sciences* 40 (3): 311–17.
Donders, F. C. 1868. "On the Speed of Mental Processes." Translated by W. G. Koster, 1969. *Acta Psychologica* 30:412–31.
Donnellan, B. 1982. "Friedrich Nietzsche and Paul Rée: Cooperation and Conflict." *Journal of the History of Ideas* 43 (4): 595–612.
Driesch, H. 1922/2016. *Geschichte des Vitalismus*. Salzwasser-Verlag.
du Bois-Reymond, E. 1847/1927. *Zwei große Naturforscher des 19. Jahrhunderts: Ein Briefwechsel zwischen Emil du Bois-Reymond und Carl Ludwig*, edited by du Bois-Reymond and Paul Diepgen. Barth.
du Bois-Reymond, E. 1848. *Untersuchungen über thierische Elektricität, Erster Band*. Verlag von G. Reimer. Berlin.
du Bois-Reymond, E. 1849. *Untersuchungen über thierische Elektricität, Zweiter Band*. Verlag von G. Reimer. Berlin.
du Bois-Reymond, E. 1875 [1886]. *Rede, Erste Folge*. Leipzig.
du Bois-Reymond, E. 1887. *Rede, Zweite Folge*. Leipzig.
Duguid, P. 2003. "Developing the Brand: The Case of Alcohol, 1800–1880." *Enterprise & Society* 4 (3): 405–41.
Dyde, S. 2015. "Cullen, a Cautionary Tale." *Medical History* 59 (2): 222–40.
Earles, M. 1961. "Studies in the Development of Experimental Pharmacology in the Eighteenth and Early Nineteenth Centuries." PhD diss., University of London.
Ebert, A., and K. J. Bär. 2010. "Emil Kraepelin: A Pioneer of Scientific Understanding of Psychiatry and Psychopharmacology." *Indian Journal of Psychiatry* 52 (2): 191–92.
Echenberg, M. 2017. *Humboldt's Mexico: In the Footsteps of the Illustrious German Scientific Traveller*. McGill-Queen's University Press.
Eder, A. 1889. *Aerztlicher Bericht der Private-Heilanstalt des Dr. Albin Eder von dem Jahre 1888*. Vienna.
Edney, M. H., and M. S. Pedley, eds. 2019. *The History of Cartography*. Vol. 4, *Cartography in the European Enlightenment*. University of Chicago Press.
Ehrhardt, W. 2006. "Goethe und Auguste Böhmer. War sie vielleicht Goethes natürliche Tochter?" In *Vernunft und Glauben. Ein philosophischer Dialog der Moderne mit dem Christentum*, edited by Steffen Dietzsch and Gian Franco Frigo. Akademie Verlag.
Elkana, Y. 1970. "'Helmholtz' 'Kraft': An Illustration of Concepts in Flux." *Historical Studies in the Physical Sciences* 2:263–98.
Emden, C. 2016. "Metaphor, Perception, and Consciousness: Nietzsche on Rhetoric and Neurophysiology." In *Nietzsche and Science*, edited by G. Moore and T. Brodjer. Routledge.
Engmann, B., and H. Steinberg. 2017. "Die Dorpater Zeit von Emil Kraepelin— Hinterließ dieser Aufenthalt Spuren in der russischen und sowjetischen Psychiatrie?" *Fortschritte der Neurologie-Psychiatrie* 85 (11): 675–82.

Engstrom, E. J. 1991. "Emil Kraepelin: Psychiatry and Public Affairs in Wilhelmine Germany." *History of Psychiatry* 2 (6): 111–32.

Engstrom, E. J. 2003. *Clinical Psychiatry in Imperial Germany: A History of Psychiatric Practice*. Cornell University Press.

Engstrom, E. J. 2007. "'On the Question of Degeneration' by Emil Kraepelin (1908)." *History of Psychiatry* 18 (3): 389–98.

Engstrom, E. J. 2015. "On Attitudes Toward Philosophy and Psychology in German Psychiatry, 1867–1917." In *Philosophical Issues in Psychiatry III: The Nature and Sources of Historical Change*, edited by K. S. Kendler and J. Parnas. Oxford University Press.

Engstrom, E. J. 2016. "Tempering Madness: Emil Kraepelin's Research on Affective Disorders." *Osiris* 31 (1): 163–80.

Engstrom, E. J., and K. S. Kendler. 2015. "Emil Kraepelin: Icon and Reality." *The American Journal of Psychiatry* 172 (12): 1190–96.

Ettmüller. 1809. *Med. Chirurg. Zeit* 3:255–56.

Exner, S. 1873. "Experimentelle Untersuchung der einfachsten psychischen Processe, Erster Abhandlung." *Pflüger, Arch.* 7:601–88.

Exner, S. 1875. "Experimentelle Untersuchung der einfachsten psychischen Processe, Dritter Abhandlung." *Pflüger, Arch.* 11:403–32.

Fahrenberg, J. 2016. "Leibniz' Einfluss auf Wundts Psychologie und Philosophie." *Psychologische Rundschau* 67 (4): 276.

Fancher, R., and A. Rutherford. 2016. "The Sensing and Perceiving Mind: From Kant Through the Gestalt Psychologists." In *Pioneers of Psychology*. Norton and Company.

Fattorusso, E., and O. Taglialatela-Scafati. 2008. *Modern Alkaloids: Structure, Isolation, Synthesis and Biology*. Wiley-VCH.

Fechner, G., ed. 1826. *Repertorium der organischen Chemie: in sechs Bänden, Band 1,Teil 1*. Leipzig.

Fechner, G., ed. 1834. *Das Hauslexikon. Vollständiges Handbuch praktischer Lebenskentnisse für alle Stände. Erster Band*. Leipzig.

Fechner, G., ed. 1841. *Das Hauslexikon. Vollständiges Handbuch praktischer Lebenskentnisse für alle Stände. Zweiter Band. Zweite Auflage*. Leipzig.

Fechner, G. 1855. *Über die physikalische und philosophische Atomlehre*. Leipzig.

Fechner, G. 1861. *Ueber die Seelenfrage. Ein Gang durch die sichtbare Welt, um die unsichtbare zu Finden*. Leipzig.

Fechner, G. 1876. *Vorschule der Aesthetik, Erster Teil*. Leipzig.

Fechner, G. 1879. *Briefwechsel Fechner-Wundt, Wundt-Nachlaß*. Universitätsarchiv Leipzig.

Fechner, G. 1888 [1860]. *Elemente der Psychophysik, Erster Theil*. 2nd ed. Leipzig.

Fechner, G. 1889 [1860]. *Elemente der Psychophysik, Zweiter Theil*. 2nd ed. Leipzig.

Fechner, G. 1901 [1851]. *Zend-Avesta oder über die Dinge des Himmels und des Jenseits: Vom Standpunkt der Naturbetrachtung*. 2nd ed. Leipzig.

Feest, U., and T. Sturm. 2011. "What (Good) Is Historical Epistemology? Editors' Introduction." *Erkenntnis (1975-)* 75 (3): 285–302.

Feuchtersleben, E. v. 1845. *Lehrbuch der ärztlichen Seelenkunde*. Vienna.

Feuchtersleben, E. v., and M. v. Schwind. [1875] 1978. *Album von Radirungen*. Harenberg Kommunikation.

Filley, C. M. 2001. *The Behavioral Neurology of White Matter*. Oxford University Press.
Finger, S. 2000. *Minds Behind the Brain: A History of the Pioneers and Their Discoveries*. Oxford University Press.
Finger, S., and M. Piccolino. 2011. *The Shocking History of Electric Fishes: From Ancient Epochs to the Birth of Modern Neurophysiology*. Oxford University Press.
Finger, S., M. Piccolino, and F. W. Stahnisch. 2013. "Alexander von Humboldt: Galvanism, Animal Electricity, and Self-Experimentation Part 2: The Electric Eel, Animal Electricity, and Later Years." *Journal of the History of the Neurosciences* 22 (4): 327–52.
Finger, S., and N. J. Wade. 2002. "The Neuroscience of Helmholtz and the Theories of Johannes Müller Part 1: Nerve Cell Structure, Vitalism, and the Nerve Impulse." *Journal of the History of the Neurosciences* 11 (2): 136–55.
Finkelstein, G. 1996. "Emil Du Bois-Reymond: The Making of a Liberal German Scientist (1818–1851)." PhD diss., Princeton University.
Finkelstein, G. 2006. "Emil du Bois-Reymond vs Ludimar Hermann." *Comptes Rendus Biologies* 329 (5–6): 340–47.
Finkelstein, G. 2013. *Emil du Bois-Reymond: Neuroscience, Self, and Society in Nineteenth-Century Germany*. MIT Press
Fischer, E. P. 1998. *Byk Gulden: Forschergeist und Unternehmermut*. Piper.
Fishman, R. S. 2010. "Darwin and Helmholtz on Imperfections of the Eye." *Arch Ophthalmol* 128 (9):1209–11.
Floyer, J. 1687. *Pharmako-Basanos: Or, the Touchstone of Medicines*. London.
Forel, A. 1935. *Rückblick auf mein Leben*. Europa-Verlag.
Foss, R. 2012. *Rum: A Global History*. Reaktion Books.
Foucault, M. 1990. *The History of Sexuality, Volume 1*. Vintage.
Foucault, M. 2005. *The Order of Things*. Routledge.
Foye, W., D. Williams, and T. Lemke. 2002. *Foye's Principles of Medicinal Chemistry*. Lippincott Williams & Wilkins.
Freckelton, I. 2012. "The Deaths of King Ludwig II of Bavaria and of His Psychiatrist, Professor von Gudden: Warnings from the Nineteenth Century." *Psychiatry, Psychology and Law* 19 (1): 1–10.
Freud, S. 1882. "Über den Bau der Nervenfaser und Nervenzellen beim Flußkrebs." *Sitzungber. Akad. Wiss. Wien* 85:9–46.
Freud, S. 1884a. "Die Struktur der Elemente des Nervensytem." *Jahrbücher für Psychiatrie* 5:221–29.
Freud, S. 1884b. "Ein Fall von Hirnblutung mit indirekten basalen Herdsymptomen bei Skorbut." *Wien. Med. Wschr* 34:244–46, 276–79.
Freud, S. 1885a. "Beitrag zur Kenntnis der Cocawirkung." *Wiener Med. Wochenschrift* 35 (5): 1–8.
Freud, S. 1885b. "Ein Fall von Muskelatrophie mit ausgebreiteten Sensibilitätsstörungen (Syringomyelie)." *Wien. med. Wschr* 35:389–92, 425–29.
Freud, S. 1885c. *Über Coca*. Vienna.
Freud, S. 1885d. "Ueber die Allgemeinwirkung des Cocains." *Zeitschrift für Therapie mit Einbeziehung der Elektro- und Hydroherapie* 3:49–51.
Freud, S. 1886. "Akute multiple Neuritis der spinalen und Hirnnerven." *Wien. Med. Wschr* 36:168–72.
Freud, S. 1891. *Zur Auffassung der Aphasien*. Leipzig and Vienna.

Freud, S. 1895. *Entwurf einer Psychologie*. Unpublished manuscript, Library of Congress.
Freud, S. 1899. *Die Traumdeutung*. Leipzig and Vienna.
Friedrich, C. 1998. "Der Fabrikant und sein Berater: Heinrich Emanuel Merck (1794–1855) und Johann Bartholomäus Trommsdorff (1770–1837)." *Pharm. Ind.* 60 (6): 508–11.
Friedrich, C. n.d. "German Pharmacopoeias." ISHPWG Germany.
Friedrich, M. 1883. "Über die Apperzeptionsdauer bei einfachen und zusammengesetzten Vorstellungen." *Philos Stud* 1:39–77
Fritsch, G., and E. Hitzig. 1870. "Über die elektrische Erregbarkeit des Großhirns." *Archiv für Anatomie, Physiologie und Wissenschaftliche Medicine* 37:300–32.
Fuhrmans, H. 1962. *F. W. J. Schelling Briefe und Dokumente*. Bonn: H. Bouvier.
Fullinwider, S. 1991. "Darwin Faces Kant: A Study in Nineteenth-Century Physiology." *The British Journal for the History of Science* 24 (1): 21–44.
Fye, W. B. 1999. "Rudolf Albrecht von Koelliker." *Clin Cardiol* 22:376–77.
Gaedcke, F. 1855. "Ueber das Erythroxylin, dargestellt aus den Blättern des in Südamerika cultivirten Strauches Erythroxylon Coca Lam." *Archiv der Pharmazie* 132 (2): 141–50.
Gambaratto, A. 2018a. "Lorenz Oken (1779–1851): Naturphilosophie and the reform of natural history." *The British Journal for the History of Science* 51 (3): 329–40.
Gambarotto, A. 2018b. "Vital Forces, Teleology and Organization." In *History, Philosophy and Theory of the Life Sciences*. Springer International Publishing.
Gantet, C. 2021. "The Dissemination of Mesmerism in Germany (1784–1815): Some Patterns of the Circulation of Knowledge." *Centaurus* 63 (4): 762–78.
Gao, M. 2020. "'Founding Its Empire on Spells of Pleasure': Brunonian Excitability, the Invigorated English Opium-Eater, and De Quincey's 'China Question.'" *Literature and Medicine* 38 (1): 1–25.
Garrett, B. 2003. "Vitalism and Teleology in the Natural Philosophy of Nehemiah Grew (1641–1712)." *The British Journal for the History of Science* 36 (1): 63–81.
Gemes, K. 2021. "The Biology of Evil: Nietzsche on Entartung and Verjüdung (Degeneration and Jewification)." *The Journal of Nietzsche Studies* 52 (1): 1–25.
Gerabek, W. E. 2001. "Lorenz Oken und die Medizin der Romantik." In *Lorenz Oken (1779–1851)*, edited by O. Breidbach, H. J. Fliedner, and K. Ries. Verlag Hermann Böhlaus Nachfolger.
Gescheider, G. 2015. *Psychophysics: The Fundamentals*. Routledge Press.
Geschwind, N. 1970. "The Organization of Language and the Brain." *Science* 170 (3961): 940–44.
Gibson, W. 1970. "The Bio-Medical Pursuits of Christopher Wren." *Medical History* 14 (4): 331–41.
Gladstone, E. 1933. "Johann Sigismund Elsholtz: Clysmatica Nova (1665): Elsholtz' Neglected Work on Intravenous Injection: Part IV." *Cal West Med* 39 (3): 190–93.
Glymour, C. 1991. "Freud's Androids." In *The Cambridge Companion to Freud*, edited by J. Neu. Cambridge University Press
Glynn, I. 2013. *Elegance in Science: The Beauty of Simplicity*. Oxford University Press.
Goethe, J. W. von. [1811] 1989. *Goethes Werke (Aus meinem Leben: Dichtung und Wahrheit, Dritte Teil, 11. Band)*. Beck.
Golinski, J. 2011. "Humphry Davy: The Experimental Self." *Eighteenth-Century Studies* 45 (1): 15–28.

Golomb, J. 2015. "Introductory Essay: Nietzsche's New Psychology." In *Nietzsche and Depth Psychology*, edited by J. Golomb, R. Lehrer, and W. Santaniello. SUNY Press.

Gordon, C. 1998. "Canguilhem: Life, Health and Death." *Economy and Society* 27 (2-3): 182-89.

Grant, I. H. 2008. *Philosophies of Nature after Schelling*. Continuum International Pub. Group.

Greenblatt, S. H. 1991. "The Development of Modern Neurological Thinking in the 1860s." *Perspectives in Biology and Medicine* 35 (1): 129-39.

Griffith, R. D., M.-A. Y. Abyaneh, L. Falto-Aizpurua, and K. Nouri. 2014. "Demystifying Merkel." *JAMA Dermatology* 150 (8): 814.

Gross, C. G. 2007. "The Discovery of Motor Cortex and its Background." *Journal of the History of the Neurosciences* 16 (3): 320-31.

Grundfest, H. 1963. "The Different Careers of Gustav Fritsch (1838-1927)." *Journal of the History of Medicine and Allied Sciences* 18 (2): 125-29.

Grzybowski, A., and J. Żołnierz. 2021. "Sigmund Freud (1856-1939)." *Journal of Neurology* 268 (6): 2299-300.

Guareschi, I. 1896. *Einführung in das Stadium der Alkaloids mit besonderer Berücksichtigung der vegetabilischen Alkaloids und der Ptomaine*. Berlin.

Gudden, B. V. 1875. "Ueber ein neues Microtom." *Archiv für Psychiatrie und Nervenkrankheiten* 5:229-34.

Guenther, K. 2015. *Localization and its Discontents: A Genealogy of Psychoanalysis and the Neurodisciplines*. University of Chicago Press.

Gwyn, J. 2001. "The Halifax Naval Yard and Mast Contractors." *The Northern Mariner/Le marin du nord* 9 (4): 1-25.

Haaz, I. 2002. "Les conceptions du corps chez Ribot et Nietzsche." *À partir des Fragments posthumes de Nietzsche, de la Revue philosophique de la France et de l'étranger et de la Recherche-Nietzsche*. L'Harmattan.

Habermas, J. 1990. *Strukturwandel der Öffentlichkeit Untersuchungen zu einer Kategorie der bürgerlichen Gesellschaft*. Suhrkamp.

Hachtmann, R. 1997. *Berlin 1848: eine Politik- und Gesellschaftsgeschichte der Revolution*. J. H. W. Dietz.

Hacking, I. 2002. *Historical Ontology*. Harvard University Press.

Hacking, I. 2009. "La Mettrie's Soul: Vertigo, Fever, Massacre, and The Natural History." *CBMH* 26 (1): 179-202.

Hagner, M. 2008. *Homo cerebralis: der Wandel vom Seelenorgan zum Gehirn*. Suhrkamp.

Hagner, M. 2012. "The Electrical Excitability of the Brain: Toward the Emergence of an Experiment." *Journal of the History of the Neurosciences* 21 (3): 237-49.

Hajdu, Steven I. 2002. "A Note from History: Introduction of the Cell Theory." *Annals of Clinical and Laboratory Science* 32 (1): 98-100.

Hakosalo, H. 2006. "The Brain under the Knife: Serial Sectioning and the Development of Late Nineteenth-Century Neuroanatomy." *Studies in History and Philosophy of Biological and Biomedical Sciences* 37 (2): 172-202.

Hänel, H. 1898. "Die psychischen Wirkungen des Trionals." *Kraepelin's Psychologische Arbeiten* 2 (2): 326-98.

Hanzlik, P. J. 1929. "125 Anniversary of the Discovery of Morphine by Sertürner." *Journal of Pharmaceutical Sciences* 18 (4): 375-84.

Hay, M. 1882/1883. "Vegetable Alkaloids, and the Methods of their Separation." *The Pharmaceutical Journal and Transactions* 3 (13): 719–21.

Hayter, A. 1968. *Opium and the Romantic Imagination*. University of California Press.

Healy, D. 2008. *Mania: A Short History of Bipolar Disorder*. Johns Hopkins University Press.

Healy, D. 2009. *The Creation of Psychopharmacology*. Harvard University Press.

Heckers, S., and K. S. Kendler. 2020. "The Evolution of Kraepelin's Nosological Principles." *World Psychiatry: Official Journal of the World Psychiatric Association* 19 (3): 381–88.

Heidelberger, M. 2003. "The Mind-Body Problem in the Origin of Logical Empiricism." In *Logical Empiricism*, translated by C. Klohr. University of Pittsburgh Press.

Heidelberger, M. 2004. *Nature from Within: Gustav Theodor Fechner and His Psychophysical Worldview*. University of Pittsburgh Press.

Heilbron, J. L. 1990. "Introductory Essay." In *The Quantifying Spirit in the 18th Century*, edited by Tore Frängsmyr et al. University of California Press.

Heine, H. 1834/87. *Zur Geschichte der Religion und Philosophie in Deutschland*. Halle.

Heine, H. 1972. *Werke und Briefe in zehn Bänden*. Band 5. Berlin und Weimar: Aufbau Verlag.

Helmholtz, H. von. 1850. "Vorläufiger Bericht über die Fortpflanzungs-Geschwindigkeit der Nervenreizung." *Archiv für Anatomie, Physiologie und wissenschaftliche Medicin*.

Helmholtz, H. von. 1867. *Handbuch der physiologischen Optik*. Leipzig.

Helmholtz, H. von. [1877] 1995. "On Thought in Medicine." In *Science and Culture*, edited and translated by David Cahan. University of Chicago Press

Helmholtz, H. v., E. du Bois-Reymond, C. Kirsten, H. Hörz, and S. Wollgast. 1986. *Dokumente einer Freundschaft: Briefwechsel zwischen Hermann von Helmholtz und Emil du Bois-Reymond, 1846–1894*. Akademie-Verlag.

Hennis, W. 1991. "The Pitiless 'Sobriety of Judgement': Max Weber between Carl Menger and Gustav von Schmoller—The Academic Politics of Value Freedom." *History of the Human Sciences* 4 (1): 27–59.

Hennock, E. P. 2007. *The Origin of the Welfare State in England and Germany, 1850–1914: Social Policies Compared*. Cambridge University Press.

Herbart, J. 1850. "Lehrbuch zur Psychologie." In *Herbarts Sämmtliche Werke* (Bande 5), edited by G. Hartenstein. Leipzig.

Hergenhahn, B., and T. Henley. 2014. *An Introduction to the History of Psychology*. Wadsworth Cengage Learning.

Hermann, L. 1864. "Ueber die physiologischen Wirkungen des Stickstoffoxydulgases." *Archiv für Anatomie, Physiologie und wissenschaftliche Medicin*, 521–536

Hermann, L. 1865. "Ueber die Wirkungen des Stickstoffoxydgases auf das Blut." *Archiv für Anatomie, Physiologie und wissenschaftliche Medicin*, 469–481.

Hermann, L. 1866. "Ueber die Wirkungsweise einer Gruppe von Giften." *Archiv für Anatomie, Physiologie und wissenschaftliche Medicin* 38:27–40.

Hermann, L. 1874. *Lehrbuch der experimentelle Toxicologie*. Berlin.

Hildebrandt, H. 1993. "Der psychologische Versuch in der Psychiatrie: Was wurde aus Kraepelins (1895) Programm?" *Psychologie und Geschichte* 5:5–30.

Hippius, H. 2008. *The University Department of Psychiatry in Munich: From Kraepelin and His Predecessors to Molecular Psychiatry*. Springer.
Hitzig, E. 1874. *Physiologische und klinische Untersuchungen über das Gehirn, Erster Teil*. Berlin.
Hlade, J. 2021. "Reconsidering 'Brain Mythology.'" *Medicina Historica* 5 (1): 1–9.
Hoff, P. 1992a. "Emil Kraepelin and Philosophy: The Implicit Philosophical Assumptions of Kraepelinian Psychiatry." In *Phenomenology, Language & Schizophrenia.*, edited by M. Spitzer, F. Uehlein, M. A. Schwartz, and C Mundt. Springer.
Hoff, P. 1992b. "Psychiatrie und Psychologie—Bemerkungen zum Hintergrund des Kraepelinschen Wissenschaftsverständnisses." In *Pharmakopsychologie— Experimentelle und klinische Aspekte*, edited by J. Oldigs-Kerber and J. P. Leonhard. Fischer.
Hoff, P. 1994. *Emil Kraepelin und die Psychiatrie als klinische Wissenschaft. Ein Beitrag zum Selbsverständnis psychiatrischer Forschung*. Springer.
Hoff, P. 2008. "Kraepelin and Degeneration Theory." *Eur Arch Psychiatry Clin Neurosci* 258 [Suppl 2]: 12–17.
Hoff, P. 2015. "The Kraepelinian Tradition." *Dialogues in Clinical Neuroscience* 17 (1): 31–41.
Hoffmann, C. 2006. *Unter Beobachtung–Naturforschung in der Zeit der Sinnesapparate*. Wallstein Verlag.
Hoffman, S. 2021. "Kant on Intoxication." In *The Court of Reason: Proceedings of the 13th International Kant Congress*, edited by B. Himmelmann and C. Serck-Hanssen. De Gruyter.
Hofmann, A. W. v. 1876. "The Life-Work of Liebig." In *Experimental and Philosophic Chemistry; with Allusions to His Influence on the Development of the Collateral Sciences and of the Useful Arts; a Discourse Delivered to the Fellows of the Chemical Society of London in the Theatre of the Royal Institution of Great Britain, on March the 18th, 1875; the Faraday Lecture for 1875*. United Kingdom.
Holmes, R. 2011. *Coleridge: Early Visions, 1772–1804*. Pantheon.
Holub, R. 2015. "The Birth of Psychoanalysis from the Spirit of Enmity: Nietzsche, Rée, and Psychology in the Nineteenth Century." In *Nietzsche and Depth Psychology*, edited by J. Golomb, R. Lehrer, and W. Santaniello. SUNY Press.
Hoppe, B. 2007. "Robin Keen: The Life and Work of Friedrich Wöhler (1800–1882)." *Isis* 98 (1): 195–96.
Hoquet, T. 2018. *Revisiting the Origin of Species: The Other Darwins*. Routledge.
Housman, B., S. S. Bellary, A. Walters, N. Mirzayan, R. Shane Tubbs, and M. Loukas. 2012. "Moritz Heinrich Romberg (1795–1873): Early Founder of Neurology." *Clinical Anatomy* 27 (2): 147–49.
Hubbard, J. A., and D. K. Binder. 2016. "History of Astrocytes." In *Astrocytes and Epilepsy*. Elsevier.
Huhle-Kreutzer, G. 1989. *Die Entwicklung arzneilicher Produktionsstätten*. Dt. Apotheker-Verl.
Hunter, C. 1865. *On the Speedy Relief of Pain and Other Nervous Activities by Means of the Hypodermic Method*. London.
Huxtable, R., and S. Schwarz. 2001. "The Isolation of Morphine—First Principles in Science and Ethics." *Molecular Interventions* 1 (4): 189–91.
Jacobson, M. 1991. *Developmental Neurobiology*. Springer.

James, K., and P. Withington. 2022. "Introduction to Intoxicants and Early Modern European Globalization." *The Historical Journal* 65 (1): 1–11.
Jaspers, K. 1913. *Allgemeine Psychopathologie. Ein Leitfaden für Studierende, Ärzte und Psychologen.* Springer.
Jay, M. 2009. "The Atmosphere of Heaven: The 1799 Nitrous Oxide Researches Reconsidered." *Notes and Records of the Royal Society of London* 63 (3): 297–309.
Jenkins, A. 2007. "Alexander von Humboldt's 'Kosmos' and the Beginnings of Ecocriticism." *Interdisciplinary Studies in Literature and Environment* 14 (2): 89–105.
Jeston, J., and P. Dick. 1823. "No IV. English Opium." *Transactions of the Society, Instituted at London, for the Encouragement of Arts, Manufactures, and Commerce* 41:17–31.
Jones, E. 1953. *Life and Work.* Vol. 1, *The Young Freud, 1856–1900*. Basic Books
Jones, R. 1994. "The Positive Science of Ethics in France: German Influences on 'De la division du travail social.'" *Sociological Forum* 9 (1): 37–57.
Jünger, E. 1970. *Annäherungen: Drogen und Rausch.* Klett.
Jurna, I. 2003. "Sertürner und Morphin—eine historische Vignette." *Der Schmerz* 17 (4): 280–83.
Kamieński, Ł. 2016. *Shooting Up: A Short History of Drugs and War.* Oxford University Press.
Kane, H. 1881. *Drugs that Enslave: The Opium, Morphine, Chloral, and Hashisch Habits.* Philadelphia.
Kant, I. 1781 [1919]. *Kritik der reinen Vernunft.* Leipzig: F. Meiner.
Kant, I. 1787. *Metaphysische Anfangsgründe der Naturwissenschaft.* Riga.
Kant, I. 1790a. "Brief 411." In *Akademieausgabe von Immanuel Kants Gesammelten Werken Bände und Verknüpfungen zu den Inhaltsverzeichnissen Band XI: Briefe, 1789–1794*.
Kant, I. 1790b. "Kritik der Urteilskraft." In *Akademieausgabe von Immanuel Kants Gesammelten Werken Bände und Verknupfungen zu den Inhaltsverzeichnissen Band V: Kritik der praktischen Vernunft, Kritik der Urteilskraft*.
Kant, I. 1797. "Metaphysik der Sitten, in zwei Teilen." In *Akademieausgabe von Immanuel Kants Gesammelten Werken Bände und Verknüpfungen zu den Inhaltsverzeichnissen Band VI: Die Religion innerhalb der Grenzen der blossen Vernunft, Die Metaphysik der Sitten*.
Kant, I. 1798. "Anthropologie in pragmatischer Hinsicht." In *Akademieausgabe von Immanuel Kants Gesammelten Werken Bände und Verknüpfungen zu den Inhaltsverzeichnissen Band VII: Der Streit der Fakultäten, Anthropologie in pragmatischer Hinsicht*.
Kant, I. [1784] 1999. "Was ist Aufklärung?" In *Was ist Aufklärung:Ausgewählte Kleine Schriften*, edited by E. Cassirer and H. D. Brandt. Meiner.
Kapoor, L. D. 1995. *Opium Poppy: Botany, Chemistry, and Pharmacology.* Food Products Press.
Karch, S. 2006. *A Brief History of Cocaine.* 2nd ed. Taylor and Francis.
Karenberg, A. 1990. "'Lehrbuch der Nervenkrankheiten des Menschen' von Moritz Heinrich Romberg (1840)." *Therapiewoche* 40:3407–10.
Kelly, A. 2012. *The Descent of Darwin: The Popularization of Darwinism in Germany, 1860–1914.* UNC Press Books.
Kennedy, J. 1985. *Coca Exotica: The Illustrated Story of Cocaine.* Fairleigh Dickinson University Press.

Kennedy, P. 1987. *The Rise and Fall of the Great Powers: Economic Change and Military Conflict from 1500 to 2000.* 1st ed. Random House.
Khalili, M. 2024. "Entity Realism Meets Perspectivism." *Acta Anal.* 39:79–95.
Kielhorn, F.-W. 1996. "The History of Alcoholism: Brühl-Cramer's Concepts and Observations." *Addiction* 91 (1): 121–28.
Kitchen, M. 2011. *A History of Modern Germany: 1800 to the Present.* John Wiley & Sons.
Klein, U., and W. Lefèvre. 2007. *Materials in Eighteenth-Century Science: A Historical Ontology.* MIT Press.
Klempe, S. H. 2021. "The Importance of Leibniz for Wundt." *Hu Arenas* 4:20–31.
Klinge, M. 2004. "Teachers." In *A History of the University in Europe.* Vol. 3, *Universities in the Nineteenth and Early Twentieth Centuries (1800–1945),* edited by W. Rüegg. Cambridge University Press.
Klockgether-Radke, A. P. 2002. "F. W. Sertürner und die Entdeckung des Morphins—200 Jahre Schmerztherapie mit Opioiden." *Ains Anästhesiologie Intensivmedizin Notfallmedizin Schmerztherapie Anasthesiol Intensivmed Notfallmed* 37 (5): 244–49.
Knabe, P.-E. 1978. Die Rezeption der französischen Aufklärung in den 'Göttingischen Gelehrten Anzeigen' (1739–1779)., Frankfurt am Main: Vittorio Klostermann.
Knight, R. J. B. 1986. "New England Forests and British Seapower: Albion Revised." *American Neptune* 46, 221–29.
Koehler, P. 2012. "Eduard Hitzig's Experiences in the Franco-Prussian War (1870–1871): The Case of Joseph Masseau." *Journal of the History of the Neurosciences* 21 (3): 250–62.
Kölliker, A. 1852. *Mikroskopische Anatomie, oder, Gewebelehre des Menschen.* Vol. 2, 1852–54, *Specielle Gewebelehre.* Leipzig.
Kölliker, A. 1859. *Handbuch der Gewebelehre des Menschen.* Leipzig.
Kölliker, A. 1880. *Grundriss der Entwicklungsgeschichte des Menschen und der höheren Thiere.* Leipzig.
Kölliker, R. A. 1889. *Handbuch der Gewebelehre des Menschens.* Leipzig.
Kolmer, W., and H. Lauber, H. 2013. *Haut und Sinnesorgane: Auge.* Springer-Verlag.
Königsberger, L. 1906. *Hermann von Helmholtz.* Dover Publications.
Kraepelin, E. 1880. *Die Abschaffung des Strafmasses—Ein Vorschlag zur Reform der heutigen Strafrechtspflege.* Stuttgart.
Kraepelin, E. 1881a. "Brief an Wilhelm Wundt, München, 1. Aug. 1881. Wundtbriefwechsel." https://home.uni-leipzig.de/wundtbriefe/viewer.htm.
Kraepelin, E. 1881b. "Ueber den Einfluss acuter Krankheiten auf die Entstehung von Geisteskrankheiten." *Archiv für Psychiatrie und Nervenkrankheiten* 11, S. 137–183.
Kraepelin, E. 1881c. "Ueber psychische Zeitmessung." *Schmidt's Jahrbücher der gesamten Medizin* 196:205–13.
Kraepelin, E. 1882. "Ueber die Dauer einfacher psychischer Vorgänge." *Biologisches Centralblatt,* Bd. 1, 1881/82, S. 654–672, S. 721–733, S. 751–766.
Kraepelin, E. 1883a. *Compendium der Psychiatrie: Zum Gebrauche für Studirende und Aerzte.* Leipzig.
Kraepelin, E. 1883b. "Ueber die Einwirkung einiger medikamentöser Stoffe auf die Dauer einfacher psychicher Vorgänge, Erste Abteilung, Ueber die Einwirkung von Amylnitrit, Aethyläther und Chloroform." *Philosophische Studien* 1:417–62.

Kraepelin, E. 1883c. "Ueber die Einwirkung einiger medicamentöser Stoffe auf die Dauer einfacher psychicher Vorgänge, Zweite Abteilung, Ueber die Einwirkung von Aethylalkohol." *Philosophische Studien* 1:573–605.
Kraepelin, E. 1885. Review of Paul Rée, "Die Entstehung des Gewissens." *Literarisches Centralblatt* 36:1697–98.
Kraepelin, E. 1886a. Review of Paul Rée, "Die Illusion der Willensfreiheit. Ihre Ursachen und ihre Folgen." *Literarisches Centralblatt* 37:41.
Kraepelin, E. 1886b. "Schlaflosigkeit und deren Behandlung durch die neueren Schlafmittel." *Jahresberichte der Gesellschaft für Natur- und Heilkunde in Dresden* 1885/86:153–55.
Kraepelin, E. 1887. *Psychiatrie: Ein kurzes Lehrbuch für Studirende und Aerzte*. 2nd revised ed. Leipzig.
Kraepelin, E. 1888. "Psychologische Forschungsmethoden." *Humboldt— Monatsschrift für die gesamten Naturwissenschaften* 7:12–14.
Kraepelin, E. 1892a. *Ueber die Beeinflussung einfacher psychischer Vorgänge durch einige Arzneimittel*. Jena.
Kraepelin, E. 1892b. "Ueber die centrale Wirkung einiger Arzneimittel." *Archiv für Psychiatrie und Nervenkrankheiten* 24: 641–642.
Kraepelin, E. 1893. *Psychiatrie: Ein kurzes Lehrbuch für Studirende und Aerzte*. 4th revised edition. Leipzig.
Kraepelin, E. 1894. "Ueber geistige Arbeit." *Neue Heidelberger Jahrbücher* 4: 31–52.
Kraepelin, E. 1895. Der psychologische Versuch in der Psychiatrie. *Psychol Arbeiten* 1:1–91.
Kraepelin, E. 1896. *Psychiatrie: Ein Lehrbuch für Studirende und Aerzte*. Fünfte, vollständig umgearbeitete Auflage. Leipzig.
Kraepelin, E. 1899. "Neuere Untersuchungen über die psychischen Wirkungen des Alkohols." *Internationale Monatsschrift zur Bekämpfung der Trinksitten* 9:321–32.
Kraepelin, E. 1903. *Ueber geistige Arbeit, Zweite*. Jena: Gustav Fischer.
Kraepelin, E. 1906a. "Das Verbrechen als soziale Krankheit." *Monatsschrift für Kriminalpsychologie und Strafrechtsreform* 3:257–279.
Kraepelin, E. 1906b. "Der Alkoholismus in München." *Münchener Medizinische Wochenschrift* 53:737–41.
Kraepelin, E. 1906c. *Über Sprachstörungen im Traume*. Engelmann Verlag [vgl. 1910].
Kraepelin, E. 1983. *Lebenserinnerungen*. Edited by H. Hippius, G. Peters, and D. Ploog. Springer.
Kraepelin, E., and A. Hoch. 1896. "Ueber die Wirkung der Theebestandtheile auf körperliche und geistige Arbeit." *Psychologische Arbeiten* 1:378–488.
Kraepelin, E., and A. Oseretzkowsky. 1901. "Ueber die Beeinflussung der Muskelleistung durch verschiedene Arbeitsbedingungen." *Psychologische Arbeiten* 3:587–690.
Kremers, E., and G. Sonnedecker. 1986. *Kremers and Urdang's History of Pharmacy*. Amer. Inst. History of Pharmacy.
Kuehn, M. 2001. *Kant: A Biography*. Cambridge University Press
Kuhn, T. 2012. *The Structure of Scientific Revolutions: 50th Anniversary Edition*. Foreword by Ian Hacking. University of Chicago Press.
La Mettrie, J. de. 1748. *L'homme machine*. Lyon.
Lampl, H. 1987. "Ex Oblivione: Das Féré-Palimpseste: Noten zur Beziehung Friedrich Nietzsche-Charles Féré (1857–1907)." *Nietzsche-Studien* 15:225–64.

Latour. B. 2000. "On the Partial Existence of Existing and Nonexisting Objects." In *Biographies of Scientific Objects*. University of Chicago Press.
Laudan, L. 1981. "The Sources of Modern Methodology: Two Models of Change." In *Science and Hypothesis*. Springer Netherlands.
Lazarsfeld, P. F., and A. R. Oberschall. 1965. "Max Weber and Empirical Social Research." *American Sociological Review* 30 (2): 185–99.
Lehrer, R. 2015. "Nietzsche and Psychology." In *Nietzsche and Depth Psychology*, edited by J. Golomb, R. Lehrer, and W. Santaniello. SUNY Press.
Lenoir, T. 1982. *The Strategy of Life: Teleology and Mechanics in Nineteenth Century German Biology*. D. Reidel.
Lentacker, A. 2019. "The Codex Nationalized: Naming People and Things in the Wake of a Revolution." In *Drugs on the Page: Pharmacopoeias and Healing Knowledge in the Early Modern Atlantic World*, edited by. M. Crawford and J. Gabriel. University of Pittsburgh Press.
Lesch, J. 1981. "Conceptual Change in an Empirical Science: The Discovery of the First Alkaloids." *Historical Studies in the Physical Sciences* 11 (2): 305–28.
Lesch, J. 1990. "Systematics and the Geometrical Spirit." In *The Quantifying Spirit in the 18th Century*, edited by Tore Frängsmyr, John L. Heilbron, and Robin E. Rider. University of California Press.
Levinger, M. 1998. "Kant and the Origins of Prussian Constitutionalism." *History of Political Thought* 19 (2): 241–63.
Levinstein, E. 1877. *Die Morphiumsucht; eine Monographie nach eignen Beobachtungen*. Berlin.
Lewontin, R. C. 2003. *Biology as Ideology*. Anansi.
Lichtenfels, R., and R. Fröhlich. 1851. "Ueber den Puls als ein Symptom, sowie als numerisches Maass der physiologischen Arzeneiwirkung." *Sitzungsber. d. Wein* 6:824–49.
Liebig, J. v. 1831. "Ueber einen neuen Apparat zur Analyse organischer Körper, und über die Zusammensetzung einiger organischen Substanzen." *Ann. Phys.* 97:1–43.
Liebig, J. v. 1832a. *Annalen der Pharmacie*. Band I. Lemgo and Heidelberg.
Liebig, J. v. 1832b. "Ueber die Verbindungen, welche durch Einwirkung des Chlors auf Alkohol, Aether, ölbildendes Gas und Essiggeist entstehen." *Annalen der Chemie* 1:182–230.
Liebig, J. v. 1842. *Die organische Chemie in ihrer Anwendung auf Physiologie und Pathologie*. Braunschweig.
Liebig, J. v. 1843. *Die Thier-Chemie oder die organische Chemie in ihrer Anwendung auf Physiologie und Pathologie*. Braunschweig.
Liebig, J. v. 1874. *Reden und Abhandlungen*. Leipzig.
Liebig, J. v., and F. Wöhler. 1838. "Untersuchungen über die Nature der Harnsäure." *Annalen der Pharmacie* 26:241–336.
Liebreich, O. 1869. *Das Chloralhydrat: ein neues Hypnoticum und Anaestheticum und dessen Anwendung in der Medicin; eine Arzneimittel-Untersuchung*. Berlin.
Liebscher, M. 2014. "Nietzsche und die Psychologie." In *Handbuch Nietzsche und die Wissenschaft*, edited by H Heit and L. Heller. De Gruyter.
Link, S. 1994. "Rediscovering the Past: Gustav Fechner and Signal Detection Theory." *Psychological Science* 5 (6): 335–40.
Link, S., D. Murray, and H. Ross. 2009. "The Wide Impact of Gustav Fechner's

Thought-Provoking Writings." *The American Journal of Psychology* 122 (3): 405–10.
Lipman, T. 1967. "Vitalism and Reductionism in Liebig's Physiological Thought." *Isis* 58 (2): 167–85.
Löbenstein-Löbel, E. 1816. *Die Anwendung und Wirkung der Weine in lebensgefährlichen Krankheiten*. Leipzig and Altenburg.
Lohff, B. 1978. "Johannes Müllers Rezeption der Zellenlehre in seinem 'Handbuch der Physiologie des Menschen.'" *Medizinhistorisches Journal* 13;247–58.
López-Muñoz, F., J. Boya, and C. Alamo. 2006. "Neuron Theory, the Cornerstone of Neuroscience, on the Centenary of the Nobel Prize Award to Santiago Ramón y Cajal." *Brain Research Bulletin* 70 (4–6): 391–405.
Louis, E. D. 2002. "Erb and Westphal: Simultaneous Discovery of the Deep Tendon Reflexes." *Seminars in Neurology* 22 (4): 385–90.
Löw, R. 1977. *Pflanzenchemie zwischen Lavoisier und Liebig*. Straubing and Munich.
Löwald, A. 1896. "Über die psychischen Wirkungen des Broms." *Kraepelins Psychol. Arbeiten* 1:489.
Luchsinger, B. 1877. "Die Wirkungen von Pilocarpin und Atropin auf die Schweissdrüsen der Katze. Ein Beitrag zur Lehre vom doppelseitigen Antagonismus zweier Gifte." *Pflüger's Archiv für Anatomie und Physiologie* 15:482–92.
Lytle, A. 1906. "Materia Medica, Pharmacy, Therapeutics." *The American Journal of Nursing* 6 (4): 217–24.
Macht, D. I., and S. Isaacs. 1917. "Action of Some Opium Alkaloids on the Psychological Reaction Time." *Psychobiology* 1 (1): 19–32.
Maehle, A. 1995. "Pharmacological Experimentation with Opium in the Eighteenth Century." In *Drugs and Narcotics in History*, edited by R. Porter and M. Teich. Cambridge University Press.
Maehle, A. 1999. *Drugs on Trial: Experimental Pharmacology and Therapeutic Innovation in the Eighteenth Century*. Amsterdam: Rodopi.
Magner, L. 2002. *A History of the Life Sciences*. CRC Press.
Marriott, J. 2010. *Pharmaceutical Compounding and Dispensing*. Pharmaceutical Press.
Martin, N. 2004. "Breeding Greeks: Nietzsche, Gobineau and Classical Theories of Race." In *Nietzsche and Antiquity: His Reaction and Response to the Classical Tradition*, edited by P. Bishop. Camden House.
Marx, K. 2010. *Marx & Engels Collected Works*. Vol. 46, *Letters 1880–83*. Lawrence and Wishart.
Marx, O. 1970. "Nineteenth-Century Medical Psychology: Theoretical Problems in the Work of Griesinger, Meynert, and Wernicke." *Isis* 61 (3): 355–70.
Marx, O. 1990. "German Romantic Psychiatry: Part 1." *History of Psychiatry* 1 (4): 351–80.
Mason, T. 2006. "An Archaeology of the Psychopath." In *Forensic Psychiatry*, edited by T. Mason. Humana Press.
Mayer, A. 2013. *Sites of the Unconscious: Hypnosis and the Emergence of the Psychoanalytic Setting*. University of Chicago Press.
McNeely, I. 2002. "Medicine on a Grand Scale." Wellcome Trust Centre for the History of Medicine at UCL.
Meischner-Metge, A. 2010. "Gustav Theodor Fechner: Life and Work in the Mirror of His Diary." *History of Psychology* 13 (4): 411–23.

Meltzer, S. 1897. "Emil du Bois-Reymond." *Science* 5 (110): 217–19.
Melville, H. 1892. *Moby Dick*. Dana Estes & Company.
Mendel, E. 1878. "Ueber den Verlauf der Fasern des Bindearms." Paper presented at the Berliner medicinisch-psychologische Gesellschaft, January 7, 1878. Abstract in *Berliner Klinische Wochenschrift* 15:402–3.
Mendelsohn, E. 1965. "Physical Models and Physiological Concepts: Explanation in Nineteenth-Century Biology." *British Journal for the History of Science* 2:203.
Merkel, F. 1875. "Tastzellen und Tastkörperchen bei den Hausthieren und beim Menschen." *Archiv für mikroskopische Anatomie* 11:636–652.
Meulders, M. 2010. "5: Helmholtz and the Understanding of Nature." In *Helmholtz: From Enlightenment to Neuroscience*, edited by Laurence Garey. MIT Press.
Meyer, M. 1901. "Über die Beeinflussung der Schrift durch Alkohol." *Kraepelins Psychol. Arbeiten* 3:535.
Meynert, T. 1865. "Bau und Funktion des Gehirns und Rückenmarkes mit Beziehung auf deren Erkrankungen." Habilitation, Wien.
Meynert, T. 1867. "Der Bau der Gross-Hirnrinde und seine örtlichen Verschiedenheiten, nebst einem pathologisch-anatomischen Corollarium, I." *Vierteljahrschrift für Psychiatrie* 1 (77–93): 198–217.
Meynert, T. 1872. "Vom Gehirne der Säugethiere." In *Handbuch der Lehre von den Geweben des Menschen und der Thiere, Zweiter Teil*. Leipzig.
Meynert, T. 1882. "Ueber die Gefühle." *Jahrb. Psychiat.* 3:165–76.
Meynert, T. 1884. *Psychiatrie: Klinik der Erkrankungen des Vorderhirns begründet auf dessen Bau, Leistungen und Ernährung*. Austria: Braumüller.
Miller, J. A., M. Sabshin, J. E. Gedo, G. H. Pollock, L. Sadow, and N. Schlessinger. 1969. "Some Aspects of Charcot's Influence on Freud." *Journal of the American Psychoanalytic Association* 17 (2): 608–23.
Miller, R., and J. Dennison. 2015. *An Outline of Psychiatry in Clinical Lectures: The Lectures of Carl Wernicke*. Springer.
Millett, D. 1998. "Illustrating a Revolution: An Unrecognized Contribution to the 'Golden Era' of Cerebral Localization." *Notes and Records of the Royal Society of London* 52 (2): 283–305.
Möbius, P. 1882. *Die Nervosität*. Leipzig.
Monakow, C. v. 1970. *Mein Leben—vita mea*. Edited by A. W. Gulser and Erwin H. Ackerknecht. Huber.
Montiel, Luis. 2009. "Une révolution manquée: Le magnétisme animal dans la médecine du romantisme allemand." *Revue d'histoire du XIXe Siècle* 38:61–77.
Monro, A. 1771. *Essays and Observations*, Vol. 3. Edinburgh.
Moon, J. 2021. "The Circle of Life: Christopher Wren and the First Intravenous Anesthetic." *Anesthesiology* 135:530.
Moore, G. 2000. "Nietzsche, Degeneration, and the Critique of Christianity." *Journal of Nietzsche Studies* 19:1–18.
Moran, B. T. 1999. [Review:] "Leben, Arbeit und Umwelt des Arztes: Johann Daniel Major (1634–1693), Eine Biographie aus dem 17. Jahrhundert, mit neuen Erkenntnissen." *Bulletin of the History of Medicine* 73 (4): 701–2.
Morel, B.-A. 1857. *Traité des dégénérescences physiques, intellectuelles et morales de l'espèce humaine*. Paris.
Morrison, R. 1997. "Opium-Eaters and Magazine Wars: De Quincey and Coleridge in 1821." *Victorian Periodicals Review* 30 (1): 27–40.

Mullen, P. 1977. "The Romantic as Scientist: Lorenz Oken." *Studies in Romanticism* 16 (3): 381–99.
Müller, J. P. 1826. *Zur vergleichenden Physiologie des Gesichtssinnes des Menschen und der Thiere, Einleitung*. Leipzig.
Müller, J. 2006. *Der Deutsche Bund 1815–1866*. De Gruyter: Oldenbourg Verlag.
Müller, J. P. 1834. *Handbuch der Physiologie des Menschen, für Vorlesungen, Band 1*. Coblenz.
Müller, J. P. 1837a. *Handbuch der Physiologie des Menschen, für Vorlesungen, Band 1*. Coblenz.
Müller, J. P. 1837b. *Handbuch der Physiologie des Menschen, für Vorlesungen, Band 2*. Coblenz.
Müller, J. P. 1840. *Handbuch der Physiologie des Menschen, für Vorlesungen, Band 2*. Coblenz.
Müller, U., P. C. Fletcher, and H. Steinberg. 2006. "The Origin of Pharmacopsychology: Emil Kraepelin's Experiments in Leipzig, Dorpat and Heidelberg (1882–1892)." *Psychopharmacology* 184 (2): 131–38.
Ndubaku, U., and de M. E. Bellard. 2008. "Glial Cells: Old Cells with New Twists." *Acta histochemica* 110 (3): 182–95.
Nencini, P. 2022. "Facts and Factoids in the Early History of the Opium Poppy." *The Social History of Alcohol and Drugs* 36 (1): 45–71.
Neubauer, J. 1967. "Dr. John Brown (1735–88) and Early German Romanticism." *Journal of the History of Ideas* 28 (3): 367–82.
Niemann, A. 1860. "Über eine neue organische Base in den Cocablättern." *Arch. Pharm. Band* 153(2–3): 129–55, 291–308.
Nietzsche, F. 1975. *Digitale Kritische Gesamtausgabe Werke und Briefe*. Edited by P. D'Iorio. Digitization of *Nietzsche Werke. Kritische Gesamtausgabe*, de Gruyter, 1967– and *Nietzsche Briefwechsel. Kritische Gesamtausgabe*, de Gruyter, 1975 (ed. G. Colli and M. Montinari). http://www.nietzschesource.org/.
Nissl, F. 1898. "Rindenbefunde bei Vergiftungen." *Neurolog. Centralblatt* 17:613–14.
Noll, R. 2004. "Historical Review: Autointoxication and Focal Infection Theories of Dementia Praecox." *The World Journal of Biological Psychiatry: The Official Journal of the World Federation of Societies of Biological Psychiatry* 5 (2): 66–72.
Nordau, M. 1892. *Entartung, Bd. 1*. Berlin.
Nordbruch, C. H. R. 1996. *Über die Pflicht: Eine Analyse des Werkes von Siegfried Lenz*. Hildesheim.
Novalis. [1800] 1988. *Hymns to the Night*. Translated by Dick Higgins. McPherson
Oken, L. 1805. *Die Zeugung*. Würzburg.
Oken, L. 1810. *Lehrbuch der Naturphilosophie. Dritter Teil, erstes and zweites Stück*. Jena.
Oken, L. 1843. *Lehrbuch der Naturphilosophie*. Zürich.
Oken, L., T. Bach, O. Breidbach, and D Engelhardt. 2007. *Gesammelte Werke*. Verlag Hermann Böhlaus Nachfolger.
Olesko, K. M. 1995. "The Meaning of Precision: The Exact Sensibility in Early Nineteenth-Century Germany." In *The Values of Precision*, edited by M. N. Wise. Princeton University Press.
Olesko, K. M., and F. L. Holmes. 1993. "Experiment, Quantification, and Discovery: Helmholtz's Early Physiological Researches, 1843–50." In *Hermann von Helm-*

holtz and the Foundations of Nineteenth-Century Science, edited by D. Cahan. University of California Press.
Ostrander, G. 1956. "The Colonial Molasses Trade." *Agricultural History* 30 (2): 77–84.
Otis, L. 2007. *Müller's Lab*. Oxford University Press.
Owen, D. 2021. "Rhetorics of Degeneration: Nietzsche, Lombroso, and Napoleon." *The Journal of Nietzsche Studies* 52 (1): 51–64.
Panhuysen, G. 1998. "The Relationship Between Somatic and Psychic Processes: Lessons from Freud's Project." *Annals of the New York Academy of Sciences* 843 (1): 20–42.
Pantalony, D. 2009. "Hermann von Helmholtz and the Sensations of Tone." In *Altered Sensations*. Springer.
Paracelsus. 1538/1965. "Die dritte Defension wegen des Schreibens der neuen Rezepte, Septem Defensiones 1538." *Werke Bd. 2*. Darmstadt: Wissenschaftliche Buchgesellschaft
Partington, J. R. 1962. "Chemistry in Germany." In *A History of Chemistry*. Palgrave.
Patil, P. 2012. *Discoveries in Pharmacological Sciences*. World Scientific.
Paul, H. W. 2001. *Bacchic Medicine: Wine and Alcohol Therapies from Napoleon to the French Paradox*. Rodopi.
Pearce, J. 2005. "Romberg and His Sign." *European Neurology* 53:210–13.
Pecere, P. 2020. "Reconsidering the Ignorabimus: Du Bois-Reymond and the Hard Problem of Consciousness." *Science in Context* 33 (1): 1–18.
Pemberton, H. 1746. *Pharmacopoeia Londinensis*. Dispensary of the Royal College of Physicians, London.
Pendergrast, M. 2010. *Uncommon Grounds: The History of Coffee and How It Transformed Our World*. Basic Books.
Perkins-McVey, M. 2022. "A Portrait of the Neurophysiologist as a Young Man: Claus, Darwin, and Sigmund Freud's Search for the Testes of the Eel (1875–1877)." *History of Psychology* 25 (4): 367–84.
Perkins-McVey, M. 2023. "Were the Scale of Excitability a Circle: Tracing the Roots of the Disease Theory of Alcoholism Through Brunonian Stimulus Dependence." *Studies in History and Philosophy of Science* 99:46–55.
Perkins-McVey, M. 2024. "Cortical Localization and the Nerve Cell: Freud's Work in Meynert's Psychiatry Clinic." *History of Psychology* 27 (4): 333–49.
Petty, R. 2019. "Pain-Killer: A 19th Century Global Patent Medicine and the Beginnings of Modern Brand Marketing." *Journal of Macromarketing* 39 (3): 287–303.
Pfaff, C. H. 1798. *John Brown's System der Heilkunde*. Copenhagen.
Phillipson, J. 2012. *Chemistry and Biology of Isoquinoline Alkaloids*. Springer-Verlag.
Pichot, P. 2013. *Psychiatry The State of the Art. Vol. 8, History of Psychiatry, National Schools, Education, and Transcultural Psychiatry*. Springer.
Pickstone, J. V. 2000. *Ways of Knowing*. Manchester University Press.
Pillmann, F., and A. Marneros. 2001. "Carl Wernicke: Wirkung und Nachwirkung. Fortschritte Der Neurologie." *Psychiatrie* 69 (10): 488–94.
Pincus, S. 2012. "Rethinking Mercantilism: Political Economy, the British Empire, and the Atlantic World in the Seventeenth and Eighteenth Centuries." *The William and Mary Quarterly* 69 (1): 3–34.
Plitt, G. L. 1870. "Aus Schellings Leben." In *Briefen. Erster Band, Zweiter Band*. Leipzig.

Porter, R. 1999. *The Greatest Benefit to Mankind: A Medical History of Humanity*. W. W. Norton.
Porter, T. M. 2010. *Karl Pearson: The Scientific Life in a Statistical Age*. Princeton University Press.
Powell, T. 1988. "Kant's Fourth Paralogism." *Philosophy and Phenomenological Research* 48 (3): 389–414.
Presner, T. 2007. *Mobile Modernity: Germans, Jews, Trains*. Columbia University Press.
Rachlin, H. 2005. "What Müller's Law of Specific Nerve Energies Says about the Mind." *Behavior and Philosophy* 33:41–54.
Radkau, J. 2009. *Max Weber: A Biography*. Polity Press.
Redwood, T. 1847. *Gray's Supplement to the Pharmacopoeia (rewrite and expansion on original text by S. Gray)*. London.
Reicheneder, J. 1988. "Cocaineuphorie und Naturwissenschaft." *Sudhoffs Archiv* 72 (2): 170–84.
Reinbacher, R. 1998. *Leben, Arbeit und Umwelt des Arztes Johann Daniel Major (1634–1693): eine Biographie aus dem 17. Jahrhundert, mit neuen Erkenntnissen*. Kroeber.
Rheinberger, H.-J. 1997. *Toward a History of Epistemic Things: Synthesizing Proteins in the Test Tube*. Stanford University Press.
Rheinberger, H.-J. 2005. "Reassessing the Historical Epistemology of Georges Canguilhem." In *Continental Philosophy of Science*, edited by G. Gutting. Blackwell Publishing.
Risse, G. 1976. "Schelling, 'Naturphilosophie' and John Brown's System of Medicine." *Bulletin of the History of Medicine* 50 (3): 321–34.
Risse, G. 2003. *Medicine in Society: Historical Essays*. Edited by Andrew Wear. Cambridge University Press.
Ritter, J. 1798. *Beweis, daß ein beständiger Galvanismus den Lebensprozeß in dem Thierreich begleite*. Weimar.
Ritvo, L. 1990. *Darwin's Influence on Freud: A Tale of Two Sciences*. Yale University Press.
Roberto, M. 2007. "Ørsted, Ritter and Magnetochemistry." In *Hans Christian Ørsted and the Romantic Legacy in Science: Ideas, Disciplines, Practices*, edited by R. M. Brain, R. S. Cohen, and O. Knudsen. Springer.
Robinson, D. 2001. "Reaction-time Experiments in Wundt's Institute and Beyond." In *Wilhelm Wundt in History: The Making of a Scientific Psychology*, edited by R. Rieber and D. Robinson. Springer Science.
Rocca, J. 2007. "William Cullen (1710–1790) and Robert Whytt (1714–1766) on the Nervous System." In *Brain, Mind and Medicine: Essays in Eighteenth-Century Neuroscience*, edited by H. A. Whitaker, C. Smith, and S. Finger. Springer.
Rocke, A. 1993. *The Quiet Revolution: Hermann Kolbe and the Science of Organic Chemistry*. University of California Press.
Rocke, A. 2023. "Origins and Spread of the 'Giessen Model' in University Science." In *The Laboratory Revolution and the Creation of the Modern University, 1830–1940*, edited by K. van Berkel and E. Homburg. Amsterdam University Press.
Roeckelein, J. E. 1998. *Dictionary of Theories, Laws, and Concepts in Psychology*. Greenwood Press.
Roelcke, V. 1999. "Laborwissenschaft und Psychiatrie: Prämissen und Implikatio-

nen bei Emil Kraepelins Neuformulierung der psychiatrischen Krankheitslehre." In *Strategien der Kausalität: Konzepte der Krankheitsverursachung im 19. und 20. Jahrhundert*, edited by Christoph Gradmann and Thomas Schlich. Pfaffenweiler: Centaurus.

Rogers, M. R. 2003. *Newtonianism for the Ladies and Other Uneducated Souls: The Popularization of Science in Leipzig, 1687–1750*. Lang.

Rokitanskys-Tilscher, U. 2018. "Im fruchtbaren Spannungsfeld von Natur- und Geisteswissenschaften—eine erste Studie zur Prägung der Persönlichkeit Carl Freiherr von Rokitanskys aus familienbiografischer Sicht." In *Strukturen und Netzwerke: Medizin und Wissenschaft in Wien 1848–1955*, edited by D. C. Angetter and P. Weindling. V&R Unipress.

Röschlaub, A. 1798. *Untersuchungen über Pathogenie oder Einleitung in die medizinische Theorie, Band I*. Andreä.

Röschlaub, A. 1799. Untitled ("An J. G. Fichte"). *Magazin zur Vervollkommnung der theoretischen und praktischen Heilkunde*. 1st edition. 2:xi–xvi.

Roth, H. L. 2002. "Finding Language in the Matter of the Brain: Origins of the Clinical Aphasia Examination." *Seminars in Neurology* 22 (4): 335–48.

Rothschuh, K. 1973. *History of Physiology*. Edited and translated by Guenter B. Risse. Robert E. Krieger.

Ruberg, W. 2019. *History of the Body*. Macmillan Education.

Rüegg, W. 2004. "Themes." In *A History of the University in Europe*. Vol. 3, *Universities in the Nineteenth and Early Twentieth Centuries (1800–1945)*, edited by W. Rüegg. Cambridge University Press.

Rumore, P. 2014. "Kant's Understanding of the Enlightenment with Reference to his Refutation of Materialism." *Con-Textos Kantianos* 1:81–97.

Rupke, N. A. 2008. *Alexander von Humboldt: A Metabiography*. University of Chicago Press.

Rynd, F. 1845. "Neuralgia—Introduction of Fluid to the Nerve." *Dublin Med Press* 13:167–68.

Sachs, A. 1995. "Humboldt's Legacy and the Restoration of Science." *World Watch* 8 (2): 28–38.

Sarikcioglu, L., and R. Y. Arican. 2006. "Wilhelm Heinrich Erb (1840–1921) and His Contributions to Neuroscience." *Journal of Neurology, Neurosurgery & Psychiatry* 78 (7): 732.

Scaff, L. 1989. *Fleeing the Iron Cage: Culture, Politics, and Modernity in the Thought of Max Weber*. University of California Press.

Schelling, F. W. 1798 [1857]. "Von der Weltseele." In *Sämtliche Werke von F. W. Schelling 1797–1798, Zweiter Band*. Stuttgart and Augsburg.

Schelling, F. W. 1799a. "Einige Bemerkungen aus Gelegenheit einer Rezension Brown'scher Schriften in der A. L. Z." *Magazin zur Vervollkommnung der theoretischen und praktischen Heilkunde* 1 (2): 255–62.

Schelling, F. W. 1799b/1858. "Erster Entwurf eines Systems der Naturphilosophie." In *Sämtliche Werke von F. W. Schelling 1799–1800, Dritter Band*. Stuttgart and Augsburg.

Scherf, C. 1799. *Dispensatorium Lippiacum, genio moderno accommodatum auctoritate Collegii Medici*. Vol. 2. 2nd edition.

Schiebinger, L. 2003. "Human Experimentation in the Eighteenth Century: Natural

Boundaries and Valid Testing." In *The Moral Authority of Nature*. University of Chicago Press.
Schiffter, R. 2007. "Lebensbilder von Romberg." *Fortschr Neurol Psychiatr* 75:160–67.
Schleich, C. L. 1936. *Those Were Good Days*. Translated by B. Miall. University of Minnesota Press.
Schmidgen, H. 2002. "Of Frogs and Men: The Origins of Psychophysiological Time Experiments, 1850–1865." *Endeavour* 26 (4): 142–48.
Schmidgen, H. 2003. "Wundt as Chemist? A Fresh Look at His Practice and Theory of Experimentation." *The American Journal of Psychology* 116 (3): 469–76.
Schmidgen, H. 2005. "The Donders Machine: Matter, Signs, and Time in a Physiological Experiment, ca. 1865." *Configurations* 13 (2): 211–56.
Schmitz, R. 1985. "Friedrich Wilhelm Sertürner and the Discovery of Morphine." *Pharmacy in History* 27 (2): 61–74.
Schultz, M. 2008. "Rudolf Virchow." *Emerging Infectious Diseases* 14 (9): 1480–81.
Scott, H. M. 1990. *Enlightened Absolutism: Reform and Reformers in Later Eighteenth-Century Europe*. Macmillan.
Seebacher, F. 2006. "The Case of Ernst Wilhelm Brücke versus Joseph Hyrtl–The Viennese Medical School Quarrel Concerning Scientific and Political Traditions." In *Controversies and Disputes in Life Sciences in the 19th and 20th Centuries*, edited by Brigitte Hoppe. E. Rauner Verlag.
Seitelberger, F. 1997. "Theodor Meynert (1833–1892), Pioneer and Visionary of Brain Research." *Journal of the History of the Neurosciences* 6 (3): 264–74.
Sertürner, F. 1805. "Säure im Opium." *Journal der Pharmacie* 14:33–57.
Sertürner, F. 1806. "Darstellung der reinen Mohnsäure (Opiumsäure) nebst einer chemischen Untersuchung des Opiums mit vorzüglicher Hinsicht auf einen darin neu entdeckten Stoff und die dahin gehörigen Bemerkungen." *Journal der Pharmacie* 14:47–93.
Sertürner, F. 1811. "Ueber das Opium und dessen krystallisirbare Substanz." *Journal der Pharmacie* 20:99–103.
Sertürner, F. 1817. "Ueber das Morphium, eine neue salzfähige Grundlage, und die Mekonsäure, als Hauptbestandtheile des Opiums." *Annalen der Physik* 55 (1): 56–92.
Seyfarth, E.-A. 2006. "Julius Bernstein (1839–1917): Pioneer Neurobiologist and Biophysicist." *Biol. Cybern* 94:2–8.
Sheehan, J. 1971. "Liberalism and the City in Nineteenth-Century Germany." *Past & Present* 51:116–37.
Shell, S. M. 2013. "Kant as 'Vitalist': The 'Principium of Life' in Anthropologie Friedländer." In *Kant's Lectures on Anthropology*, edited by A. Cohen. Cambridge University Press.
Shepherd, G. M. 1991. *Foundations of the Neuron Doctrine*. Oxford University Press.
Shorter, E. 1997. *A History of Psychiatry*. John Wiley & Sons.
Shrady, G., and T. Stedman. 1885. *Medical Record*. Vol. 28. New York.
Siemerling, E. 1890. "Dr. Carl Westphal." *Journal of Mental Science* 36 (153): 312–13.
Simpson, D. 2005. "Phrenology and the Neurosciences: Contributions of F. J. Gall and J. G. Spurzheim." *ANZ Journal of Surgery* 75 (6): 475–82.

Sloan, P. 1986. "Darwin, Vital Matter, and the Transformism of Species." *Journal of the History of Biology* 19 (3): 369–445.
Sloane, H. 1721. *Pharmacopoeia Londinensis*. 4th ed.
Smith, D. 2008. "Hail Mariani: The Transformation of Vin Mariani from Medicine to Food in American Culture, 1886–1910." *Social History of Alcohol and Drugs* 23 (1): 42–57.
Snelders, H. 1970. "Romanticism and Naturphilosophie and the Inorganic Natural Sciences 1797–1840: An Introductory Survey." *Studies in Romanticism* 9 (3): 193–215.
Snelders, H. 1983. "Sertürner: Opium en 'Naturphilosophie.'" *Bull Cercle Benelux Hist Pharm*. 65:1–4.
Snelders, S., C. Kaplan, and T. Pieters. 2006. "On Cannabis, Chloral Hydrate, and Career Cycles of Psychotropic Drugs in Medicine." *Bulletin of the History of Medicine* 80 (1): 95–114.
Springer, A. 2002. "Kokain, Freud und die Psychoanalyse." *Suchttherapie* 3 (1): 18–23.
Squire, L. R. 2003. *Fundamental Neuroscience*. Academic Press.
Staal, F. 2001. "How a Psychoactive Substance Becomes a Ritual: The Case of Soma." *Social Research* 68 (3): 745–78.
Steigerwald, J. 2016. "The Subject as Instrument: Galvanic Experiments, Organic Apparatus and Problems of Calibration." In *Human Experimentation*, edited by L. Stewart and E. Dyck. Brill.
Steinberg, H. 2000. "Emil Kraepelins Rezensionen für das 'Literarische Centralblatt für Deutschland.'" *Psychiatrische Praxis* 27 (3): 119–26.
Steinberg, H., and M. C. Angermeyer. 2001. "Emil Kraepelin's Years at Dorpat as Professor of Psychiatry in Nineteenth-Century Russia." *History of Psychiatry* 12 (47 pt. 3): 297–327.
Steinberg, H., and H. Himmerich. 2013. "Emil Kraepelin's Habilitation and His Thesis: A Pioneer Work for Modern Systematic Reviews, Psychoimmunological Research and Categories of Psychiatric Diseases." *The World Journal of Biological Psychiatry: The Official Journal of the World Federation of Societies of Biological Psychiatry* 14 (4): 248–57.
Steinberg, H., and U. Müller. 2005. "Emil Kraepelin 1882/83 in Leipzig und seine frühen pharmakopsychologischen Arbeiten im Licht der aktuellen Forschung." In *200 Jahre Psychiatrie an der Universität Leipzig*, edited by M. C. Angermeyer and H. Steinberg. Springer.
Steinkamp, F. 2002. "Schelling's 'Clara': Editors' Obscurity." *The Journal of English and Germanic Philology* 101 (4): 478–96.
Steinke, H. 2005. *Irritating Experiments: Haller's Concept and the European Controversy on Irritability and Sensibility, 1750–90*. Brill.
Steudel, J. 1963. *Le Physiologiste Johannes Müller*. Paris: Université de Paris.
Strickland, S. 1998. "The Ideology of Self-Knowledge and the Practice of Self-Experimentation." *Eighteenth-Century Studies* 31 (4): 453–41.
Strohl, J. 1935. "Okens Stellung zu Paracelsus und zur Geschichte der Naturwissenschaften überhaupt." In *Verhandlungen der Schweizerischen Gesellschaft für Geschichte der Medizin und der Naturwissenschaften* 116:395–97.
Stuckrad, K. 2014. *The Scientification of Religion: An Historical Study of Discursive Change, 1800–2000*. De Gruyter.

Sturm, T. 2009. "Die Kritik an der empirischen Psychologie." In *Kant und die Wissenschaften vom Menschen*. Brill.
Tannahill, R. 1973. *Food in History*. Stein and Day.
Tasaki, I. 2014. *Physiology and Electrochemistry of Nerve Fibres*. Elsevier Science.
Tattersall, I. 1995. *The Fossil Trail: How We Know What We Think We Know About Human Evolution*. Oxford University Press.
Timmermann, J. 2006. "Kantian Duties to the Self, Explained and Defended." *Philosophy* 81(3): 505–30.
Titchener, E. 1921. "Wilhelm Wundt." *The American Journal of Psychology* 32 (2): 161–78.
Tlusty, B. A. 1998. "Water of Life, Water of Death: The Controversy over Brandy and Gin in Early Modern Augsburg." *Central European History* 31 (1–2): 1–30.
Trede, K. 2007. "150 Years of Freud-Kraepelin Dualism." *The Psychiatric Quarterly* 78 (3): 237–40.
Tsouyopoulos, N. 1982. *Andreas Röschlaub und die romantische Medizin: die philosophischen Grundlagen der modernen Medizin*. Stuttgart: G. Fischer.
Tsouyopoulos, N. 1988. "The Influence of John Brown's Ideas in Germany." *Medical History Supplement* 8:63–74.
Tsouyopoulos, N. 1990. "Doctors Contra Clysters and Feudalism: The Consequences of a Romantic Revolution Romanticism and the Sciences." In *Romanticism and the Sciences*, edited by A. Cunningham. Cambridge University Press.
Turner, R. 1993. "Vision Studies in Germany: Helmholtz versus Hering." *Osiris* 8:80–103.
Urdang, G. 1946. "Pharmacopoeias as Witnesses of World History." *Journal of the History of Medicine and Allied Sciences* 1 (1): 46–70
Van den Berg, H. 2009. "Kant's Conception of Proper Science." *Synthese* 183 (1): 7–26.
Van Wyhe, J. 2002. "The Authority of Human Nature: The 'Schädellehre' of Franz Joseph Gall." *The British Journal for the History of Science* 35 (1): 17–42.
Vanzo, A. 2012. "Kant on Experiment." In *Rationis Defensor*, edited by J. Maclaurin. Springer.
Vartanian, A. 1983. "La Mettrie and Diderot Revisited: An Intertextual Encounter." *Diderot Studies* 21:155–97.
Verplaetse, J. 2009. *Localizing the Moral Sense: Neuroscience and the Search for the Cerebral Seat of Morality, 1800–1930*. Springer.
Vickers, N. 1997. "Coleridge, Thomas Beddoes and Brunonian Medicine." *European Romantic Review* 8 (1): 47–94.
Vigilantius, J. 1997. "Part IV—Kant on the Metaphysics of Morals: Vigilantius's Lecture Notes." In *Lectures on Ethics*, edited by P. Heath and J. Schneewind. Cambridge University Press.
Vintschgau, M. v., and Hönigschmied, J. 1875. "Versuche über die Reactionszeit einer Geschmacksempfindung." *Pflüger Archiv* 10:1–48.
Virchow, R. 1848. *Mittheilungen über die in Oberschlesien herrschende Typhus-Epidemie*. Berlin.
Virchow, R. 1858. *Die Cellularpathologie in ihrer Begründung auf physiologische und pathologische Gewebelehre*. Berlin.
Virchow, R. 1860. *Handbuch der speciellen Pathologie und Therapie*. Erlangen.
Vogt, K. 1855. *Köhlergluabe und Wissenschaft*. Gießen.

Volhard, J. 1898. *Justus v. Liebig: Vortrag in der Hauptversammlung des Vereins deutscher Chemiker*. Berlin and Heidelberg.
vom Scheidt, J. 1973. "Sigmund Freud and Cocaine." *Psyche: Zeitschrift für Psychoanalyse und ihre Anwendungen* 27 (5): 385–430.
von Engelhardt, D. 1994. "Paracelsus im Urteil der Naturforschung und Medizin der Romantik." *NTM* (n.s.) 2:97–116.
Vucinich, A. 1988. *Darwin in Russian Thought*. University of California Press.
Wackenroder, H. 1853. "Chemische Prüfung der Coca-Blätter." *Archiv der Pharmacie* 75:24–27.
Waldenburg, L. 1864. *Die Inhalationen der zerstäubten Flüssigkeiten sowie der Dämpfe und Gase in ihrer Wirkung auf die Krankheiten der Athmungsorgane: Lehrbuch der respiratorischen Therapie*. Berlin.
Walter, E., and M. Scott. 2017. "The Life and Work of Rudolf Virchow, 1821–1902: Cell Theory, Thrombosis and the Sausage Duel." *Journal of the Intensive Care Society* 18 (3): 234–35.
Wagner, L. E. 1938. "Colridge's Use of Laudanum and Opium." *The Psychoanalytic Review* (1913–1957) 25: 309.
Warren, M. E. 1992. "Max Weber's Nietzschean Conception of Power." *History of the Human Sciences* 5 (3): 19–37.
Wasianski, E. 1804. *Immanuel Kant in seinen letzten Lebensjahren, Dritter Band*. Königsberg.
Wassmann, C. 2009. "Physiological Optics, Cognition and Emotion: A Novel Look at the Early Work of Wilhelm Wundt." *J Hist Med Allied Sci.* 64 (2): 213–49.
Watson, P. 2010. *The German Genius: Europe's Third Renaissance, the Second Scientific Revolution and the Twentieth Century*. Simon and Schuster.
Weatherby, L. 2016. *Transplanting the Metaphysical Organ: German Romanticism between Leibniz and Marx*. Fordham University Press.
Weber, M. 1917 [1994]. *Wissenschaft als Beruf 1917/1919/Politik als Beruf, 1919 Studienausgabe*. J.C.B. Mohr (Paul Siebeck).
Weber, M. 1995. "Zur Psychophysik der industriellen Arbeit." In *Max Weber Gesamtausgabe (Band 11): Zur Psychophysik der industriellen Arbeit. Schriften und Reden, 1908–1912.*, edited by W. Schluchter and S. Frommer. J. C. B. Mohr.
Weber, M. 2016. "Die protestantische Ethik und der Geist des Kapitalismus." In *Max Weber Gesamtausgabe (Band 18): Die protestantische Ethik und der Geist des Kapitalismus / Die protestantischen Sekten und der Geist des Kapitalismus Schriften 1904–1920*, edited by H. Baier et al. J.C.B. Mohr.
Weber, M. 2018. "Die 'Objektivität' sozialwissenschaftlicher und sozialpolitischer Erkenntnis." In *Max Weber Gesamtausgabe (Band 7): Zur Logik und Methodik der Sozialwissenschaften Schriften 1900–1907*, edited by H. Baier et al. J.C.B. Mohr.
Weber, M. M., and E. Engstrom. 1997. "Kraepelin's 'Diagnostic Cards': The Confluence of Clinical Research and Preconceived Categories." *History of Psychiatry* 8 (31 pt. 3): 375–85.
Weikart, R. 2016. *From Darwin to Hitler: Evolutionary Ethics, Eugenics and Racism in Germany*. Palgrave Macmillan.
Weindling, P. 1993. *Health, Race and German Politics Between National Unification and Nazism, 1870–1945*. Cambridge University Press.

Wellman, K. A. 1992. *La Mettrie: Medicine, Philosophy, and Enlightenment*. Duke University Press.
Werner, P., and F. Holmes. 2002. "Justus Liebig and the Plant Physiologists." *Journal of the History of Biology* 35 (3): 421–41.
Wernicke, K. 1874. *Der aphasische Symptomencomplex: Eine psychologische Studie auf anatomischer Basis*. Breslau.
Wernicke, K. 1896. *Grundriss der Psychiatrie in klinischen Vorlesungen*. Leipzig
Wernicke, K. 1900. *Grundriss der Psychiatrie in klinischen Vorlesungen*. Leipzig.
Westphal, C. F. O. 1875. "Ueber einige durch mechanische Einwirkung auf Sehnen und Muskeln hervorgebrachte Bewegungs-Erscheinungen (Knie-, Fussphänomen)." *Archiv für Psychiatrie und Nervenkrankheiten* 5:803–34.
Westphal, C. F. O. 1883. "Über eine dem Bilde der cerebrospinalen grauen Degeneration ähnliche Erkrankung des centralen Nervensystems ohne anatomischen Befund, nebst einigen Bemerkungen über paradoxe Contraction." *Archiv für Psychiatrie und Nervenkrankheiten* 14:87–134.
Wiesing, U. 1989. "Der Tod der Auguste Böhmer. Chronik eines medizinischen Skandals, seine Hintergründe und seine historische Bedeutung" ["Auguste Böhner's Death: Chronicle of a Medical Scandal, Its Background and Its Historical Significance"]. *History and Philosophy of the Life Sciences* 11 (2): 275–95.
Wilhelmy, P. 1989. *Der Berliner Salon im 19. Jahrhundert*. Walter de Gruyter.
Winfield, R. 2011. "Hegel's Solution to the Mind-Body Problem." In *A Companion to Hegel*. Wiley-Blackwell.
Wolpert, L. 1995. "Evolution of the Cell Theory." *Philosophical Transactions: Biological Sciences* 349 (1329): 227–33.
Wood, P. 1996 [1974]. *Black Majority: Negroes in Colonial South Carolina from 1670 through the Stono Rebellion*. Norton.
Woycke, J. 1992. "Patent Medicines in Imperial Germany." *CBMH/BCHM* 9:41–56.
Wundt, W. 1857. "Ueber die Elasticität der thierischen Gewebe." *Heidelberger Jahrbücher der Literatur* 50 (16): 249–50.
Wundt, W. 1858a. *Die Lehre von der Muskelbewegung. Nach eigenen Untersuchungen bearbeitet*. Braunschweig.
Wundt, W. 1858b. "Ueber das Gesetz der Zuckungen und die Modifikation der Erregbarkeit durch geschlossene Ketten." *Archiv für physiologische Heilkunde* 2, 354–400.
Wundt, W. 1862. *Beiträge zur Theorie der Sinneswahrnehmung*. Leipzig and Heidelberg.
Wundt, W. 1874a. *Grundzüge der physiologischen Psychologie*. Leipzig.
Wundt, W. 1874b. "Ueber die Aufgabe der Philosophie in der Gegenwart." *Akademische Antrittsrede in Zürich*. Leipzig.
Wundt, W. 1876. *Ueber den Einfluss der Philosophie auf die Erfahrungswissenschaften*. Leipzig.
Wundt, W. 1909. *Grundriss der Psychologie*. 9th ed. Engelmann.
Wundt, W. 2013. *Erlebtes und Erkanntes*. Dogma Publishing.
Yakira, E. 2010. "The Anachronism of Parallelism: A Chapter in Historical Irony." *Iyyun: The Jerusalem Philosophical Quarterly* 59:99–114.
Youngquist, P. 1999. "De Quincey's Crazy Body." *PMLA* 114 (3): 346–58.

Zieger, S. 2005. "'How Far Am I Responsible?': Women and Morphinomania in Late-Nineteenth-Century Britain." *Victorian Studies* 48 (1): 59–81.
Zimmer, H.-G. 2006. "Profiles in Cardiology Johannes Müller." *Clin. Cardiol.* 29:327–28.
Zwirn, S. 2015. "Butterflies of the Soul: Cajal's Neuron Theory and Art." *The Journal of Aesthetic Education* 49 (4): 105–19.

Index

Page numbers in italics refer to figures.

active principle concept, 20, 73, 82
addiction, 6
"Aktenstücke eines Psychologen" ("From the Case Files of a Psychologist") (Nietzsche), 234
alcohol: availability as an intoxicant, 165; Brown and, 28; Brunonian ideas about studies & treatments for, 43–44; colonial trade and, 21–22; Kant and, 56–57; Kraepelin's reaction time trials of, 198–200, *201*, *203*, *205*, 217–18. *See also* intoxicants
Aldini, Giovanni, 110
alkaloidal chemistry, 82
alkaloids, 87
Allgemeine Literatur-Zeitung (journal), 62
Also Sprach Zarathustra (Nietzsche), 237
Annäherungen: Drogen und Rausch (Jünger), 1
Annalen der Pharmacie (scientific journal), 169–70
Annalen der Physik, 87
"Annalen für das Universalsystem der Elemente" (Sertürner), 81
Anthropologie in pragmatischer Hinsicht abgefaßt (Kant), 53, 55–56
Anton, Gabriel, 246
Araujo, Saulo de Freitas, 151
Archiv der Pharmazie (scientific journal), 170–71

Archiv für Anatomie, Physiologie und wissenschaftliche Medicin (Archive for Anatomy, Physiology, and Scientific Medicine) (Müller), 96
Archiv für Psychiatrie und Nervenkrankheiten (Journal for Psychiatry and Nervous Diseases), 120
Atomlehre (Fechner), 141

Bacon, Francis, 18, 34
Barfoot, Michael, 25
Baxt, Nikolay, 159
Bayer AG, 174
Beddoes, Thomas, 46–47
"Beitrag zur Kenntnis der Cocawirkung" (Freud), 249
Beiträge zur Theorie der Sinneswahrnehmung (Wundt), 150–55
Bérard, Jacques Étienne, 144
Berlin (University of), 92–93, 137
Berlin, Rudolf, 130
Berlin Medical Society, 177
Bernard, Claude, 93
Bernstein, Julius, 107, 135–37, 153–54, 233
Bessel, Friedrich, 158
Beweis, daß ein beständiger Galvanismus den Lebensprozeß in dem Thierreich begleite (Ritter), 58
biologism (biological subject), 3–4, 6–7, 72, 88, 137–38, 230–31, 243, 263–66
Black, Sara, 180

Blumenbach, Johann, 58
Böhmer, Auguste, 61–62
Borowski, Ludwig, 37
Boyle, Robert, 19–20, 178
brain localization (theory of), 115–17
Brobjer, Thomas H., 233
Broca, Paul, 124
Brock, William Hodson, 168
Brown, John: Brunonian system of, 25–26, 29–32; opium and, 24, 269n2. *See also* Brunonianism; excitability (theory of); therapeutics
Brücke, Ernst Wilhelm von, 103, 120, 129, 136
Brühl-Cramer, Carl Wilhelm von, 43, 56
Brunonianism: alcohol, 43–44; discovery of nitrous oxide and, 47–48; Fechner and, 143–44; influence on English Romantics, 46–48; Müller and, 100; psychiatry and, 44; Röschlaub and, 40–43, 264; Schelling and, 41, 62–63. *See also* Brown, John; Kant, Immanuel; Röschlaub, Andreas; Schelling, Friedrich W.
Buchheim, Rudolf, 169
Buchholz, Christian Friedrich, 80–81
Bucquet, Jean-Baptiste Michel, 75
Byk, Heinrich, 180, 272n11

Cadet de Gassicourt, Louis-Claude, 82
caffeine: coffeehouses and, 177; Dehio and, 214–16; discovery of, 87
Canales, Jimena, 107
Canstatt, Karl Friedrich, 44
Cauchy, Augustin-Louis, 141
Caventou, Joseph-Bienaimé, 74, 87
Charcot, Jean-Martin, 246–47
Chemische Untersuchung über das Fleisch (Liebig), 172
Chevreul, Michel-Eugène, 76
Chirurgia Infusoria (Major), 178
chloral hydrate, 127–28, 166, 169, 179–80, 218–19
chloroform, 136–37
Clysmatica Nova (Elsholtz), 178
cocaine: cocaine tonic, 176–77; discovery of, 170–73; Freud studies on, 246–52; pharmaceutical manufacturing of, 176–77
Cocteau, Jean, 263
Coleridge, Samuel Taylor, 46–47, 269n2
Compendium der Psychiatrie (Kraepelin), 209, 229
Confessions of an English Opium-Eater (De Quincey), 46–47
Cullen, William, 18–20, 25

Danzinger, Kurt, 159
"Darstellung der reinen Mohnsäure" (Sertürner), 74, 79
Das Chloralhydrat ein neues Hypnoticum und Anaestheticum und dessen Anwendung in der Medicin (Liebreich), 179
"Das Verbrechen als soziale Krankheit" (Kraepelin), 253
"Das Verhalten der physiologischen Reactionszeit unter dem Einfluss von Morphium, Caffee und Wein" (Dietl & Vintschgau), 187
Davy, Humphry, 47–48, 135
De glandularum secernetium (Müller), 96
De Quincey, Thomas, 46–47, 175, 269n3, 272n6
Dehio, Heinrich, 214–17, 221–22
"Der aphasische Symptomencomplex Eine psychologische Studie auf anatomischer Basis" (The Aphasic Symptom Complex) (Wernicke), 123–24
Der Ursprung der moralischen Empfindungen (Rée), 235
Derosne, Jean-François, 77–80, 82–84, 84
Descartes, René, 12, 34–35, 37. *See also* dualism; Kant, Immanuel
Die Abschaffung des Strafmasses—Ein Vorschlag zur Reform der heutigen Strafrechtspflege (Kraepelin), 253
Die Evangelisch-sozialen Kurse in Berlin im Herbst dieses Jahres (Weber), 258
Die Fröhliche Wissenschaft (Nietzsche), 238

Die fünf Sinne des Menschens (Bernstein), 233
Die Geburt der Tragödie aus dem Geiste der Musik (Nietzsche), 238
Die Lehre von der Muskelbewegung (Wundt), 149, 153
"Die Metamorphose der Pflanzen" (Goethe), 58
Die Metaphysik der Sitten in zwei Teilen (Kant), 53
Die Nervosität (Möbius), 191
Die "Objektivität" sozialwissenschaftlicher und sozialpolitischer Erkenntnis (Weber), 257
Die organische Chemie in ihrer Anwendung auf Agricultur und Physiologie (Liebig), 172
Die protestantische Ethik und der Geist des Kapitalismus (Weber), 258
"Die Psychologie des Alkohols" (Kraepelin), 254
"Die Schildknappen des Weinkapitals an der Arbeit" (Kraepelin), 253-54
"Die Struktur der Elemente des Nervensytem" (Freud), 246
Die Thier-Chemie, oder die organische Chemie in ihrer Anwendung auf Physiologie und Pathologie (Liebig), 167-68, 172
Die Zeugung (Oken), 67
Dietl, Michael, 187-89, 198-99, 206-7
Dilthey, Wilhelm, 12
Dispensatorium Lippiacum, 17
Donders, Franz, 158-60
Driesch, Hans, 68
Drugs that Enslave: The Opium, Morphine, Chloral, and Hashisch Habit (Kane), 128
du Bois-Reymond, Emil: Darwinism and, 104; electrophysiology and, 104-6, 108; Hermann and, 135-36; Kantianism and, 104; La Mettrie and, 109; nerve galvanometer of, 106; Nietzsche and, 233; as a student of Müller, 103; use of vital substances by, 109-12; Wundt and, 149
dualism, 12, 34-35

Ecce Homo (Nietzsche), 238
Ehrhardt, Walter, 61
"Ein Fall von Hirnblutung mit indirekten basalen Herdsymptomen bei Skorbut" (Freud), 246
electrophysiology, 103-5, 108. *See also* du Bois-Reymond, Emil; Helmholtz, Hermann Ludwig Ferdinand von
Elementa Medicinae (Brown), 25-26, 46, 269-70n1
Elemente der Psychophysik (Fechner), 142, 145
Elsholtz, Johann Sigismund, 178
Emminghaus, Hermann, 209, 213
endogenous psychoses, 191
Engel-Apotheke, 174
Engstrom, Eric J., 223
Erb, Wilhelm Heinrich, 209
Erster Entwurf eines Systems der Naturphilosophie (Schelling), 41, 59
Ettmüller, Michael, 178
excitability (theory of), 4, 26-28, 27, 102, 264-65. *See also* Brown, John; Röschlaub, Andreas; Schelling, Friedrich W.
Exner, Sigmund, 159, 187-89, 198-99, 206-7, 249, 256
"Experimentelle Untersuchung der einfachsten psychischen Processe, Erster Abhandlung" (Exner), 159, 187

Fechner, Gustav: atomism and, 141-42; criticism of, 146; doctrine of parallelism, 5, 141, 143; as editor of the *Hauslexikon*, 143-44; intoxication and, 145-46; Kantianism and, 139; at Leipzig, 138; materialism and, 143; mechanism and, 142; *Naturphilosophie* and, 140; Oken and, 140; *Psychophysik* and, 137-38; publications of, 140-41; spiritism and, 156-57; vital substances and, 143-45; Weber and, 140, 142; Wundt and, 152-53
Feuchtersleben, Ernst von, 44-46
Fichte, Johann G., 42-43, 92
Fiedler, Karl, 179
Flechsig, Paul, 186

Floyer, John, 19–20
Fontana, Felice, 40
Forel, Auguste, 130
Fourcroy, Antoine, 74–76
Fresnel, Augustin-Jean, 141
Freud, Sigmund: cocaine and, 176, 245–52; Exner and, 250; experimental psychology and, 251–52, 273n8; Kraepelin and, 245, 250–51; Meynert and, 246–47; on Wernicke, 128
Friedrich, Max, 160
Friedrich II, King of Prussia, 12
Fritsch, Gustav, 116–19, 123
Fröhlich, Rudolf, 145, 271n2

Gaedcke, Friedrich, 170–71
Galenism, 17, 19–20
Gall, Franz, 115–16, 125
galvanism, 58–59, 64–65, 69–70, 81, 104, 109–10
galvanometer, 106–8
Gay-Lussac, Joseph Louis, 87
Gemes, Ken, 240
Geschichte der Materialismus (Lange), 233
Geschichte des Vitalismus (Driesch), 68
Gierke, Hans, 184
Giessen (University of), 168–69
Gilbert, Ludwig Wilhelm, 86
Glymour, Clark, 251
Goethe, Wolfgang von, 58, 142
Golgi, Camillo, 130
Götzen-Dämmerung (Nietzsche), 242
Griesinger, Wilhelm, 120, 123, 209
Grundriss der Naturphilosophie, der Theorie der Sinne, mit der darauf gegründeten Classification der Thier (Oken), 66
Grundriss der Psychiatrie (Wernicke), 125–28
Grundzüge der physiologischen Psychologie (Wundt), 150, 153, 155–57, 237
Gudden, Bernhard von, 130, 185, 213

Hacking, Ian, 12, 31
Haeckel, Ernst, 118
Hakosalo, Heini, 129–30

Hall, A. Rupert, 168
Haller, Albrecht von, 13, 25
Handbuch der Lehre von den Geweben des Menschen und der Thiere (Meynert), 123
Handbuch der Physiologie (Handbook of Physiology) (Müller), 96, 99
Handbuch der physiologischen Optik (Helmholtz), 152
Hänel, Hans, 229
Hardenberg, Karl August von, 33
Hartmann, Eduard, 236
Hauslexikon: Vollständiges Handbuch praktischer Lebenskenntnisse für alle Stände, 143
Hegel, Georg Wilhelm Friedrich, 59, 141
Heidelberger, Michael, 143
Heilbron, John L., 74
Heine, Heinrich, 33, 178, 272n6
Helmholtz, Hermann Ludwig Ferdinand von: criticism of vitalism, 108, 112; Darwinism and, 104; electrophysiology and, 104, 106–7; Exner and, 159; Fechner and, 152; Kantianism and, 104; on the state of physics and physiology, 115; as a student of Müller, 103; Wundt and, 148, 150
Herbart, Johann, 139, 149–50, 152, 157
Hermann, Ludimar, 135–36
Hildebrandt, Helmut, 223
Hipp, Matthäus, 158–59
Hirsch, Adolph, 158
His, Wilhelm, 233
Hitzig, Eduard, 116–19, 123, 129
Hoch, August, 229
Hoffman, Friedrich, 19
Hogarth, William, 28
Holbach, Baron d,' 142
homme machine, L' (La Mettrie), 12–13
Hönigschmied, J., 188
Hooke, Robert, 24
Hufeland, Cristoph, 43
Humboldt, Alexander, 92–94, 100, 105, 109, 168
Humboldt, Wilhelm von, 92–94
Hunter, Charles, 178–79
hypodermic method, 179

Ideen zu einer Philosophie der Natur als Einleitung in das Studium dieser Wissenschaft (Schelling), 59
Immanuel Kant in seinen letzten Lebensjahren (Wasianski), 50
intoxicants: Brunonian system and, 29–32; definition of, 11; degeneration and, 239–40, 254; diseases of, 6, 56–57; excitability and, 48; fatigue studies, 255–56; history of the biological subject and, 266; imagination and, 54–56; impact of colonial trade on, 21–24; model psychosis from, 190–91; Müller and, 100–102; Nietzsche and, 238–39; Oken and, 70–71; public availability of, 165, 180, 265; reaction time experiments, 190–200, *191*, *193*, *195*, *197*, *201*, *203*, *205*, *227*; as therapeutics, 29–30; upper-tier theories of, 86; uses of, 11–12; as a way of knowing, 2–3, 31–32, 260–61, 263–64. *See also* alcohol; opium; therapeutics; vital substances
intravenous injection, 178–79
irritability (theory of), 25

Jahrbücher der Medizin als Wissenschaft, Die (Schelling/Marcus), 41
Jaspers, Karl, 120, 129, 273n6
Jenseits von Gut und Böse (Nietzsche), 237
Jünger, Ernst, 1

Kane, Harold, 128, 271n5
Kant, Immanuel: Brunonian theory and, 49–52; Cartesian dualism and, 34–35; definition of scientificity/*eigentliche Wissenschaft*, 34, 38–39; disease of intoxicants and, 56–57; empirical psychology, 37; experimentation and, 38; influence of, 33; medicine and, 39–40; mind-body relationship and, 36–37; Müller and, 98; publications on intoxicants by, 53–54; Schelling and, 60; scientific epistemology of, 37
Kastner, Karl, 168

Kielmeyer, Karl, 58
Klein, Ursula, 75
Klempe, Sven Hroar, 149
Koehlerglaube und Wissenschaf (Vogt), 141
Kölliker, Albert von, 151, 184
König, Joseph, 87
Kosmos (Humboldt, A.), 93
Kraepelin, Emil: amyl nitrate trials by, 192–94, *193*, *195*; apparatus of, *191*, 192; background of, 184–85; biologism and, 230; chloral hydrate trials by, 219; chloroform trials by, 196, *197*; criminality and, 253; degeneration and, 254; diagnostic cards (*Zählkarte*) of, 223–24; diagnostic techniques and, 211–12; Dorpat Psychological Society, 213–14, 273n3; effects of studies by, 204, 206–7; Exner and, 187–88, 272n3; experiential science and, 5–6, 201–2, 224–26; experimental pharmapsychology and, 216–17; fatigue studies by, 254–56; Flechsig and, 186, 209; Freud and, 245, 250–51; Gudden and, 213; at Heidelberg, 216–17; intoxication as a way of knowing for, 224–26, 265; intoxication as model psychosis and, 190–91, 221–23, 273n5; morphine and, 183–84, 219; Nietzsche and, 243–45; nosology of intoxication, 203–4, 227; paraldehyde trials by, 218–19; psychiatric nosology by, 209–12, 228–30, 273n6; publications of, 209–10; reaction time/alcohol experiments, 2, 198–200, *201*, *203*, *205*, 217–18, 272n5; reaction time/inhalation experiments by, 190–92, 196, 198; Weber and, 256–57; Wundt and, 184, 189–90, 220, 272–73n2, 272n1
Kraepelins Psychologische Arbeiten (Kraepelin), 229
Krafft-Ebing, Richard von, 209
Kritik der reinen Vernunft (Kant), 34, 46, 98
Kritik der Urteilskraft (Kant), 46, 54

308   INDEX

Krueger, Louis, 184
Kuhn, Thomas, 79, 270n1

La Mettrie, Julien Offrey de, 12–13, 36, 109
Lähr, Johann Sigismund, 179
Lange, Friedrich, 157, 233
Latour, Bruno, 82
Laudan, Larry, 37–38
Lavoisier, Antoine, 75
Lefèvre, Wolfgang, 75
Lehmann, Alfred, 218
Lehrbuch der ärztlichen Seelenkunde (Feuchtersleben), 44
Lehrbuch der Naturphilosophie (Oken), 67, 70
Leibniz, Gottfried Wilhelm, 141, 149
Leipzig (University of), 137
Lenoir, Timothy, 58, 270n4
Lesch, John, 75, 79–81
Leuckart, Rudolf, 184
Levinstein, Eduard, 179
Lichtenfels, Rudolf, 145, 271n2
Liebig, Justus von: agricultural chemistry and, 172–73; chloral hydrate and, 169; industrial pharmacies and, 170, 174; legacy in organic chemistry, 166–67; materialism and, 168; *Naturphilosophie* and, 167–68; at University of Giessen, 168–69; as a vitalist, 167–68, 271n1; Wöhler and, 166, 170, 271n2
Liebreich, Oskar, 169
Linnaeus, Carl, 19–20
Lippian Pharmacopoeia, 17
Literarische Centralblatt für Deutschland (periodical), 244
Löbenstein-Löbel, Eduard, 43–44, 48
Lotze, Rudolf Hermann, 151
Löw, Reinhard, 79
Löwald, Arnold, 229
Luchsinger, Balthasar, 250
Ludwig, Carl, 146, 209

Magazin für Literatur des Auslands (periodical), 146
Magazin zur Vervollkommnung der theoretischen und praktischen Heilkunde (Röschlaub), 41

Magnan, Valentin, 191
Major, Daniel, 178
Marcus, Adalbert, 40–41
Mariani, Angelo, 176–77
Marx, Karl, 128
Masseau, Joseph, 119
materia medica, 15–16
materialism, 12–13
Matteucci, Carlo, 105–6, 110
measurement, 108–9
Mechanik der Nerven (Wundt), 154
mechanistic physiology, 12–13
Meissner, Carl Friedrich Wilhelm, 87
Mendel, Emmanuel, 130–31
Merck, Friedrich Jacob, 174
Merck, Heinrich Emanuel, 174
Metaphysische Anfangsgründe der Naturwissenschaft (Kant), 38–39
Meyer, Martin, 229
Meynert, Theodor: *Abfaserung* technique of, 129–30; background of, 120, 271n3; downfall of, 129–32; Freud and, 246–47; Hitzig & Fritsch and, 123; intoxication and, 127–28; mental illnesses and, 120–21; neurophysiology and, 116–17; theories of cortical localization, 121, 122
Möbius, Paul Julius, 191, 224
Monakow, Constantin von, 131
Monro, Alexander Secundus, 20–21
Morgenröte: Gedanken über die moralischen Vorurteile (Nietzsche), 235
morphine: discovery of, 73–74, 81–83, 84; English morphine business, 175; hypodermic method for, 178–79, 272n8; Kraepelin and, 183–84, 219; Merck and, 174; morphinism, 179, 272n10; patent medicine manufacturers and, 175, 272n6; as a *principium somniferum*, 87
Morson, Thomas, 175
motor cortex (study of), 117–19, 123
Müller, Johannes: background information on, 95–96; Brunonianism and, 100; discoveries in body function by, 96; du Bois-Reymond and, 111; Humboldt brothers and, 94; intoxicants and, 100–102; Kant and, 98; law of

specific sense energies, 97–98; opium and, 102–3; publications of, 96; as a pupil of Oken, 72, 264; as a reformer, 97; significance of, 94; as a teacher, 91; as a vitalist, 4–5, 95, 98–103; Wundt and, 148

Munk, Hermann, 149

Nature (scientific journal), 139

Naturphilosophie: Fechner and, 140; Liebig and, 167–68; Oken and, 65–66; Schelling and, 57, 59, 64–65, 270n8; Sertürner and, 81; vital substances and, 4. See also Oken, Lorenz; Schelling, Friedrich W.

Naumann, Constantin Georg, 236

neomechanism: definition of, 91; difficulties of, 131–32; impact of, 133; Meynert & Wernicke's development of, 131; vitalism and, 5, 104–5

Newtonianism, 26

Niemann, Albert, 87, 171–73, 271n3–4

Nietzsche, Friedrich: concept of will, 237, 240–41, 243; degeneration and, 239–44, 273n4; du Bois and, 233; experimentation and, 235; intoxication and, 238–39, 242–43; Kraepelin and, 243–45; physiology and, 233; as a psychiatrist, 241–42; as a psychologist, 234–36; Rée and, 232–35, 273n2; science reading list of, 233–34; Wundtian thinking and, 236–38

nitrous oxide, 135–36

Novalis, 42, 58

Novatore, Renzo, 240

Novum Organum (Bacon), 18

Oken, Lorenz: animal classification and, 66; excitability in organisms, 67–69; Fechner and, 140; galvanism and, 69–70; on intoxication, 70–71; *Naturphilosophie* and, 65–66, 264; opium and, 69–70; Schelling and, 68–69

Olesko, Kathy, 107

"On the Speedy Relief of Pain and Other Nervous Activities by Means of the Hypodermic Method" (Hunter), 179

opium: active principle concept and, 20; Brown and, 24, 28; Cocteau and, 263; colonial trade and, 22–23; English opium-eaters, 46–47; La Mettrie and, 13; Oken and, 69–70; Schelling's principle of causation and, 63–64. See also intoxicants

Opium: Journal d'une désintoxication (Cocteau), 263

organic principle analysis, 74–75

Oseretskowsky, A., 229, 255

Owen, David, 240

paraldehyde, 218–19

Pasteur, Louis, 93

patent medicine industry, 175–78

Pelletier, Pierre-Joseph, 74, 87

Pfizer, 174

pharmacopoeia, 14–18, 264

Pharmacopoeia Edinburgensis, 17

Pharmacopoeia Londinensis, 17

pharmacy, 15, 73, 173–81, 271–72n5

Physiologische Psychologie (Wundt), 184

Physiologische und klinische Untersuchungen über das Gehirn, 119

Pickstone, John, 31

plant principle analysis, 74–79

Platen, Karl Friedrich von, 168

polarity, 79

Posselt, Wilhelm Heinrich, 87

Pouillet, Claude, 107

proximate principle analysis, 75

Psychiatrie: Ein kurzes Lehrbuch für Studirende und Aerzte (Kraepelin), 221–23, 229

Psychiatrie: Klinik der Erkrankungen des Vorderhirns begründet auf dessen Bau, Leistungen und Ernährung (Meynert), 122

psychiatry: brain, 116–17, 131; Brunonianism and, 44; French, 239; Freud and, 248, 252; Kraepelin and, 184–85, 210–11, 253–54; Meynert and, 119–21; Wernicke and, 125–26

Psychologische Beobachtungen (Rée), 235

"Psychologische Forschungsmethoden" (Kraepelin), 228

psychosis, 44

rational psychology, 37
Rée, Paul, 232–36
"Refutation of Idealism" (Kant), 35
Rehm, Ernst, 218
Reil, Johann, 58
Reimann, Karl Ludwig, 87
Repertorium der organischen Chemie (Fechner), 143
Reubein, Joseph, 62
Richter, F. A., 177
Riedel, J. D., 174–75
Rinecker, Franz von, 183–84
Risse, Guenter, 25
Ritter, Johann, 58–59, 110, 270n5
Robinson, David, 160
Robiquet, Pierre-Jean, 87
Roelcke, Voelker, 224
Rohde, Erwin, 245
Rokitansky, Carl von, 120
Romantic science, 58–59
Romberg, Moritz, 117, 271n1
Röschlaub, Andreas: Brunonianism and, 40–43, 264; English influence of, 46; Schelling and, 59. *See also* Schelling, Friedrich W.
Rudolphi, Carl Asmund, 96
Rüegg, Walter, 92
Runge, Friedlieb Ferdinand, 87
Rush, Benjamin, 43
Rynd, Francis, 178

"Säure im Opium" (Sertürner), 80
Saussure, Nicolas-Théodore de, 144
Scaff, Lawrence, 260
Schädellehre (Gall), 115
Scheele, Carl Wilhelm, 75
Schelling, Friedrich W.: atomism and, 141; Böhmer and, 61–62; Brunonianism and, 41, 62–63; English influence of, 46; excitability in organisms, 67; Fichte's philosophy and, 60; Hegel and, 59; Kant and, 60; *Naturphilosophie* and, 57, 59, 64–65, 270n8; Oken and, 68–69; opium and, 63–64, 264, 270n6; Ritter and, 58–59; Röschlaub and, 59
Scherf, Johann Christian Friedrich, 17
Schering, Emil, 174

Scherzer, Karl von, 171
Schlager, Ludwig, 120
Schlegel, August Wilhelm and Caroline, 61
Schleiermacher, Ferdinand, 92
Schmidgen, Henning, 189
Schmidt, Alexander, 214
Schopenhauer, Arthur, 232
Schüle, Heinrich, 209
Schwind, Mortiz von, 45
Sechenov, Ivan, 150
Seguin, Armand, 78
Sel Nacrotigue de Derosne (Derosne), 78
self-experimentation: Kraepelin and, 201–2; neomechanism and, 112–13; Sertürner and, 83, 85, 264
Sertürner, Friedrich W.: chemistry and, 80, 82; Derosne and, 77–78, 80–84, 84; discovery of morphine by, 4, 73–74; morphine self-experimentation and, 83, 85, 264; *Naturphilosophie* and, 81; Niemann and, 172; polarity and, 79; resistance against, 86–87; as a Romantic, 80–81; use of intoxicants for knowledge by, 2
Shakespeare, William, 1
Sloane, Hans, 17
sobriety, 252–53, 257–60
Spinoza, Baruch, 141
Spurzheim, Gaspar, 116
Stahl, Georg Ernst, 12
Steigerwald, Joan, 112, 270n3
Stein, Karl vom, 33
Structure of Scientific Revolutions, The (Kuhn), 79
"System der chem. Physik" (Sertürner), 81
Système de la nature (Holbach), 142

Temkin, Oswei, 91
Thénard, Louis Jacques, 76
therapeutics: plant- and animal-based, 14–15, 18; stimulant/Brunonian, 28; theory of therapeutic action, 18–21. *See also* intoxicants; vital substances
Thomson, John, 25
Thomson, Thomas, 75–77

Trommsdorff, Johann Bartholomäus, 80, 174
Trotter, Thomas, 43
Tsouyopoulos, Nelly, 41–42

Über Coca (Freud), 176, 245–47, 251
"Über den Bau der Nervenfaser und Nervenzellen beim Flußkrebs" (Freud), 246
"Über die Apperzeptionsdauer bei einfachen und zusammengesetzten Vorstellungen" (Friedrich), 160–61
Über die Natur der Kometen: Beiträge zur Geschichte und Theorie der Erkenntnis (Zöllner), 233
"Über Wahrheit und Lüge im außermoralischen Sinn" (Nietzsche), 235
"Ueber das Bedürfnis der Physiologie nach einer philosophischen Naturbetrachtung" (On the Need of Physiology for a Philosophical Contemplation of Nature) (Müller), 97
"Ueber das Erythroxylin, dargestellt aus den Blättern des in Südamerika cultivirten Strauches Erythroxylon Coca Lam" (Gaedcke), 170
"Ueber das Morphium als Hauptbestandteil des Opium" (Sertürner), 82
"Ueber den Einfluss acuter Krankheiten auf die Entstehung von Geisteskrankheiten" (Kraepelin), 185, 209
"Ueber den Missbrauch der Morphium-Injectionen" (Fiedler), 179
"Ueber die Allgemeinwirkung des Cocains" (Freud), 248
Ueber die Beeinflussung (Kraepelin), 221, 254–55
Ueber die Beeinflussung einfacher psychischer Vorgänge durch einige Arzneimittel (Kraepelin), 216, 222
"Ueber die Dauer einfacher psychischer Vorgänge" (Kraepelin), 209
"Ueber die Einwirkung einiger medicamentöser Stoffe auf die Dauer einfacher psychicher Vorgänge" (Kraepelin), 190, 209, 244
"Ueber die physiologische Wirkung des Chloroforms" (Bernstein), 135
"Ueber die physiologischen Wirkungen des Stickstoffoxydulgases" (Hermann), 135
Ueber die Trunksucht und eine rationelle Heilmethode derselben (Brühl-Cramer), 43, 56
"Ueber eine neue organische Base in den Cocablätter" (Niemann), 171
Ueber geistige Arbeit (Kraepelin), 255
"Ueber Missbrauch mit Morphium-Injectionen" (Lähr), 179
"Ueber psychische Zeitmessung" (Kraepelin), 215
Unsere Körperform und das physiologische Problem ihrer Entstehung (His), 233
"Untersuchungen über den Einfluss des Coffeins und Thees auf der Dauer einfacher psychischer Vorgänge" (Dehio), 214
Untersuchungen über den Erregungsvorgang im Nerven- und Muskelsysteme (Bernstein), 154
Untersuchungen über die Pathogenie oder Einleitung in die medicinische Theorie (Röschlaub), 40
Untersuchungen über Thierische Elektricität (du Bois-Reymond), 108, 111
Untersuchungen zur Mechanik der Nerven und Nervencentren (Wundt), 154

Vanzo, Alberto, 37–38
Vauquelin, Nicolas-Louis, 87
"Versuche die Zerlegung des Opiums" (Buchholz), 80
Vin Mariani, 176–77
Vintschgau, Maximilian von, 187–89, 198–99, 206–7
Virchow, Rudolf, 130, 271n2
vital substances: Brown and, 25, 31–32; definition of, 4, 14; du Bois-Reymond and, 110–11; Fechner and, 143–45; morphine as a, 85; Schelling and, 62–63. See also intoxicants; Müller, Johannes
vitalism: importance of, 265; Oken and, 69; opposition of, 5; Stahl and, 12. See also Brown, John

Vogt, Karl, 141
Volta, Alessandro, 59
"Vom Nutzen und Nachtheil der Historie für das Leben" (Nietzsche), 235
Vorlesungen über die Menschen- und Thier-Seele (Wundt), 184
Vorschule der Aesthetik (Fechner), 145

Wackenroder, Heinrich, 170–71
Wagner, Richard, 238
Waitz, Theodor, 151
Waldeyer, Wilhelm, 130
Was ist Aufklärun (Kant), 52
Wasianski, Ehregott Andreas Christoph, 49–52, 56–57
Wassman, Christian, 148–49
Watson, Peter, 58
Weber, Ernst, 104, 138–40, 151, 153, 223
Weber, Max: intoxication as a way of knowing for, 260–61; Kraepelin and, 256–61; sobriety and, 252–53, 257–60
Weikard, Melchoir Adam, 50
Weindling, Paul, 118
Weiner Logik (Kant), 38
Weltseele, Von der (Schelling), 41, 59
Wepfer, Johann Jakob, 19–20
Wernicke, Karl: aphasia and, 124–25; applied neurophysiology of, 125–27, 271n4; downfall of, 129; intoxicants and, 128; Meynertian neurophysiology and, 123–24; neurophysiology and, 116–17
Westphal, Karl, 125
Wiedemann, Gustav, 184
Wislicenus, Johannes, 184
Wissenschaftslehre (Fichte), 42

Wöhler, Friedrich, 166–67, 170, 271n2
Wood, Scot Alexander, 178–79
Woskresensky, Aleksandr, 87
Wren, Christopher, 178
Wundt, Wilhelm: background of, 148; *Beiträge* and, 150–55; Bernstein and, 154; development of scientific psychology of, 147–48; du Bois-Reymond and, 149; Fechner and, 138, 152; *Grundzüge*, 155–57; Helmholtz and, 150, 152; Herbart and, 150, 157; Kraepelin and, 184, 189–90, 220, 272–73n2; Leibniz and, 149; Müller and, 148; Nietzsche and, 236–38; physical-psychology of, 156; physiology and, 148–49, 154–55; psychological parallelism, 5, 153, 204, 206; publications of, 148; reaction time and, 158–60, 271n3; spiritism and, 156–57; Weber and, 151, 153

Youngquist, Paul, 47

Zend-Avesta (Fechner), 141–42
Zöllner, Karl, 233
Zur Genealogie der Moral (Nietzsche), 236
"Zur Physiologie der Kunst" (Nietzsche), 242
"Zur Theorie des Fechner'schen Gesetzes der Empfindung" (Bernstein), 154
Zur vergleichenden Physiologie des Gesichtssinnes des Menschen und der Thiere (On the Comparative Physiology of Vision in Men and Animals) (Müller), 97

www.ingramcontent.com/pod-product-compliance
Lightning Source LLC
Chambersburg PA
CBHW022035290426
44109CB00014B/864